中 外 物 理 学 精 品 书 系
本书出版得到"国家出版基金"资助

国家出版基金项目
NATIONAL PUBLICATION FOUNDATION

中外物理学精品书系

引进系列·63

An Introduction to Non-Abelian Discrete Symmetries for Particle Physicists

粒子物理学家用非阿贝尔离散对称导论

（影印版）

〔日〕石森一（H. Ishimori）
〔日〕小林达夫（T. Kobayashi）
〔日〕大木洋（H. Ohki）　　　著
〔日〕冈田宽（H. Okada）
〔日〕清水勇介（Y. Shimizu）
〔日〕谷本盛光（M. Tanimoto）

北京大学出版社
PEKING UNIVERSITY PRESS

著作权合同登记号　图字：01-2014-4298

图书在版编目(CIP)数据

粒子物理学家用非阿贝尔离散对称导论 = An introduction to non-Abelian discrete symmetries for particle physicists：英文/(日)石森一等著.—影印本.—北京：北京大学出版社,2014.12

(中外物理学精品书系)

ISBN 978-7-301-25184-3

Ⅰ.①粒…　Ⅱ.①石…　Ⅲ.①粒子物理学—对称—研究—英文　Ⅳ.①O572.2

中国版本图书馆 CIP 数据核字(2014)第 278927 号

Reprint from English language edition：
An Introduction to Non-Abelian Discrete Symmetries for Particle Physicists
by Hajime Ishimori, Tatsuo Kobayashi ,Hiroshi Ohki, Hiroshi Okada, Yusuke Shimizu, Morimitsu Tanimoto
Copyright © 2012 Springer Berlin Heidelberg
Springer Berlin Heidelberg is a part of Springer Science＋Business Media
All Rights Reserved

"This reprint has been authorized by Springer Science & Business Media for distribution in China Mainland only and not for export therefrom."

书　　　名：	An Introduction to Non-Abelian Discrete Symmetries for Particle Physicists(粒子物理学家用非阿贝尔离散对称导论)(影印版)
著作责任者：	〔日〕石森一(H. Ishimori)　〔日〕小林达夫(T. Kobayashi)　〔日〕大木洋(H. Ohki)　〔日〕冈田宽(H. Okada)　〔日〕清水勇介(Y. Shimizu)　〔日〕谷本盛光(M. Tanimoto)　著
责 任 编 辑：	刘　啸
标 准 书 号：	ISBN 978-7-301-25184-3/O・1055
出 版 发 行：	北京大学出版社
地　　　址：	北京市海淀区成府路 205 号　100871
网　　　址：	http://www.pup.cn
新 浪 微 博：	@北京大学出版社
电 子 信 箱：	zpup@pup.cn
电　　　话：	邮购部 62752015　发行部 62750672　编辑部 62752038　出版部 62754962
印 刷 者：	北京中科印刷有限公司
经 销 者：	新华书店
	730 毫米×980 毫米　16 开本　19 印张　362 千字
	2014 年 12 月第 1 版　2014 年 12 月第 1 次印刷
定　　　价：	51.00 元

未经许可,不得以任何方式复制或抄袭本书之部分或全部内容。
版权所有,侵权必究
举报电话：010-62752024　电子信箱：fd@pup.pku.edu.cn

"中外物理学精品书系"
编 委 会

主　任：王恩哥
副主任：夏建白
编　委：（按姓氏笔画排序，标*号者为执行编委）

王力军	王孝群	王　牧	王鼎盛	石　兢
田光善	冯世平	邢定钰	朱邦芬	朱　星
向　涛	刘　川*	许宁生	许京军	张　酣*
张富春	陈志坚*	林海青	欧阳钟灿	周月梅*
郑春开*	赵光达	聂玉昕	徐仁新*	郭　卫*
资　剑	龚旗煌	崔　田	阎守胜	谢心澄
解士杰	解思深	潘建伟		

秘　书：陈小红

序　　言

物理学是研究物质、能量以及它们之间相互作用的科学。她不仅是化学、生命、材料、信息、能源和环境等相关学科的基础，同时还是许多新兴学科和交叉学科的前沿。在科技发展日新月异和国际竞争日趋激烈的今天，物理学不仅囿于基础科学和技术应用研究的范畴，而且在社会发展与人类进步的历史进程中发挥着越来越关键的作用。

我们欣喜地看到，改革开放三十多年来，随着中国政治、经济、教育、文化等领域各项事业的持续稳定发展，我国物理学取得了跨越式的进步，做出了很多为世界瞩目的研究成果。今日的中国物理正在经历一个历史上少有的黄金时代。

在我国物理学科快速发展的背景下，近年来物理学相关书籍也呈现百花齐放的良好态势，在知识传承、学术交流、人才培养等方面发挥着无可替代的作用。从另一方面看，尽管国内各出版社相继推出了一些质量很高的物理教材和图书，但系统总结物理学各门类知识和发展，深入浅出地介绍其与现代科学技术之间的渊源，并针对不同层次的读者提供有价值的教材和研究参考，仍是我国科学传播与出版界面临的一个极富挑战性的课题。

为有力推动我国物理学研究、加快相关学科的建设与发展，特别是展现近年来中国物理学者的研究水平和成果，北京大学出版社在国家出版基金的支持下推出了"中外物理学精品书系"，试图对以上难题进行大胆的尝试和探索。该书系编委会集结了数十位来自内地和香港顶尖高校及科研院所的知名专家学者。他们都是目前该领域十分活跃的专家，确保了整套丛书的权威性和前瞻性。

这套书系内容丰富，涵盖面广，可读性强，其中既有对我国传统物理学发展的梳理和总结，也有对正在蓬勃发展的物理学前沿的全面展示；既引进和介绍了世界物理学研究的发展动态，也面向国际主流领域传播中国物理的优秀专著。可以说，"中外物理学精品书系"力图完整呈现近现代世界和中国物理科学发展的全貌，是一部目前国内为数不多的兼具学术价值和阅读乐趣的经典物理丛书。

"中外物理学精品书系"另一个突出特点是，在把西方物理的精华要义"请进来"的同时，也将我国近现代物理的优秀成果"送出去"。物理学科在世界范围内的重要性不言而喻，引进和翻译世界物理的经典著作和前沿动态，可以满足当前国内物理教学和科研工作的迫切需求。另一方面，改革开放几十年来，我国的物理学研究取得了长足发展，一大批具有较高学术价值的著作相继问世。这套丛书首次将一些中国物理学者的优秀论著以英文版的形式直接推向国际相关研究的主流领域，使世界对中国物理学的过去和现状有更多的深入了解，不仅充分展示出中国物理学研究和积累的"硬实力"，也向世界主动传播我国科技文化领域不断创新的"软实力"，对全面提升中国科学、教育和文化领域的国际形象起到重要的促进作用。

值得一提的是，"中外物理学精品书系"还对中国近现代物理学科的经典著作进行了全面收录。20世纪以来，中国物理界诞生了很多经典作品，但当时大都分散出版，如今很多代表性的作品已经淹没在浩瀚的图书海洋中，读者们对这些论著也都是"只闻其声，未见其真"。该书系的编者们在这方面下了很大工夫，对中国物理学科不同时期、不同分支的经典著作进行了系统的整理和收录。这项工作具有非常重要的学术意义和社会价值，不仅可以很好地保护和传承我国物理学的经典文献，充分发挥其应有的传世育人的作用，更能使广大物理学人和青年学子切身体会我国物理学研究的发展脉络和优良传统，真正领悟到老一辈科学家严谨求实、追求卓越、博大精深的治学之美。

温家宝总理在2006年中国科学技术大会上指出，"加强基础研究是提升国家创新能力、积累智力资本的重要途径，是我国跻身世界科技强国的必要条件"。中国的发展在于创新，而基础研究正是一切创新的根本和源泉。我相信，这套"中外物理学精品书系"的出版，不仅可以使所有热爱和研究物理学的人们从中获取思维的启迪、智力的挑战和阅读的乐趣，也将进一步推动其他相关基础科学更好更快地发展，为我国今后的科技创新和社会进步做出应有的贡献。

"中外物理学精品书系"编委会 主任
中国科学院院士，北京大学教授
王恩哥
2010年5月于燕园

Hajime Ishimori · Tatsuo Kobayashi ·
Hiroshi Ohki · Hiroshi Okada · Yusuke Shimizu ·
Morimitsu Tanimoto

An Introduction to Non-Abelian Discrete Symmetries for Particle Physicists

Preface

The purpose of these lecture notes is to introduce the basic framework of non-Abelian discrete symmetries, and to present some important applications in particle physics. Discrete non-Abelian groups have in fact played an important role in particle physics. However, they may not be so familiar to particle physicists as continuous non-Abelian symmetries. These lecture notes are written for particle physicists and differ in this respect from standard books on group theory. However, preliminary knowledge of group theory is not required to understand the non-Abelian discrete symmetries.

We hope our lecture notes will serve as a handbook for serious learners, and also as a helpful reference book for experts, as well perhaps as triggering future research.

It is pleasure to acknowledge fruitful discussions with H. Abe, T. Araki, K.S. Choi, Y. Daikoku, K. Hashimoto, J. Kubo, H.P. Nilles, F. Ploger, S. Raby, S. Ramos-Sanchez, M. Ratz, and P.K.S. Vaudrevange.

Hajime Ishimori
Tatsuo Kobayashi
Hiroshi Ohki
Hiroshi Okada
Yusuke Shimizu
Morimitsu Tanimoto

Contents

1	**Introduction**	1
	References	3
2	**Basics of Finite Groups**	13
	References	20
3	S_N	21
	3.1 S_3	21
	3.1.1 Conjugacy Classes	21
	3.1.2 Characters and Representations	22
	3.1.3 Tensor Products	22
	3.2 S_4	25
	3.2.1 Conjugacy Classes	27
	3.2.2 Characters and Representations	27
	3.2.3 Tensor Products	29
	References	30
4	A_N	31
	4.1 A_4	31
	4.2 A_5	34
	4.2.1 Conjugacy Classes	35
	4.2.2 Characters and Representations	35
	4.2.3 Tensor Products	37
	References	41
5	T'	43
	5.1 Conjugacy Classes	43
	5.2 Characters and Representations	44
	5.3 Tensor Products	47
6	D_N	51
	6.1 D_N with N Even	51
	6.1.1 Conjugacy Classes	52

	6.1.2	Characters and Representations	52
	6.1.3	Tensor Products	54
6.2	D_N with N Odd		56
	6.2.1	Conjugacy Classes	56
	6.2.2	Characters and Representations	56
	6.2.3	Tensor Products	57
6.3	D_4		58
6.4	D_5		59

7 Q_N — 61
- 7.1 Q_N with $N = 4n$ — 61
  - 7.1.1 Conjugacy Classes — 62
  - 7.1.2 Characters and Representations — 62
  - 7.1.3 Tensor Products — 62
- 7.2 Q_N with $N = 4n + 2$ — 64
  - 7.2.1 Conjugacy Classes — 64
  - 7.2.2 Characters and Representations — 64
  - 7.2.3 Tensor Products — 65
- 7.3 Q_4 — 66
- 7.4 Q_6 — 67

8 QD_{2N} — 69
- 8.1 Generic Aspects — 69
  - 8.1.1 Conjugacy Classes — 70
  - 8.1.2 Characters and Representations — 70
  - 8.1.3 Tensor Products — 71
- 8.2 QD_{16} — 72

9 $\Sigma(2N^2)$ — 75
- 9.1 Generic Aspects — 75
  - 9.1.1 Conjugacy Classes — 75
  - 9.1.2 Characters and Representations — 76
  - 9.1.3 Tensor Products — 77
- 9.2 $\Sigma(18)$ — 78
- 9.3 $\Sigma(32)$ — 80
- 9.4 $\Sigma(50)$ — 84

10 $\Delta(3N^2)$ — 87
- 10.1 $\Delta(3N^2)$ with $N/3 \neq$ Integer — 87
  - 10.1.1 Conjugacy Classes — 88
  - 10.1.2 Characters and Representations — 89
  - 10.1.3 Tensor Products — 89
- 10.2 $\Delta(3N^2)$ with $N/3$ Integer — 91
  - 10.2.1 Conjugacy Classes — 91
  - 10.2.2 Characters and Representations — 92
  - 10.2.3 Tensor Products — 93
- 10.3 $\Delta(27)$ — 94
- References — 95

Contents

11 T_N 97
    11.1 Generic Aspects 97
        11.1.1 Conjugacy Classes 98
        11.1.2 Characters and Representations 99
        11.1.3 Tensor Products 99
    11.2 T_7 100
    11.3 T_{13} 102
    11.4 T_{19} 104
    References 108

12 $\Sigma(3N^3)$ 109
    12.1 Generic Aspects 109
        12.1.1 Conjugacy Classes 110
        12.1.2 Characters and Representations 111
        12.1.3 Tensor Products 112
    12.2 $\Sigma(81)$ 113
    References 121

13 $\Delta(6N^2)$ 123
    13.1 $\Delta(6N^2)$ with $N/3 \neq$ Integer 123
        13.1.1 Conjugacy Classes 123
        13.1.2 Characters and Representations 126
        13.1.3 Tensor Products 128
    13.2 $\Delta(6N^2)$ with $N/3$ Integer 131
        13.2.1 Conjugacy Classes 131
        13.2.2 Characters and Representations 133
        13.2.3 Tensor Products 134
    13.3 $\Delta(54)$ 138
        13.3.1 Conjugacy Classes 138
        13.3.2 Characters and Representations 139
        13.3.3 Tensor Products 141
    References 145

14 Subgroups and Decompositions of Multiplets 147
    14.1 S_3 147
        14.1.1 $S_3 \to Z_3$ 148
        14.1.2 $S_3 \to Z_2$ 148
    14.2 S_4 149
        14.2.1 $S_4 \to S_3$ 150
        14.2.2 $S_4 \to A_4$ 151
        14.2.3 $S_4 \to \Sigma(8)$ 151
    14.3 A_4 152
        14.3.1 $A_4 \to Z_3$ 152
        14.3.2 $A_4 \to Z_2 \times Z_2$ 153
    14.4 A_5 153
        14.4.1 $A_5 \to A_4$ 153

	14.4.2	$A_5 \to D_5$	153
	14.4.3	$A_5 \to S_3 \simeq D_3$	154
14.5	T'		154
	14.5.1	$T' \to Z_6$	154
	14.5.2	$T' \to Z_4$	155
	14.5.3	$T' \to Q_4$	155
14.6	General D_N		155
	14.6.1	$D_N \to Z_2$	156
	14.6.2	$D_N \to Z_N$	157
	14.6.3	$D_N \to D_M$	157
14.7	D_4		158
	14.7.1	$D_4 \to Z_4$	158
	14.7.2	$D_4 \to Z_2 \times Z_2$	159
	14.7.3	$D_4 \to Z_2$	159
14.8	General Q_N		159
	14.8.1	$Q_N \to Z_4$	160
	14.8.2	$Q_N \to Z_N$	161
	14.8.3	$Q_N \to Q_M$	161
14.9	Q_4		162
	14.9.1	$Q_4 \to Z_4$	162
14.10	QD_{2N}		162
	14.10.1	$QD_{2N} \to Z_2$	163
	14.10.2	$QD_{2N} \to Z_N$	163
	14.10.3	$QD_{2N} \to D_{N/2}$	163
14.11	General $\Sigma(2N^2)$		164
	14.11.1	$\Sigma(2N^2) \to Z_{2N}$	164
	14.11.2	$\Sigma(2N^2) \to Z_N \times Z_N$	164
	14.11.3	$\Sigma(2N^2) \to D_N$	165
	14.11.4	$\Sigma(2N^2) \to Q_N$	166
	14.11.5	$\Sigma(2N^2) \to \Sigma(2M^2)$	166
14.12	$\Sigma(32)$		167
14.13	General $\Delta(3N^2)$		168
	14.13.1	$\Delta(3N^2) \to Z_3$	169
	14.13.2	$\Delta(3N^2) \to Z_N \times Z_N$	169
	14.13.3	$\Delta(3N^2) \to T_N$	170
	14.13.4	$\Delta(3N^2) \to \Delta(3M^2)$	170
14.14	$\Delta(27)$		172
	14.14.1	$\Delta(27) \to Z_3$	172
	14.14.2	$\Delta(27) \to Z_3 \times Z_3$	172
14.15	General T_N		173
	14.15.1	$T_N \to Z_3$	173
	14.15.2	$T_N \to Z_N$	173
14.16	T_7		174
	14.16.1	$T_7 \to Z_3$	174
	14.16.2	$T_7 \to Z_7$	175

Contents   xi

  14.17 General $\Sigma(3N^3)$ 175
    14.17.1 $\Sigma(3N^2) \to Z_N \times Z_N \times Z_N$ 175
    14.17.2 $\Sigma(3N^3) \to \Delta(3N^2)$ 175
    14.17.3 $\Sigma(3N^3) \to \Sigma(3M^3)$ 176
  14.18 $\Sigma(81)$ 176
    14.18.1 $\Sigma(81) \to Z_3 \times Z_3 \times Z_3$ 177
    14.18.2 $\Sigma(81) \to \Delta(27)$ 177
  14.19 General $\Delta(6N^2)$ 178
    14.19.1 $\Delta(6N^2) \to \Sigma(2N^2)$ 179
    14.19.2 $\Delta(6N^2) \to \Delta(3N^2)$ 180
    14.19.3 $\Delta(6N^2) \to \Delta(6M^2)$ 180
  14.20 $\Delta(54)$ 181
    14.20.1 $\Delta(54) \to S_3 \times Z_3$ 182
    14.20.2 $\Delta(54) \to \Sigma(18)$ 182
    14.20.3 $\Delta(54) \to \Delta(27)$ 183

15 Anomalies 185
  15.1 Generic Aspects 185
  15.2 Explicit Calculations 189
    15.2.1 S_3 189
    15.2.2 S_4 190
    15.2.3 A_4 190
    15.2.4 A_5 191
    15.2.5 T' 192
    15.2.6 D_N (N Even) 193
    15.2.7 D_N (N Odd) 194
    15.2.8 Q_N ($N = 4n$) 194
    15.2.9 Q_N ($N = 4n + 2$) 195
    15.2.10 QD_{2N} 196
    15.2.11 $\Sigma(2N^2)$ 197
    15.2.12 $\Delta(3N^2)$ ($N/3 \neq$ Integer) 198
    15.2.13 $\Delta(3N^2)$ ($N/3$ Integer) 199
    15.2.14 T_N 200
    15.2.15 $\Sigma(3N^3)$ 201
    15.2.16 $\Delta(6N^2)$ ($N/3 \neq$ Integer) 202
    15.2.17 $\Delta(6N^2)$ ($N/3$ Integer) 203
  15.3 Comments on Anomalies 203
  References 204

16 Non-Abelian Discrete Symmetry in Quark/Lepton Flavor Models .. 205
  16.1 Neutrino Flavor Mixing and Neutrino Mass Matrix 205
  16.2 A_4 Flavor Symmetry 207
    16.2.1 Realizing Tri-Bimaximal Mixing of Flavors 207
    16.2.2 Breaking Tri-Bimaximal Mixing 209
  16.3 S_4 Flavor Model 211
  16.4 Alternative Flavor Mixing 219

16.5	Comments on Other Applications	222
16.6	Comment on Origins of Flavor Symmetries	223
References		224

Appendix A Useful Theorems 229
    References 235

Appendix B Representations of S_4 in Different Bases 237
    B.1   Basis I 237
    B.2   Basis II 238
    B.3   Basis III 240
    B.4   Basis IV 242
    References 244

Appendix C Representations of A_4 in Different Bases 245
    C.1   Basis I 245
    C.2   Basis II 245
    References 246

Appendix D Representations of A_5 in Different Bases 247
    D.1   Basis I 247
    D.2   Basis II 253
    References 259

Appendix E Representations of T' in Different Bases 261
    E.1   Basis I 262
    E.2   Basis II 263
    References 264

Appendix F Other Smaller Groups 265
    F.1 $Z_4 \rtimes Z_4$ 265
    F.2 $Z_8 \rtimes Z_2$ 268
    F.3 $(Z_2 \times Z_4) \rtimes Z_2$ (I) 270
    F.4 $(Z_2 \times Z_4) \rtimes Z_2$ (II) 272
    F.5 $Z_3 \rtimes Z_8$ 275
    F.6 $(Z_6 \times Z_2) \rtimes Z_2$ 277
    F.7 $Z_9 \rtimes Z_3$ 281
    References 283

Index 285

Chapter 1
Introduction

These lecture notes aim to provide a pedagogical review of non-Abelian discrete groups and show some applications to physical issues. Symmetry constitutes a very important principle in physics. In particular, it has played an essential role in constructing the framework of particle physics. For example, continuous (and local) symmetries such as Lorentz, Poincaré, and gauge symmetries are crucial to understand several phenomena, such as the strong, weak, and electromagnetic interactions among particles. On the other hand, discrete symmetries such as C, P, and T are also vital concepts in particle physics. Abelian discrete symmetries, Z_N, are also often imposed in order to control allowed couplings for particle physics, in particular model-building beyond the standard model. In addition to Abelian discrete symmetries, non-Abelian discrete symmetries have also been applied for model-building in particle physics, in particular to understand the three-generation flavor structure.

There are many free parameters in the standard model, including its extension with neutrino mass terms. Most of them are Yukawa couplings of quarks and leptons to the Higgs boson. The quark and lepton sector is called the flavor sector. Flavor physics is a challenging aspect of the construction of the theory beyond the standard model. If a symmetry is imposed on the flavor sector, one can control the Yukawa couplings in the three generations, although the origin of the generations remains unknown. Therefore, quark masses and mixing angles have been studied from the standpoint of flavor symmetries.

In addition, the discovery of neutrino masses and neutrino mixing [1, 2] has stimulated work on flavor symmetries. Experiments on neutrino oscillations are now going into a new phase of precise determination of mixing angles and mass squared differences [3–7]. In particular, the recent long baseline neutrino experiment T2K is reaching the last neutrino mixing angle, so called θ_{13} [8]. The Double Chooz collaboration has also reported indications of non-zero θ_{13} [9]. Reactor neutrino experiments, Reno and Daya Bay are also attempting to observe it. Global analyses of neutrino data indicate the special neutrino mixing pattern, which is called tri-bimaximal mixing for three flavors in the lepton sector [10–13]. These large mixing angles are completely different from the quark mixing ones. Therefore, it is very important to find a natural model that leads to these mixing patterns of quarks and leptons with good accuracy.

Non-Abelian discrete symmetries are considered to be the most attractive choice for the flavor sector. Model builders have tried to derive experimental values of quark/lepton masses and mixing angles by assuming non-Abelian discrete flavor symmetries of quarks and leptons. In particular, lepton mixing has been intensively discussed in the context of non-Abelian discrete flavor symmetries, as seen, e.g., in the reviews [14, 15].

Particle physicists may be interested in the origin of the non-Abelian discrete symmetry for flavors. One of the most interesting is a higher dimensional spacetime symmetry. After it has been broken down to the 4D Poincaré symmetry through compactification, e.g., via orbifolding, a remnant symmetry appears in the flavor sector. This remnant symmetry is often a non-Abelian symmetry. Actually, it has been shown how the flavor symmetry A_4 (or S_4) can arise if the three fermion generations are taken to live on the fixed points of a specific 2D orbifold [16]. Further non-Abelian discrete symmetries can arise in a similar setup [17] (see also [18]).

Superstring theory is a promising candidate for a unified theory including gravity. Certain string modes correspond to gauge bosons, quarks, leptons, Higgs bosons, and gravitons as well as their superpartners. Superstring theory predicts six extra dimensions. Certain classes of discrete symmetries can be derived from superstring theories. A combination among geometrical symmetries of a compact space and stringy selection rules for couplings enhances discrete flavor symmetries. For example, D_4 and $\Delta(54)$ flavor symmetries can be obtained in heterotic orbifold models [19–21]. In addition to these flavor symmetries, the $\Delta(27)$ flavor symmetry can be derived from magnetized/intersecting D-brane models [22–24].

There is another possibility, namely that non-Abelian discrete groups may originate from the breaking of continuous (gauge) flavor symmetries [25–27].

Thus, a non-Abelian discrete symmetry can arise from the underlying theory, e.g., string theory or compactification via orbifolding. In addition, non-Abelian discrete symmetries are interesting tools for controlling flavor structure in model building using the bottom-up approach. Hence, non-Abelian flavor symmetries could provide a bridge between the low-energy physics and the underlying theory. It is thus quite important to understand the properties of non-Abelian groups for particle physics.

Continuous non-Abelian groups are well-known, and of course there are several good reviews and books. On the other hand, discrete non-Abelian symmetries may not be so familiar to particle physicists as continuous non-Abelian symmetries. However, discrete non-Abelian symmetries have become important tools for model building, as discussed above, in particular in the context of flavor physics. The purpose of these lecture notes is therefore to provide a pedagogical review of non-Abelian discrete groups with particle phenomenology in mind, and to exhibit the group-theoretical aspects of many concrete groups explicitly, including, for example, representations and their tensor products [15, 28–34]. We present these aspects in detail for the groups S_N [35–132], A_N [133–243], T' [33, 244–263], D_N [264–285], Q_N [286–300], QD_{2N}, $\Sigma(2N^2)$ [301], $\Delta(3N^2)$ [302–313], T_N [302–304, 312, 314–323], $\Sigma(3N^3)$ [315, 324], and $\Delta(6N^2)$ [302–304, 312, 325–330].

We explain pedagogically how to derive conjugacy classes, characters, representations, and tensor products for these groups (with a finite number) when algebraic relations are given. Thus, it will be straightforward for readers to apply this to other groups.

In applications to particle physics, the breaking patterns of discrete groups and decompositions of multiplets are often required to understand low energy phenomena. Such aspects are given in these notes.

Symmetries at the tree level are not always symmetries in quantum theory. If symmetries are anomalous, breaking terms are induced by quantum effects. Such anomalies are important in applications for particle physics. Here, we study such anomalies for discrete symmetries [331–344] and show anomaly-free conditions explicitly for the above concrete groups. If flavor symmetries are stringy symmetries, these anomalies may also be controlled by string dynamics, i.e., anomaly cancellation.

We also present flavor models with non-Abelian discrete symmetry as typical examples. One can see how to use the non-Abelian discrete symmetry for flavors. A lot of references are available to understand the model building here.

On the other hand, discrete subgroups of $SU(3)$ would also be interesting from the standpoint of phenomenological applications to flavor physics [345–349]. Here, most of them are shown for subgroups including doublets or triplets as the largest dimensional irreducible representations (for other groups see [29, 31, 34, 241, 350–354]).

The book is organized as follows. In Chap. 2, we summarize the basic group-theoretical aspects used in subsequent chapters, and also present some examples to provide a more concrete understanding. Readers familiar with group theory can skip Chap. 2. In Chaps. 3 to 13, we present the non-Abelian discrete groups S_N, A_N, T', D_N, Q_N, QD_{2N}, $\Sigma(2N^2)$, $\Delta(3N^2)$, T_N, $\Sigma(3N^3)$, and $\Delta(6N^2)$. In each chapter, groups with specific values of N are also discussed for typical examples. Chapter 14 discusses the breaking patterns of the non-Abelian discrete groups. In Chap. 15, we review the anomalies of non-Abelian flavor symmetries, which is an important topic in particle physics, and exhibit the anomaly-free conditions explicitly for the above concrete groups. Chapter 16 presents typical flavor models with the non-Abelian discrete symmetries A_4 and S_4.

Appendix A gives some useful theorems on finite group theory, while Appendices B, C, D, and E provide the representation bases of S_4, A_4, A_5, and T', which are different from those in Chaps. 3, 4, and 5. Appendix F presents other smaller groups in detail.

Note Added in Proof Finally, θ_{13} has been observed by Daya Bay [355] and Reno [356].

References

1. Pontecorvo, B.: Sov. Phys. JETP **6**, 429 (1957) [Zh. Eksp. Teor. Fiz. **33**, 549 (1957)]
2. Maki, Z., Nakagawa, M., Sakata, S.: Prog. Theor. Phys. **28**, 870 (1962)
3. Schwetz, T., Tortola, M.A., Valle, J.W.F.: New J. Phys. **10**, 113011 (2008). arXiv:0808.2016 [hep-ph]

4. Fogli, G.L., Lisi, E., Marrone, A., Palazzo, A., Rotunno, A.M.: Phys. Rev. Lett. **101**, 141801 (2008). arXiv:0806.2649 [hep-ph]
5. Fogli, G.L., Lisi, E., Marrone, A., Palazzo, A., Rotunno, A.M.: Nucl. Phys. Proc. Suppl. **188**, 27 (2009)
6. Gonzalez-Garcia, M.C., Maltoni, M., Salvado, J.: arXiv:1001.4524 [hep-ph]
7. Schwetz, T., Tortola, M., Valle, J.W.F.: New J. Phys. **13**, 063004 (2011). arXiv:1103.0734 [hep-ph]
8. Abe, K., et al. (T2K Collaboration): Phys. Rev. Lett. **107**, 041801 (2011). arXiv:1106.2822 [hep-ex]
9. Abe, Y., et al. (Double Chooz Collaboration): arXiv:1112.6353 [hep-ex]
10. Harrison, P.F., Perkins, D.H., Scott, W.G.: Phys. Lett. B **530**, 167 (2002). arXiv:hep-ph/0202074
11. Harrison, P.F., Scott, W.G.: Phys. Lett. B **535**, 163 (2002). arXiv:hep-ph/0203209
12. Harrison, P.F., Scott, W.G.: Phys. Lett. B **557**, 76 (2003). arXiv:hep-ph/0302025
13. Harrison, P.F., Scott, W.G.: arXiv:hep-ph/0402006
14. Altarelli, G., Feruglio, F.: arXiv:1002.0211 [hep-ph]
15. Ishimori, H., Kobayashi, T., Ohki, H., Shimizu, Y., Okada, H., Tanimoto, M.: Prog. Theor. Phys. Suppl. **183**, 1–163 (2010). arXiv:1003.3552 [hep-th]
16. Altarelli, G., Feruglio, F., Lin, Y.: Nucl. Phys. B **775**, 31 (2007). arXiv:hep-ph/0610165
17. Adulpravitchai, A., Blum, A., Lindner, M.: J. High Energy Phys. **0907**, 053 (2009). arXiv:0906.0468 [hep-ph]
18. Abe, H., Choi, K.-S., Kobayashi, T., Ohki, H., Sakai, M.: Int. J. Mod. Phys. A **26**, 4067–4082 (2011). arXiv:1009.5284 [hep-th]
19. Kobayashi, T., Raby, S., Zhang, R.J.: Nucl. Phys. B **704**, 3 (2005). arXiv:hep-ph/0409098
20. Kobayashi, T., Nilles, H.P., Ploger, F., Raby, S., Ratz, M.: Nucl. Phys. B **768**, 135 (2007). arXiv:hep-ph/0611020
21. Ko, P., Kobayashi, T., Park, J.H., Raby, S.: Phys. Rev. D **76**, 035005 (2007) [Erratum ibid. D **76**, 059901 (2007)]. arXiv:0704.2807 [hep-ph]
22. Abe, H., Choi, K.S., Kobayashi, T., Ohki, H.: Nucl. Phys. B **820**, 317 (2009). arXiv:0904.2631 [hep-ph]
23. Abe, H., Choi, K.S., Kobayashi, T., Ohki, H.: Phys. Rev. D **80**, 126006 (2009). arXiv:0907.5274 [hep-th]
24. Abe, H., Choi, K.S., Kobayashi, T., Ohki, H.: arXiv:1001.1788 [hep-th]
25. de Medeiros Varzielas, I., King, S.F., Ross, G.G.: Phys. Lett. B **644**, 153 (2007). arXiv:hep-ph/0512313
26. Adulpravitchai, A., Blum, A., Lindner, M.: J. High Energy Phys. **0909**, 018 (2009). arXiv:0907.2332 [hep-ph]
27. Frampton, P.H., Kephart, T.W., Rohm, R.M.: Phys. Lett. B **679**, 478 (2009). arXiv:0904.0420 [hep-ph]
28. Ramond, P.: Group Theory: A Physicist's Survey. Cambridge University Press, Cambridge (2010)
29. Miller, G.A., Dickson, H.F., Blichfeldt, L.E.: Theory and Applications of Finite Groups. Wiley, New York (1916)
30. Hamermesh, M.: Group Theory and Its Application to Physical Problems. Addison-Wesley, Reading (1962)
31. Fairbairn, W.M., Fulton, T., Klink, W.H.: J. Math. Phys. **5**, 1038 (1964)
32. Georgi, H.: Front. Phys. **54**, 1 (1982)
33. Frampton, P.H., Kephart, T.W.: Int. J. Mod. Phys. A **10**, 4689 (1995). arXiv:hep-ph/9409330
34. Ludl, P.O.: arXiv:0907.5587 [hep-ph]
35. Pakvasa, S., Sugawara, H.: Phys. Lett. B **73**, 61 (1978)
36. Tanimoto, M.: Phys. Rev. D **41**, 1586 (1990)
37. Lee, C.E., Lin, C.L., Yang, Y.W.: The minimal extension of the standard model with $S(3)$ symmetry. Prepared for 2nd International Spring School on Medium and High-Energy Nuclear Physics, Taipei, 8–12 May 1990

References

38. Kang, K., Kim, J.E., Ko, P.: Z. Phys. C **72**, 671 (1996). arXiv:hep-ph/9503436
39. Kang, K., Kang, S.K., Kim, J.E., Ko, P.W.: Mod. Phys. Lett. A **12**, 1175 (1997). arXiv:hep-ph/9611369
40. Fukugita, M., Tanimoto, M., Yanagida, T.: Phys. Rev. D **57**, 4429 (1998). arXiv:hep-ph/9709388
41. Tanimoto, M., Watari, T., Yanagida, T.: Phys. Lett. B **461**, 345 (1999). arXiv:hep-ph/9904338
42. Tanimoto, M.: Phys. Lett. B **483**, 417 (2000). arXiv:hep-ph/0001306
43. Tanimoto, M.: Phys. Rev. D **59**, 017304 (1999). arXiv:hep-ph/9807283
44. Tanimoto, M., Yanagida, T.: Phys. Lett. B **633**, 567 (2006). arXiv:hep-ph/0511336
45. Honda, M., Tanimoto, M.: Phys. Rev. D **75**, 096005 (2007). arXiv:hep-ph/0701083
46. Dong, P.V., Long, H.N., Nam, C.H., Vien, V.V.: arXiv:1111.6360 [hep-ph]
47. Xing, Z.Z., Yang, D., Zhou, S.: Phys. Lett. B **690**, 304–310 (2010). arXiv:1004.4234 [hep-ph]
48. Dicus, D.A., Ge, S.-F., Repko, W.W.: Phys. Rev. D **82**, 033005 (2010). arXiv:1004.3266 [hep-ph]
49. Meloni, D., Morisi, S., Peinado, E.: J. Phys. G **38**, 015003 (2011). arXiv:1005.3482 [hep-ph]
50. Jora, R., Schechter, J., Shahid, M.N.: Phys. Rev. D **82**, 053006 (2010). arXiv:1006.3307 [hep-ph]
51. Bhattacharyya, G., Leser, P., Pas, H.: Phys. Rev. D **83**, 011701 (2011). arXiv:1006.5597 [hep-ph]
52. Watanabe, A., Yoshioka, K.: Prog. Theor. Phys. **125**, 129–148 (2011). arXiv:1007.1527 [hep-ph]
53. Kaneko, T., Sugawara, H.: Phys. Lett. B **697**, 329–332 (2011). arXiv:1011.5748 [hep-ph]
54. Adulpravitchai, A., Batell, B., Pradler, J.: Phys. Lett. B **700**, 207–216 (2011). arXiv:1103.3053 [hep-ph]
55. Teshima, T., Okumura, Y.: Phys. Rev. D **84**, 016003 (2011). arXiv:1103.6127 [hep-ph]
56. Dev, S., Gupta, S., Gautam, R.R.: Phys. Lett. B **702**, 28–33 (2011). arXiv:1106.3873 [hep-ph]
57. Zhou, S.: Phys. Lett. B **704**, 291–295 (2011). arXiv:1106.4808 [hep-ph]
58. Koide, Y.: Phys. Rev. D **60**, 077301 (1999). arXiv:hep-ph/9905416
59. Chalut, K., Cheng, H., Frampton, P.H., Stowe, K., Yoshikawa, T.: Mod. Phys. Lett. A **17**, 1513 (2002). arXiv:hep-ph/0204074
60. Kubo, J., Mondragon, A., Mondragon, M., Rodriguez-Jauregui, E.: Prog. Theor. Phys. **109**, 795 (2003) [Erratum ibid. **114**, 287 (2005)]. arXiv:hep-ph/0302196
61. Kobayashi, T., Kubo, J., Terao, H.: Phys. Lett. B **568**, 83 (2003). arXiv:hep-ph/0303084
62. Kubo, J.: Phys. Lett. B **578**, 156 (2004) [Erratum ibid. B **619**, 387 (2005)]. arXiv:hep-ph/0309167
63. Chen, S.L., Frigerio, M., Ma, E.: Phys. Rev. D **70**, 073008 (2004) [Erratum ibid. D **70**, 079905 (2004)]. arXiv:hep-ph/0404084
64. Choi, K.Y., Kajiyama, Y., Lee, H.M., Kubo, J.: Phys. Rev. D **70**, 055004 (2004). arXiv:hep-ph/0402026
65. Kubo, J., Okada, H., Sakamaki, F.: Phys. Rev. D **70**, 036007 (2004). arXiv:hep-ph/0402089
66. Lavoura, L., Ma, E.: Mod. Phys. Lett. A **20**, 1217 (2005). arXiv:hep-ph/0502181
67. Grimus, W., Lavoura, L.: J. High Energy Phys. **0508**, 013 (2005). arXiv:hep-ph/0504153
68. Morisi, S., Picariello, M.: Int. J. Theor. Phys. **45**, 1267 (2006). arXiv:hep-ph/0505113
69. Teshima, T.: Phys. Rev. D **73**, 045019 (2006). arXiv:hep-ph/0509094
70. Koide, Y.: Phys. Rev. D **73**, 057901 (2006). arXiv:hep-ph/0509214
71. Kimura, T.: Prog. Theor. Phys. **114**, 329 (2005)
72. Araki, T., Kubo, J., Paschos, E.A.: Eur. Phys. J. C **45**, 465 (2006). arXiv:hep-ph/0502164
73. Kubo, J., Mondragon, A., Mondragon, M., Rodriguez-Jauregui, E., Felix-Beltran, O., Peinado, E.: J. Phys. Conf. Ser. **18**, 380 (2005)
74. Haba, N., Yoshioka, K.: Nucl. Phys. B **739**, 254 (2006). arXiv:hep-ph/0511108
75. Kim, J.E., Park, J.C.: J. High Energy Phys. **0605**, 017 (2006). arXiv:hep-ph/0512130

76. Kaneko, S., Sawanaka, H., Shingai, T., Tanimoto, M., Yoshioka, K.: Prog. Theor. Phys. **117**, 161 (2007). arXiv:hep-ph/0609220
77. Mohapatra, R.N., Nasri, S., Yu, H.B.: Phys. Lett. B **639**, 318 (2006). arXiv:hep-ph/0605020
78. Picariello, M.: Int. J. Mod. Phys. A **23**, 4435 (2008). arXiv:hep-ph/0611189
79. Koide, Y.: Eur. Phys. J. C **50**, 809 (2007). arXiv:hep-ph/0612058
80. Haba, N., Watanabe, A., Yoshioka, K.: Phys. Rev. Lett. **97**, 041601 (2006). arXiv:hep-ph/0603116
81. Morisi, S.: Int. J. Mod. Phys. A **22**, 2921 (2007)
82. Chen, C.Y., Wolfenstein, L.: Phys. Rev. D **77**, 093009 (2008). arXiv:0709.3767 [hep-ph]
83. Mitra, M., Choubey, S.: Phys. Rev. D **78**, 115014 (2008). arXiv:0806.3254 [hep-ph]
84. Feruglio, F., Lin, Y.: Nucl. Phys. B **800**, 77 (2008). arXiv:0712.1528 [hep-ph]
85. Jora, R., Schechter, J., Naeem Shahid, M.: Phys. Rev. D **80**, 093007 (2009). arXiv:0909.4414 [hep-ph]
86. Beltran, O.F., Mondragon, M., Rodriguez-Jauregui, E.: J. Phys. Conf. Ser. **171**, 012028 (2009)
87. Yamanaka, Y., Sugawara, H., Pakvasa, S.: Phys. Rev. D **25**, 1895 (1982) [Erratum ibid. D **29**, 2135 (1984)]
88. Brown, T., Pakvasa, S., Sugawara, H., Yamanaka, Y.: Phys. Rev. D **30**, 255 (1984)
89. Brown, T., Deshpande, N., Pakvasa, S., Sugawara, H.: Phys. Lett. B **141**, 95 (1984)
90. Ma, E.: Phys. Lett. B **632**, 352 (2006). arXiv:hep-ph/0508231
91. Zhang, H.: Phys. Lett. B **655**, 132 (2007). arXiv:hep-ph/0612214
92. Hagedorn, C., Lindner, M., Mohapatra, R.N.: J. High Energy Phys. **0606**, 042 (2006). arXiv:hep-ph/0602244
93. Cai, Y., Yu, H.B.: Phys. Rev. D **74**, 115005 (2006). arXiv:hep-ph/0608022
94. Caravaglios, F., Morisi, S.: Int. J. Mod. Phys. A **22**, 2469 (2007). arXiv:hep-ph/0611078
95. Koide, Y.: J. High Energy Phys. **0708**, 086 (2007). arXiv:0705.2275 [hep-ph]
96. Parida, M.K.: Phys. Rev. D **78**, 053004 (2008). arXiv:0804.4571 [hep-ph]
97. Lam, C.S.: Phys. Rev. D **78**, 073015 (2008). arXiv:0809.1185 [hep-ph]
98. Bazzocchi, F., Morisi, S.: Phys. Rev. D **80**, 096005 (2009). arXiv:0811.0345 [hep-ph]
99. Ishimori, H., Shimizu, Y., Tanimoto, M.: Prog. Theor. Phys. **121**, 769 (2009). arXiv:0812.5031 [hep-ph]
100. Bazzocchi, F., Merlo, L., Morisi, S.: Nucl. Phys. B **816**, 204 (2009). arXiv:0901.2086 [hep-ph]
101. Bazzocchi, F., Merlo, L., Morisi, S.: Phys. Rev. D **80**, 053003 (2009). arXiv:0902.2849 [hep-ph]
102. Altarelli, G., Feruglio, F., Merlo, L.: J. High Energy Phys. **0905**, 020 (2009). arXiv:0903.1940 [hep-ph]
103. Grimus, W., Lavoura, L., Ludl, P.O.: J. Phys. G **36**, 115007 (2009). arXiv:0906.2689 [hep-ph]
104. Ding, G.J.: Nucl. Phys. B **827**, 82 (2010). arXiv:0909.2210 [hep-ph]
105. Merlo, L.: arXiv:0909.2760 [hep-ph]
106. Daikoku, Y., Okada, H.: arXiv:0910.3370 [hep-ph]
107. Meloni, D.: arXiv:0911.3591 [hep-ph]
108. Morisi, S., Peinado, E.: arXiv:1001.2265 [hep-ph]
109. de Adelhart Toorop, R., Bazzocchi, F., Merlo, L.: J. High Energy Phys. **1008**, 001 (2010). arXiv:1003.4502 [hep-ph]
110. Hagedorn, C., King, S.F., Luhn, C.: J. High Energy Phys. **1006**, 048 (2010). arXiv:1003.4249 [hep-ph]
111. Ishimori, H., Saga, K., Shimizu, Y., Tanimoto, M.: Phys. Rev. D **81**, 115009 (2010). arXiv:1004.5004 [hep-ph]
112. Ahn, Y.H., Kang, S.K., Kim, C.S., Nguyen, T.P.: Phys. Rev. D **82**, 093005 (2010). arXiv:1004.3469 [hep-ph]
113. Ding, G.-J.: Nucl. Phys. B **846**, 394–428 (2011). arXiv:1006.4800 [hep-ph]
114. Daikoku, Y., Okada, H.: arXiv:1008.0914 [hep-ph]

115. Dong, P.V., Long, H.N., Soa, D.V., Vien, V.V.: Eur. Phys. J. C **71**, 1544 (2011). arXiv:1009.2328 [hep-ph]
116. Ishimori, H., Shimizu, Y., Tanimoto, M., Watanabe, A.: Phys. Rev. D **83**, 033004 (2011). arXiv:1010.3805 [hep-ph]
117. Parida, M.K., Sahu, P.K., Bora, K.: Phys. Rev. D **83**, 093004 (2011). arXiv:1011.4577 [hep-ph]
118. Daikoku, Y., Okada, H., Toma, T.: arXiv:1010.4963 [hep-ph]
119. Ding, G.-J., Pan, D.-M.: Eur. Phys. J. C **71**, 1716 (2011). arXiv:1011.5306 [hep-ph]
120. Ishimori, H., Tanimoto, M.: Prog. Theor. Phys. **125**, 653–675 (2011). arXiv:1012.2232 [hep-ph]
121. Stech, B.: arXiv:1012.6028 [hep-ph]
122. Park, N.W., Nam, K.H., Siyeon, K.: Phys. Rev. D **83**, 056013 (2011). arXiv:1101.4134 [hep-ph]
123. Ishimori, H., Kajiyama, Y., Shimizu, Y., Tanimoto, M.: arXiv:1103.5705 [hep-ph]
124. Yang, R.-Z., Zhang, H.: Phys. Lett. B **700**, 316–321 (2011). arXiv:1104.0380 [hep-ph]
125. Morisi, S., Peinado, E.: Phys. Lett. B **701**, 451–457 (2011). arXiv:1104.4961 [hep-ph]
126. Zhao, Z.-h.: Phys. Lett. B **701**, 609–613 (2011). arXiv:1106.2715 [hep-ph]
127. Ishimori, H., Kobayashi, T.: arXiv:1106.3604 [hep-ph]
128. Hagedorn, C., Serone, M.: arXiv:1106.4021 [hep-ph]
129. Daikoku, Y., Okada, H., Toma, T.: arXiv:1106.4717 [hep-ph]
130. Meloni, D.: arXiv:1107.0221 [hep-ph]
131. Morisi, S., Patel, K.M., Peinado, E.: arXiv:1107.0696 [hep-ph]
132. Bhupal Dev, P.S., Mohapatra, R.N., Severson, M.: arXiv:1107.2378 [hep-ph]
133. Ma, E., Rajasekaran, G.: Phys. Rev. D **64**, 113012 (2001). arXiv:hep-ph/0106291
134. Babu, K.S., Enkhbat, T., Gogoladze, I.: Phys. Lett. B **555**, 238 (2003). arXiv:hep-ph/0204246
135. Babu, K.S., Ma, E., Valle, J.W.F.: Phys. Lett. B **552**, 207 (2003). arXiv:hep-ph/0206292
136. Ma, E.: Mod. Phys. Lett. A **17**, 2361 (2002). arXiv:hep-ph/0211393
137. Babu, K.S., Kobayashi, T., Kubo, J.: Phys. Rev. D **67**, 075018 (2003). arXiv:hep-ph/0212350
138. Hirsch, M., Romao, J.C., Skadhauge, S., Valle, J.W.F., Villanova del Moral, A.: Phys. Rev. D **69**, 093006 (2004). arXiv:hep-ph/0312265
139. Ma, E.: Phys. Rev. D **70**, 031901 (2004). arXiv:hep-ph/0404199
140. Altarelli, G., Feruglio, F.: Nucl. Phys. B **720**, 64 (2005). arXiv:hep-ph/0504165
141. Chen, S.L., Frigerio, M., Ma, E.: Nucl. Phys. B **724**, 423 (2005). arXiv:hep-ph/0504181
142. Zee, A.: Phys. Lett. B **630**, 58 (2005). arXiv:hep-ph/0508278
143. Ma, E.: Mod. Phys. Lett. A **20**, 2601 (2005). arXiv:hep-ph/0508099
144. Ma, E.: Phys. Lett. B **632**, 352 (2006). arXiv:hep-ph/0508231
145. Ma, E.: Phys. Rev. D **73**, 057304 (2006). arXiv:hep-ph/0511133
146. Altarelli, G., Feruglio, F.: Nucl. Phys. B **741**, 215 (2006). arXiv:hep-ph/0512103
147. He, X.G., Keum, Y.Y., Volkas, R.R.: J. High Energy Phys. **0604**, 039 (2006). arXiv:hep-ph/0601001
148. Adhikary, B., Brahmachari, B., Ghosal, A., Ma, E., Parida, M.K.: Phys. Lett. B **638**, 345 (2006). arXiv:hep-ph/0603059
149. Ma, E., Sawanaka, H., Tanimoto, M.: Phys. Lett. B **641**, 301 (2006). arXiv:hep-ph/0606103
150. Valle, J.W.F.: J. Phys. Conf. Ser. **53**, 473 (2006). arXiv:hep-ph/0608101
151. Adhikary, B., Ghosal, A.: Phys. Rev. D **75**, 073020 (2007). arXiv:hep-ph/0609193
152. Lavoura, L., Kuhbock, H.: Mod. Phys. Lett. A **22**, 181 (2007). arXiv:hep-ph/0610050
153. King, S.F., Malinsky, M.: Phys. Lett. B **645**, 351 (2007). arXiv:hep-ph/0610250
154. Hirsch, M., Joshipura, A.S., Kaneko, S., Valle, J.W.F.: Phys. Rev. Lett. **99**, 151802 (2007). arXiv:hep-ph/0703046
155. Bazzocchi, F., Kaneko, S., Morisi, S.: J. High Energy Phys. **0803**, 063 (2008). arXiv:0707.3032 [hep-ph]
156. Grimus, W., Kuhbock, H.: Phys. Rev. D **77**, 055008 (2008). arXiv:0710.1585 [hep-ph]
157. Honda, M., Tanimoto, M.: Prog. Theor. Phys. **119**, 583 (2008). arXiv:0801.0181 [hep-ph]

158. Brahmachari, B., Choubey, S., Mitra, M.: Phys. Rev. D **77**, 073008 (2008) [Erratum ibid. D **77**, 119901 (2008)]. arXiv:0801.3554 [hep-ph]
159. Adhikary, B., Ghosal, A.: Phys. Rev. D **78**, 073007 (2008). arXiv:0803.3582 [hep-ph]
160. Fukuyama, T.: arXiv:0804.2107 [hep-ph]
161. Lin, Y.: Nucl. Phys. B **813**, 91 (2009). arXiv:0804.2867 [hep-ph]
162. Frampton, P.H., Matsuzaki, S.: arXiv:0806.4592 [hep-ph]
163. Feruglio, F., Hagedorn, C., Lin, Y., Merlo, L.: Nucl. Phys. B **809**, 218 (2009). arXiv:0807.3160 [hep-ph]
164. Morisi, S.: Nuovo Cimento B **123**, 886 (2008). arXiv:0807.4013 [hep-ph]
165. Ishimori, H., Kobayashi, T., Omura, Y., Tanimoto, M.: J. High Energy Phys. **0812**, 082 (2008). arXiv:0807.4625 [hep-ph]
166. Ma, E.: Phys. Lett. B **671**, 366 (2009). arXiv:0808.1729 [hep-ph]
167. Bazzocchi, F., Frigerio, M., Morisi, S.: Phys. Rev. D **78**, 116018 (2008). arXiv:0809.3573 [hep-ph]
168. Hirsch, M., Morisi, S., Valle, J.W.F.: Phys. Rev. D **79**, 016001 (2009). arXiv:0810.0121 [hep-ph]
169. Merlo, L.: arXiv:0811.3512 [hep-ph]
170. Baek, S., Oh, M.C.: arXiv:0812.2704 [hep-ph]
171. Morisi, S.: Phys. Rev. D **79**, 033008 (2009). arXiv:0901.1080 [hep-ph]
172. Ciafaloni, P., Picariello, M., Torrente-Lujan, E., Urbano, A.: Phys. Rev. D **79**, 116010 (2009). arXiv:0901.2236 [hep-ph]
173. Merlo, L.: J. Phys. Conf. Ser. **171**, 012083 (2009). arXiv:0902.3067 [hep-ph]
174. Chen, M.C., King, S.F.: J. High Energy Phys. **0906**, 072 (2009). arXiv:0903.0125 [hep-ph]
175. Branco, G.C., Gonzalez Felipe, R., Rebelo, M.N., Serodio, H.: Phys. Rev. D **79**, 093008 (2009). arXiv:0904.3076 [hep-ph]
176. Hayakawa, A., Ishimori, H., Shimizu, Y., Tanimoto, M.: Phys. Lett. B **680**, 334 (2009). arXiv:0904.3820 [hep-ph]
177. Altarelli, G., Meloni, D.: J. Phys. G **36**, 085005 (2009). arXiv:0905.0620 [hep-ph]
178. Urbano, A.: arXiv:0905.0863 [hep-ph]
179. Hirsch, M., Morisi, S., Valle, J.W.F.: Phys. Lett. B **679**, 454 (2009). arXiv:0905.3056 [hep-ph]
180. Lin, Y.: Nucl. Phys. B **824**, 95 (2010). arXiv:0905.3534 [hep-ph]
181. Hirsch, M.: Pramana **72**, 183 (2009)
182. Hagedorn, C., Molinaro, E., Petcov, S.T.: J. High Energy Phys. **0909**, 115 (2009). arXiv:0908.0240 [hep-ph]
183. Tamii, A., et al.: Mod. Phys. Lett. A **24**, 867 (2009)
184. Ma, E.: arXiv:0908.3165 [hep-ph]
185. Burrows, T.J., King, S.F.: arXiv:0909.1433 [hep-ph]
186. Albaid, A.: Phys. Rev. D **80**, 093002 (2009). arXiv:0909.1762 [hep-ph]
187. Merlo, L.: Nucl. Phys. Proc. Suppl. **188**, 345 (2009)
188. Ciafaloni, P., Picariello, M., Torrente-Lujan, E., Urbano, A.: arXiv:0909.2553 [hep-ph]
189. Feruglio, F., Hagedorn, C., Merlo, L.: arXiv:0910.4058 [hep-ph]
190. Morisi, S., Peinado, E.: arXiv:0910.4389 [hep-ph]
191. Berger, J., Grossman, Y.: arXiv:0910.4392 [hep-ph]
192. Hagedorn, C., Molinaro, E., Petcov, S.T.: arXiv:0911.3605 [hep-ph]
193. Feruglio, F., Hagedorn, C., Lin, Y., Merlo, L.: arXiv:0911.3874 [hep-ph]
194. Ding, G.J., Liu, J.F.: arXiv:0911.4799 [hep-ph]
195. Barry, J., Rodejohann, W.: arXiv:1003.2385 [hep-ph]
196. Cooper, I.K., King, S.F., Luhn, C.: Phys. Lett. B **690**, 396–402 (2010). arXiv:1004.3243 [hep-ph]
197. Albright, C.H., Dueck, A., Rodejohann, W.: Eur. Phys. J. C **70**, 1099–1110 (2010). arXiv:1004.2798 [hep-ph]
198. Riva, F.: Phys. Lett. B **690**, 443–450 (2010). arXiv:1004.1177 [hep-ph]
199. Kadosh, A., Pallante, E.: J. High Energy Phys. **1008**, 115 (2010). arXiv:1004.0321 [hep-ph]

200. Feruglio, F., Paris, A.: Nucl. Phys. B **840**, 405–423 (2010). arXiv:1005.5526 [hep-ph]
201. Fukuyama, T., Sugiyama, H., Tsumura, K.: Phys. Rev. D **82**, 036004 (2010). arXiv:1005.5338 [hep-ph]
202. Antusch, S., King, S.F., Spinrath, M.: Phys. Rev. D **83**, 013005 (2011). arXiv:1005.0708 [hep-ph]
203. Ahn, Y.H.: arXiv:1006.2953 [hep-ph]
204. Ma, E.: Phys. Rev. D **82**, 037301 (2010). arXiv:1006.3524 [hep-ph]
205. Hirsch, M., Morisi, S., Peinado, E., Valle, J.W.F.: Phys. Rev. D **82**, 116003 (2010). arXiv:1007.0871 [hep-ph]
206. Esteves, J.N., Joaquim, F.R., Joshipura, A.S., Romao, J.C., Tortola, M.A., Valle, J.W.F.: Phys. Rev. D **82**, 073008 (2010). arXiv:1007.0898 [hep-ph]
207. Burrows, T.J., King, S.F.: Nucl. Phys. B **842**, 107–121 (2011). arXiv:1007.2310 [hep-ph]
208. Haba, N., Kajiyama, Y., Matsumoto, S., Okada, H., Yoshioka, K.: Phys. Lett. B **695**, 476–481 (2011). arXiv:1008.4777 [hep-ph]
209. Araki, T., Mei, J., Xing, Z.-Z.: Phys. Lett. B **695**, 165–168 (2011). arXiv:1010.3065 [hep-ph]
210. Meloni, D., Morisi, S., Peinado, E.: Phys. Lett. B **697**, 339–342 (2011). arXiv:1011.1371 [hep-ph]
211. Machado, A.C.B., Montero, J.C., Pleitez, V.: Phys. Lett. B **697**, 318–322 (2011). arXiv:1011.5855 [hep-ph]
212. Carone, C.D., Lebed, R.F.: Phys. Lett. B **696**, 454–458 (2011). arXiv:1011.6379 [hep-ph]
213. de Medeiros Varzielas, I., Merlo, L.: J. High Energy Phys. **1102**, 062 (2011). arXiv:1011.6662 [hep-ph]
214. de Adelhart Toorop, R., Bazzocchi, F., Merlo, L., Paris, A.: J. High Energy Phys. **1103**, 035 (2011). arXiv:1012.1791 [hep-ph]
215. de Adelhart Toorop, R., Bazzocchi, F., Merlo, L., Paris, A.: J. High Energy Phys. **1103**, 040 (2011). arXiv:1012.2091 [hep-ph]
216. Fukuyama, T., Sugiyama, H., Tsumura, K.: Phys. Rev. D **83**, 056016 (2011). arXiv:1012.4886 [hep-ph]
217. de Medeiros Varzielas, I., Gonzalez Felipe, R., Serodio, H.: Phys. Rev. D **83**, 033007 (2011). arXiv:1101.0602 [hep-ph]
218. Boucenna, M.S., Hirsch, M., Morisi, S., Peinado, E., Taoso, M., Valle, J.W.F.: J. High Energy Phys. **1105**, 037 (2011). arXiv:1101.2874 [hep-ph]
219. Kadosh, A., Pallante, E.: J. High Energy Phys. **1106**, 121 (2011). arXiv:1101.5420 [hep-ph]
220. Ahn, Y.H., Cheng, H.-Y., Oh, S.: Phys. Rev. D **83**, 076012 (2011). arXiv:1102.0879 [hep-ph]
221. Ahn, Y.H., Kim, C.S., Oh, S.: arXiv:1103.0657 [hep-ph]
222. Morisi, S., Peinado, E., Shimizu, Y., Valle, J.W.F.: Phys. Rev. D **84**, 036003 (2011). arXiv:1104.1633 [hep-ph]
223. Toorop, R.d.A., Bazzocchi, F., Morisi, S.: arXiv:1104.5676 [hep-ph]
224. Shimizu, Y., Tanimoto, M., Watanabe, A.: Prog. Theor. Phys. **126**, 81–90 (2011). arXiv:1105.2929 [hep-ph]
225. Barry, J., Rodejohann, W., Zhang, H.: J. High Energy Phys. **1107**, 091 (2011). arXiv:1105.3911 [hep-ph]
226. Ma, E., Wegman, D.: Phys. Rev. Lett. **107**, 061803 (2011). arXiv:1106.4269 [hep-ph]
227. Albaid, A.: arXiv:1106.4070 [hep-ph]
228. Adulpravitchai, A., Takahashi, R.: arXiv:1107.3829 [hep-ph]
229. King, S.F., Luhn, C.: J. High Energy Phys. **1109**, 042 (2011). arXiv:1107.5332 [hep-ph]
230. Machado, A.C.B., Montero, J.C., Pleitez, V.: arXiv:1108.1767 [hep-ph]
231. Ding, G.-J., Meloni, D.: arXiv:1108.2733 [hep-ph]
232. Aristizabal Sierra, D., Bazzocchi, F.: arXiv:1110.3781 [hep-ph]
233. Cooper, I.K., King, S.F., Luhn, C.: arXiv:1110.5676 [hep-ph]
234. Barry, J., Rodejohann, W., Zhang, H.: arXiv:1110.6382 [hep-ph]
235. Ferreira, P.M., Lavoura, L.: arXiv:1111.5859 [hep-ph]
236. King, S.F., Luhn, C.: arXiv:1112.1959 [hep-ph]
237. Ding, G.-J., Everett, L.L., Stuart, A.J.: arXiv:1110.1688 [hep-ph]

238. Chen, C.-S., Kephart, T.W., Yuan, T.-C.: J. High Energy Phys. **1104**, 015 (2011). arXiv:1011.3199 [hep-ph]
239. Feruglio, F., Paris, A.: J. High Energy Phys. **1103**, 101 (2011). arXiv:1101.0393 [hep-ph]
240. Shirai, K.: J. Phys. Soc. Jpn. **61**, 2735 (1992)
241. Luhn, C., Nasri, S., Ramond, P.: J. Math. Phys. **48**, 123519 (2007). arXiv:0709.1447 [hep-th]
242. Luhn, C., Ramond, P.: J. Math. Phys. **49**, 053525 (2008). arXiv:0803.0526 [hep-th]
243. Everett, L.L., Stuart, A.J.: Phys. Rev. D **79**, 085005 (2009). arXiv:0812.1057 [hep-ph]
244. Aranda, A., Carone, C.D., Lebed, R.F.: Phys. Lett. B **474**, 170 (2000). arXiv:hep-ph/9910392
245. Aranda, A., Carone, C.D., Lebed, R.F.: Phys. Rev. D **62**, 016009 (2000). arXiv:hep-ph/0002044
246. Carr, P.D., Frampton, P.H.: arXiv:hep-ph/0701034
247. Feruglio, F., Hagedorn, C., Lin, Y., Merlo, L.: Nucl. Phys. B **775**, 120 (2007). arXiv:hep-ph/0702194
248. Chen, M.C., Mahanthappa, K.T.: Phys. Lett. B **652**, 34 (2007). arXiv:0705.0714 [hep-ph]
249. Frampton, P.H., Kephart, T.W.: J. High Energy Phys. **0709**, 110 (2007). arXiv:0706.1186 [hep-ph]
250. Aranda, A.: Phys. Rev. D **76**, 111301 (2007). arXiv:0707.3661 [hep-ph]
251. Ding, G.J.: Phys. Rev. D **78**, 036011 (2008). arXiv:0803.2278 [hep-ph]
252. Frampton, P.H., Kephart, T.W., Matsuzaki, S.: Phys. Rev. D **78**, 073004 (2008). arXiv:0807.4713 [hep-ph]
253. Eby, D.A., Frampton, P.H., Matsuzaki, S.: Phys. Lett. B **671**, 386 (2009). arXiv:0810.4899 [hep-ph]
254. Frampton, P.H., Matsuzaki, S.: Phys. Lett. B **679**, 347 (2009). arXiv:0902.1140 [hep-ph]
255. Chen, M.C., Mahanthappa, K.T., Yu, F.: arXiv:0909.5472 [hep-ph]
256. Frampton, P.H., Ho, C.M., Kephart, T.W., Matsuzaki, S.: Phys. Rev. D **82**, 113007 (2010). arXiv:1009.0307 [hep-ph]
257. Aranda, A., Bonilla, C., Ramos, R., Rojas, A.D.: arXiv:1011.6470 [hep-ph]
258. BenTov, Y., Zee, A.: arXiv:1101.1987 [hep-ph]
259. Eby, D.A., Frampton, P.H., He, X.-G., Kephart, T.W.: Phys. Rev. D **84**, 037302 (2011). arXiv:1103.5737 [hep-ph]
260. Chen, M.-C., Mahanthappa, K.T.: arXiv:1107.3856 [hep-ph]
261. Merlo, L., Rigolin, S., Zaldivar, B.: arXiv:1108.1795 [hep-ph]
262. Chen, M.-C., Mahanthappa, K.T., Meroni, A., Petcov, S.T.: arXiv:1109.0731 [hep-ph]
263. Eby, D.A., Frampton, P.H.: arXiv:1112.2675 [hep-ph]
264. Bergshoeff, E., Janssen, B., Ortin, T.: Class. Quantum Gravity **13**, 321 (1996). arXiv:hep-th/9506156
265. Frampton, P.H., Kephart, T.W.: Phys. Rev. D **64**, 086007 (2001). arXiv:hep-th/0011186
266. Grimus, W., Lavoura, L.: Phys. Lett. B **572**, 189 (2003). arXiv:hep-ph/0305046
267. Grimus, W., Joshipura, A.S., Kaneko, S., Lavoura, L., Tanimoto, M.: J. High Energy Phys. **0407**, 078 (2004). arXiv:hep-ph/0407112
268. Grimus, W., Joshipura, A.S., Kaneko, S., Lavoura, L., Sawanaka, H., Tanimoto, M.: Nucl. Phys. B **713**, 151 (2005). arXiv:hep-ph/0408123
269. Blum, A., Mohapatra, R.N., Rodejohann, W.: Phys. Rev. D **76**, 053003 (2007). arXiv:0706.3801 [hep-ph]
270. Ishimori, H., Kobayashi, T., Ohki, H., Omura, Y., Takahashi, R., Tanimoto, M.: Phys. Lett. B **662**, 178 (2008). arXiv:0802.2310 [hep-ph]
271. Ishimori, H., Kobayashi, T., Ohki, H., Omura, Y., Takahashi, R., Tanimoto, M.: Phys. Rev. D **77**, 115005 (2008). arXiv:0803.0796 [hep-ph]
272. Adulpravitchai, A., Blum, A., Hagedorn, C.: J. High Energy Phys. **0903**, 046 (2009). arXiv:0812.3799 [hep-ph]
273. Hagedorn, C., Ziegler, R.: Phys. Rev. D **82**, 053011 (2010). arXiv:1007.1888 [hep-ph]
274. Meloni, D., Morisi, S., Peinado, E.: Phys. Lett. B **703**, 281–287 (2011). arXiv:1104.0178 [hep-ph]

References

275. Hagedorn, C., Lindner, M., Plentinger, F.: Phys. Rev. D **74**, 025007 (2006). arXiv:hep-ph/0604265
276. Kajiyama, Y., Kubo, J., Okada, H.: Phys. Rev. D **75**, 033001 (2007). arXiv:hep-ph/0610072
277. Blum, A., Hagedorn, C., Lindner, M.: Phys. Rev. D **77**, 076004 (2008). arXiv:0709.3450 [hep-ph]
278. Hagedorn, C., Lindner, M., Plentinger, F.: Phys. Rev. D **74**, 025007 (2006). arXiv:hep-ph/0604265
279. Blum, A., Hagedorn, C., Hohenegger, A.: J. High Energy Phys. **0803**, 070 (2008). arXiv:0710.5061 [hep-ph]
280. Blum, A., Hagedorn, C.: Nucl. Phys. B **821**, 327 (2009). arXiv:0902.4885 [hep-ph]
281. Adulpravitchai, A., Blum, A., Rodejohann, W.: New J. Phys. **11**, 063026 (2009). arXiv:0903.0531 [hep-ph]
282. Kim, J.E., Seo, M.-S.: J. High Energy Phys. **1102**, 097 (2011). arXiv:1005.4684 [hep-ph]
283. Kajiyama, Y., Okada, H., Toma, T.: Eur. Phys. J. C **71**, 1688 (2011). arXiv:1104.0367 [hep-ph]
284. Kim, J.E., Seo, M.-S.: arXiv:1106.6117 [hep-ph]
285. Kajiyama, Y., Okada, H., Toma, T.: arXiv:1109.2722 [hep-ph]
286. Babu, K.S., Kubo, J.: Phys. Rev. D **71**, 056006 (2005). arXiv:hep-ph/0411226
287. Kajiyama, Y., Itou, E., Kubo, J.: Nucl. Phys. B **743**, 74 (2006). arXiv:hep-ph/0511268
288. Itou, E., Kajiyama, Y., Kubo, J.: AIP Conf. Proc. **903**, 389 (2007). arXiv:hep-ph/0611052
289. Kawashima, K., Kubo, J., Lenz, A.: Phys. Lett. B **681**, 60 (2009). arXiv:0907.2302 [hep-ph]
290. Volkov, G.: arXiv:1006.5627 [math-ph]
291. Kubo, J., Lenz, A.: Phys. Rev. D **82**, 075001 (2010). arXiv:1007.0680 [hep-ph]
292. Hackett, J., Kauffman, L.: arXiv:1010.2979 [math-ph]
293. Furui, S.: AIP Conf. Proc. **1343**, 533–535 (2011). arXiv:1011.3086 [hep-ph]
294. Kaburaki, Y., Konya, K., Kubo, J., Lenz, A.: Phys. Rev. D **84**, 016007 (2011). arXiv:1012.2435 [hep-ph]
295. Babu, K.S., Kawashima, K., Kubo, J.: Phys. Rev. D **83**, 095008 (2011). arXiv:1103.1664 [hep-ph]
296. Araki, T., Li, Y.F.: arXiv:1112.5819 [hep-ph]
297. Frigerio, M., Kaneko, S., Ma, E., Tanimoto, M.: Phys. Rev. D **71**, 011901 (2005). arXiv:hep-ph/0409187
298. Frigerio, M., Ma, E.: Phys. Rev. D **76**, 096007 (2007). arXiv:0708.0166 [hep-ph]
299. Dev, S., Verma, S.: Mod. Phys. Lett. A **25**, 2837–2848 (2010). arXiv:1005.4521 [hep-ph]
300. Aranda, A., Bonilla, C., Ramos, R., Rojas, A.D.: Phys. Rev. D **84**, 016009 (2011). arXiv:1105.6373 [hep-ph]
301. Ma, E.: arXiv:0705.0327 [hep-ph]
302. Bovier, A., Luling, M., Wyler, D.: J. Math. Phys. **22**, 1536 (1981)
303. Bovier, A., Luling, M., Wyler, D.: J. Math. Phys. **22**, 1543 (1981)
304. Fairbairn, W.M., Fulton, T.: J. Math. Phys. **23**, 1747 (1982)
305. Branco, G.C., Gerard, J.M., Grimus, W.: Phys. Lett. B **136**, 383 (1984)
306. Ma, E.: Mod. Phys. Lett. A **21**, 1917 (2006). arXiv:hep-ph/0607056
307. Luhn, C., Nasri, S., Ramond, P.: J. Math. Phys. **48**, 073501 (2007). arXiv:hep-th/0701188
308. de Medeiros Varzielas, I., King, S.F., Ross, G.G.: Phys. Lett. B **648**, 201 (2007). arXiv:hep-ph/0607045
309. Ma, E.: Phys. Lett. B **660**, 505 (2008). arXiv:0709.0507 [hep-ph]
310. Grimus, W., Lavoura, L.: J. High Energy Phys. **0809**, 106 (2008). arXiv:0809.0226 [hep-ph]
311. Howl, R., King, S.F.: J. High Energy Phys. **0805**, 008 (2008). arXiv:0802.1909 [hep-ph]
312. King, S.F., Luhn, C.: J. High Energy Phys. **0910**, 093 (2009). arXiv:0908.1897 [hep-ph]
313. King, S.F.: J. High Energy Phys. **1009**, 114 (2010). arXiv:1006.5895 [hep-ph]
314. Luhn, C., Nasri, S., Ramond, P.: Phys. Lett. B **652**, 27 (2007). arXiv:0706.2341 [hep-ph]
315. Hagedorn, C., Schmidt, M.A., Smirnov, A.Y.: Phys. Rev. D **79**, 036002 (2009). arXiv:0811.2955 [hep-ph]

316. Cao, Q.-H., Khalil, S., Ma, E., Okada, H.: Phys. Rev. Lett. **106**, 131801 (2011). arXiv:1009.5415 [hep-ph]
317. Ma, E.: Mod. Phys. Lett. A **26**, 377–385 (2011). arXiv:1101.4972 [hep-ph]
318. Cao, Q.-H., Khalil, S., Ma, E., Okada, H.: arXiv:1108.0570 [hep-ph]
319. Kajiyama, Y., Okada, H.: Nucl. Phys. B **848**, 303–313 (2011). arXiv:1011.5753 [hep-ph]
320. Parattu, K.M., Wingerter, A.: Phys. Rev. D **84**, 013011 (2011). arXiv:1012.2842 [hep-ph]
321. Ding, G.-J.: Nucl. Phys. B **853**, 635–662 (2011). arXiv:1105.5879 [hep-ph]
322. Hartmann, C., Zee, A.: Nucl. Phys. B **853**, 105–124 (2011). arXiv:1106.0333 [hep-ph]
323. Hartmann, C.: arXiv:1109.5143 [hep-ph]
324. Ishimori, H., Kobayashi, T.: arXiv:1201.3429 [hep-ph]
325. Escobar, J.A., Luhn, C.: J. Math. Phys. **50**, 013524 (2009). arXiv:0809.0639 [hep-th]
326. Ishimori, H., Kobayashi, T., Okada, H., Shimizu, Y., Tanimoto, M.: J. High Energy Phys. **0904**, 011 (2009). arXiv:0811.4683 [hep-ph]
327. Ishimori, H., Kobayashi, T., Okada, H., Shimizu, Y., Tanimoto, M.: J. High Energy Phys. **0912**, 054 (2009). arXiv:0907.2006 [hep-ph]
328. Escobar, J.A.: arXiv:1102.1649 [hep-ph]
329. Varzielas, I.d.M., Emmanuel-Costa, D.: arXiv:1106.5477 [hep-ph]
330. Toorop, R.d.A., Feruglio, F., Hagedorn, C.: Phys. Lett. B **703**, 447–451 (2011). arXiv:1107.3486 [hep-ph]
331. Krauss, L.M., Wilczek, F.: Phys. Rev. Lett. **62**, 1221 (1989)
332. Ibáñez, L.E., Ross, G.G.: Phys. Lett. B **260**, 291–295 (1991)
333. Banks, T., Dine, M.: Phys. Rev. D **45**, 1424–1427 (1992). arXiv:hep-th/9109045
334. Dine, M., Graesser, M.: J. High Energy Phys. **01**, 038 (2005). arXiv:hep-th/0409209
335. Csaki, C., Murayama, H.: Nucl. Phys. B **515**, 114–162 (1998). arXiv:hep-th/9710105
336. Ibáñez, L.E., Ross, G.G.: Nucl. Phys. B **368**, 3–37 (1992)
337. Ibáñez, L.E.: Nucl. Phys. B **398**, 301–318 (1993). arXiv:hep-ph/9210211
338. Babu, K.S., Gogoladze, I., Wang, K.: Nucl. Phys. B **660**, 322–342 (2003). arXiv:hep-ph/0212245
339. Dreiner, H.K., Luhn, C., Thormeier, M.: Phys. Rev. D **73**, 075007 (2006). arXiv:hep-ph/0512163
340. Araki, T.: Prog. Theor. Phys. **117**, 1119–1138 (2007). arXiv:hep-ph/0612306
341. Araki, T., Choi, K.S., Kobayashi, T., Kubo, J., Ohki, H.: Phys. Rev. D **76**, 066006 (2007). arXiv:0705.3075 [hep-ph]
342. Araki, T., Kobayashi, T., Kubo, J., Ramos-Sanchez, S., Ratz, M., Vaudrevange, P.K.S.: Nucl. Phys. B **805**, 124 (2008). arXiv:0805.0207 [hep-th]
343. Luhn, C., Ramond, P.: J. High Energy Phys. **0807**, 085 (2008). arXiv:0805.1736 [hep-ph]
344. Luhn, C.: Phys. Lett. B **670**, 390 (2009). arXiv:0807.1749 [hep-ph]
345. Merle, A., Zwicky, R.: arXiv:1110.4891 [hep-ph]
346. Grimus, W., Ludl, P.O.: J. Phys. A **43**, 445209 (2010). arXiv:1006.0098 [hep-ph]
347. Ludl, P.O.: J. Phys. A **43**, 395204 (2010). arXiv:1006.1479 [math-ph]
348. Ludl, P.O.: J. Phys. A **44**, 255204 (2011). arXiv:1101.2308 [math-ph]
349. Luhn, C.: J. High Energy Phys. **1103**, 108 (2011). arXiv:1101.2417 [hep-ph]
350. King, S.F., Luhn, C.: Nucl. Phys. B **820**, 269 (2009). arXiv:0905.1686 [hep-ph]
351. King, S.F., Luhn, C.: Nucl. Phys. B **832**, 414 (2010). arXiv:0912.1344 [hep-ph]
352. Everett, L.L., Stuart, A.J.: Phys. Lett. B **698**, 131–139 (2011). arXiv:1011.4928 [hep-ph]
353. Hashimoto, K., Okada, H.: arXiv:1110.3640 [hep-ph]
354. Chen, C.-S., Kephart, T.W., Yuan, T.-C.: arXiv:1110.6233 [hep-ph]
355. An, F.P., et al. (Daya Bay Collaboration): Phys. Rev. Lett. **108**, 171803 (2012). arXiv:1203.1669 [hep-ex]
356. Ahn, J.K., et al. (Reno Collaboration): Phys. Rev. Lett. **108**, 191802 (2012). arXiv:1204.0626 [hep-ex]

Chapter 2
Basics of Finite Groups

We start by introducing the basics of group theory, considering in particular finite groups. For pedagogical purposes, we shall use several theorems without proof, although proofs of useful theorems are given in Appendix A. (See also, e.g., [1–6].) On the other hand, we shall present several examples in order to obtain a clear understanding of these basic theorems.

A group G is a set with a product satisfying the following properties:

1. **Closure**
   If a and b are elements of the group G, then $c = ab$ is also an element of G.
2. **Associativity**
   $(ab)c = a(bc)$ for all $a, b, c \in G$.
3. **Identity**
   The group G includes an identity element e, which satisfies $ae = ea = a$ for any element $a \in G$.
4. **Inverse**
   The group G includes an inverse element a^{-1} for any element $a \in G$, such that $aa^{-1} = a^{-1}a = e$.

Let us present some simple examples.

Example (Cyclic Group Z_N) Discrete rotations of a complex plane form a group. Let us denote the $\exp[2\pi i/N]$ rotation by a. Then the $\exp[2\pi i m/N]$ rotation for $m =$ integer can be written a^m. The multiplication rule is defined such that $a^m a^n = a^{m+n}$. The operator a^N corresponds to the identity, $a^N = e$, and the inverse of a^m is obtained as a^{N-m}. Thus, the set

$$\{e, a, a^2, \ldots, a^{N-1}\} \qquad (2.1)$$

forms a group. Its closure and associativity should be obvious. This group is called the cyclic group Z_N.

Example (S_3 and S_N) All possible permutations among three objects, (x_1, x_2, x_3), form a group denoted by S_3. There are six permutations:

$$\begin{aligned} e &: (x_1, x_2, x_3) \rightarrow (x_1, x_2, x_3), \\ a_1 &: (x_1, x_2, x_3) \rightarrow (x_2, x_1, x_3), \\ a_2 &: (x_1, x_2, x_3) \rightarrow (x_3, x_2, x_1), \\ a_3 &: (x_1, x_2, x_3) \rightarrow (x_1, x_3, x_2), \\ a_4 &: (x_1, x_2, x_3) \rightarrow (x_3, x_1, x_2), \\ a_5 &: (x_1, x_2, x_3) \rightarrow (x_2, x_3, x_1). \end{aligned} \quad (2.2)$$

The element e is clearly the identity. Their products form a closed algebra, e.g.,

$$\begin{aligned} a_1 a_2 &: (x_1, x_2, x_3) \rightarrow (x_2, x_3, x_1), \\ a_2 a_1 &: (x_1, x_2, x_3) \rightarrow (x_3, x_1, x_2), \\ a_4 a_2 &: (x_1, x_2, x_3) \rightarrow (x_1, x_3, x_2), \end{aligned} \quad (2.3)$$

whence

$$a_1 a_2 = a_5, \quad a_2 a_1 = a_4, \quad a_4 a_2 = a_2 a_1 a_2 = a_3. \quad (2.4)$$

It is straightforward to check the closure rule for other products, as well as associativity and the presence of an inverse for each element.

Using the multiplication rules, one can write all six elements in terms of two proper elements and their products. For example, by defining $a_1 = a, a_2 = b$, all elements are written as

$$\{e, a, b, ab, ba, bab\}. \quad (2.5)$$

Note that $aba = bab$. The group S_3 is the symmetry group of an equilateral triangle, as shown in Fig. 2.1. The elements a and ab correspond to a reflection and the $2\pi/3$ rotation, respectively.

Similarly, all possible permutations among N objects x_i with $i = 1, \ldots, N$,

$$(x_1, \ldots, x_N) \rightarrow (x_{i_1}, \ldots, x_{i_N}), \quad (2.6)$$

form a group. This is denoted by S_N and contains $N!$ elements. It is often called the symmetric group.

The *order* of a group G is the number of elements in G. Obviously, the order of a finite group is finite. For example, the order of the group Z_N is N, while the order of the group S_N is $N!$.

A group G is said to be *Abelian* if all its elements commute with each other, i.e., $ab = ba$ for any elements a and b in G. If not all pairs of elements satisfy commutativity, the group is said to be *non-Abelian*. The group Z_N is Abelian, but S_3 and S_N ($N \geq 3$) are non-Abelian. For example, for S_3, we see that $a_1 a_2 \neq a_2 a_1$ in the above notation (2.2).

If a subset H of a group G is also a group, H is said to be a *subgroup* of G. The order of the subgroup H is always a divisor of the order of G. This is *Lagrange's theorem* (see Appendix A). If a subgroup N of G satisfies $g^{-1}Ng = N$ for any element $g \in G$, the subgroup N is called a *normal subgroup* or an *invariant subgroup*.

Fig. 2.1 The S_3 symmetry of an equilateral triangle

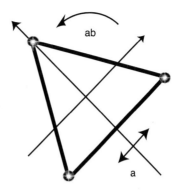

Any subgroup H and normal subgroup N of G satisfy $HN = NH$, where HN denotes

$$\{h_i n_j | h_i \in H, n_j \in N\}, \tag{2.7}$$

and NH has a similar meaning. Furthermore, HN is a subgroup of G.

Example For example, the three elements $\{e, ab, ba\}$ form a subgroup of S_3. Indeed, these elements correspond to even permutations, while the other elements, $\{a, b, bab\}$, correspond to odd permutations. This subgroup is nothing but the Z_3 group, because $(ab)^2 = ba$ and $(ab)^3 = e$. Lagrange's theorem implies that the order of any subgroup of S_3 must be equal to 1, 2, or 3, because the order of S_3 is 6 ($= 3 \times 2$). The subgroup of order 3 corresponds to the above Z_3. In addition, the S_3 group includes three subgroups of order 2, viz., $\{e, a\}$, $\{e, b\}$, and $\{e, bab\}$. These subgroups are Z_2 groups. Furthermore, it can be shown that the above Z_3 is a normal subgroup of S_3.

When $a^h = e$ for an element $a \in G$ and h is the smallest positive integer for which this is so, the number h is called the *order* of a. The elements $\{e, a, a^2, \ldots, a^{h-1}\}$ form a subgroup, which is the Abelian group Z_h of order h.

The elements $g^{-1}ag$ for $g \in G$ are called elements conjugate to the element a. The set containing all elements conjugate to an element a of G, i.e., $\{g^{-1}ag, \forall g \in G\}$, is called a *conjugacy class*. All elements in a conjugacy class have the same order since

$$(gag^{-1})^h = ga(g^{-1}g)a(g^{-1}g)\cdots ag^{-1} = ga^h g^{-1} = geg^{-1} = e. \tag{2.8}$$

The conjugacy class containing the identity e consists of the single element e.

Example All the elements of S_3 are classified into three conjugacy classes:

$$C_1 : \{e\}, \quad C_2 : \{ab, ba\}, \quad C_3 : \{a, b, bab\}. \tag{2.9}$$

Here, the subscript n of C_n denotes the number of elements in the conjugacy class C_n.

We consider two groups, G and G', and a map f of G into G'. This map is *homomorphic* if and only if it preserves the multiplicative structure, that is,

$$f(a)f(b) = f(ab), \tag{2.10}$$

for all $a, b \in G$. Furthermore, the map is *isomorphic* when it is a one-to-one correspondence.

A *representation* D of G is a homomorphic map of elements of G onto matrices $D(g)$ for $g \in G$. The representation matrices then satisfy $D(a)D(b) = D(c)$ if $ab = c$ for $a, b, c \in G$. The vector space V on which the representation matrices act is called a *representation space*, with $D(g)_{ij} v_j$, $(j = 1, \ldots, n)$, for $v \in V$ with components v_j relative to some basis $\{e_1, \ldots, e_n\}$. The dimension n of the vector space V is called the *dimension* of the representation.

A subspace in the representation space is said to be an *invariant subspace* if, for any vector v in the subspace and any element $g \in G$, $D(g)_{ij} v_j$ also corresponds to a vector in the same subspace. If there is an invariant subspace, such a representation is said to be *reducible*. In contrast, a representation is *irreducible* if it has no invariant subspace. In particular, a representation is said to be *completely reducible* if, for every $g \in G$, $D(g)$ can be written in the following block diagonal form:

$$\begin{pmatrix} D_1(g) & 0 & & \\ 0 & D_2(g) & & \\ & & \ddots & \\ & & & D_r(g) \end{pmatrix}, \tag{2.11}$$

where each $D_\alpha(g)$ is irreducible for $\alpha = 1, \ldots, r$. We then say that the reducible representation $D(g)$ is the direct sum of the $D_\alpha(g)$:

$$\bigoplus_{\alpha=1}^{r} D_\alpha(g). \tag{2.12}$$

Every (reducible) representation of a finite group is completely reducible. Furthermore, every representation of a finite group is equivalent to a unitary representation (see Appendix A). The simplest (irreducible) representation is just $D(g) = 1$ for all elements g, that is, a trivial 1D representation or *singlet*. The matrix representations satisfy the following orthogonality relation:

$$\sum_{g \in G} D_\alpha(g)_{i\ell} D_\beta(g^{-1})_{mj} = \frac{N_G}{d_\alpha} \delta_{\alpha\beta} \delta_{ij} \delta_{\ell m}, \tag{2.13}$$

where N_G is the order of G and d_α is the dimension of $D_\alpha(g)$ for each α (see Appendix A).

The *character* $\chi_D(g)$ of a representation $D(g)$ is the trace of the representation matrix:

$$\chi_D(g) = \mathrm{tr}\, D(g) = \sum_{i=1}^{d_\alpha} D(g)_{ii}. \tag{2.14}$$

2 Basics of Finite Groups

The elements conjugate to a have the same character because of the following property of the trace:

$$\operatorname{tr} D(g^{-1}ag) = \operatorname{tr}\left[D(g^{-1})D(a)D(g)\right] = \operatorname{tr} D(a). \tag{2.15}$$

That is, the characters are constant in a conjugacy class. The characters satisfy the following orthogonality relation:

$$\sum_{g \in G} \chi_{D_\alpha}(g)^* \chi_{D_\beta}(g) = N_G \delta_{\alpha\beta}, \tag{2.16}$$

where N_G denotes the order of a group G (see Appendix A). That is, the characters of different irreducible representations are orthogonal and different from each other. Furthermore, it can be shown that *the number of irreducible representations must be equal to the number of conjugacy classes* (see Appendix A). In addition, they satisfy the following orthogonality relation:

$$\sum_\alpha \chi_{D_\alpha}(g_i)^* \chi_{D_\alpha}(g_j) = \frac{N_G}{n_i} \delta_{C_i C_j}, \tag{2.17}$$

where C_i denotes the conjugacy class of g_i and n_i denotes the number of elements in the conjugacy class C_i (see Appendix A). The right-hand side is equal to N_G/n_i if g_i and g_j belong to the same conjugacy class, and otherwise it must vanish. A trivial singlet, $D(g) = 1$ for any $g \in G$, must always be included. Thus, the corresponding character satisfies $\chi_1(g) = 1$ for any $g \in G$.

Suppose that there are m_n n-dimensional irreducible representations, that is, with $D(g)$ represented by $(n \times n)$ matrices. The identity e is always represented by the $(n \times n)$ identity matrix. Clearly, the character $\chi_{D_\alpha}(C_1)$ for the conjugacy class $C_1 = \{e\}$ is just $\chi_{D_\alpha}(C_1) = n$ for an n-dimensional representation. The orthogonality relation (2.17) then requires

$$\sum_\alpha [\chi_\alpha(C_1)]^2 = \sum_n m_n n^2 = m_1 + 4m_2 + 9m_3 + \cdots = N_G, \tag{2.18}$$

where $m_n \geq 0$. Furthermore, m_n must satisfy

$$\sum_n m_n = \text{number of conjugacy classes}, \tag{2.19}$$

because the number of irreducible representations is equal to the number of conjugacy classes. Equations (2.18) and (2.19), together with (2.16) and (2.17), will often be used in the following sections to determine characters.

Example Let us study the irreducible representations of S_3. The number of irreducible representations must be equal to three, because there are three conjugacy classes. We assume that there are m_n n-dimensional representations, that is, with $D(g)$ represented by $(n \times n)$ matrices. Here, m_n must satisfy $\sum_n m_n = 3$. Furthermore, the orthogonality relation (2.18) requires

$$\sum_\alpha [\chi_\alpha(C_1)]^2 = \sum_n m_n n^2 = m_1 + 4m_2 + 9m_3 + \cdots = 6, \tag{2.20}$$

Table 2.1 Characters of S_3 representations

	h	χ_1	$\chi_{1'}$	χ_2
C_1	1	1	1	2
C_2	3	1	1	−1
C_3	2	1	−1	0

where $m_n \geq 0$. This equation has only two possible solutions, $(m_1, m_2) = (2, 1)$ and $(6, 0)$, but only the former $(m_1, m_2) = (2, 1)$ satisfies $m_1 + m_2 = 3$. Thus, irreducible representations of S_3 include two singlets **1** and **1'**, and a doublet **2**. We denote their characters by $\chi_1(g)$, $\chi_{1'}(g)$, and $\chi_2(g)$, respectively. Clearly, $\chi_1(C_1) = \chi_{1'}(C_1) = 1$ and $\chi_2(C_1) = 2$. Furthermore, one of the singlet representations corresponds to a trivial singlet, that is, $\chi_1(C_2) = \chi_1(C_3) = 1$.

The characters, which are not fixed at this stage, are $\chi_{1'}(C_2)$, $\chi_{1'}(C_3)$, $\chi_2(C_2)$, and $\chi_2(C_3)$. Now let us determine them. For a non-trivial singlet **1'**, the representation matrices are nothing but the characters, $\chi_{1'}(C_2)$ and $\chi_{1'}(C_3)$. They must satisfy

$$[\chi_{1'}(C_2)]^3 = 1, \qquad [\chi_{1'}(C_3)]^2 = 1. \tag{2.21}$$

Thus, $\chi_{1'}(C_2)$ is one of 1, ω, and ω^2, where $\omega = \exp[2\pi i/3]$, and $\chi_{1'}(C_3)$ is 1 or −1. On top of that, the orthogonality relation (2.16) requires

$$\sum_g \chi_1(g)\chi_{1'}(g) = 1 + 2\chi_{1'}(C_2) + 3\chi_{1'}(C_3) = 0. \tag{2.22}$$

Its unique solution is $\chi_{1'}(C_2) = 1$ and $\chi_{1'}(C_3) = -1$. Furthermore, the orthogonality relations (2.16) and (2.17) require

$$\sum_g \chi_1(g)\chi_2(g) = 2 + 2\chi_2(C_2) + 3\chi_2(C_3) = 0, \tag{2.23}$$

$$\sum_\alpha \chi_\alpha(C_1)^*\chi_\alpha(C_2) = 1 + \chi_{1'}(C_2) + 2\chi_2(C_2) = 0. \tag{2.24}$$

Their solution is $\chi_2(C_2) = -1$ and $\chi_2(C_3) = 0$. These results are shown in Table 2.1.

Next, we figure out the representation matrices $D(g)$ of S_3 using the character Table 2.1. For singlets, their characters are nothing but representation matrices. Let us consider representation matrices $D(g)$ for the doublet, where $D(g)$ are (2×2) unitary matrices. Obviously, $D_2(e)$ is the (2×2) identity matrix. Since $\chi_2(C_3) = 0$, we can diagonalize one element of the conjugacy class C_3. Here we choose, e.g., a in C_3, as the diagonal element:

$$a = \begin{pmatrix} 1 & 0 \\ 0 & -1 \end{pmatrix}. \tag{2.25}$$

The other elements in C_3, as well as those in C_2, are non-diagonal matrices. Recalling that $b^2 = e$, we can write

$$b = \begin{pmatrix} \cos\theta & \sin\theta \\ \sin\theta & -\cos\theta \end{pmatrix}, \qquad bab = \begin{pmatrix} \cos 2\theta & \sin 2\theta \\ \sin 2\theta & -\cos 2\theta \end{pmatrix}. \tag{2.26}$$

Then, we can write elements in C_2 as

$$ab = \begin{pmatrix} \cos\theta & \sin\theta \\ -\sin\theta & \cos\theta \end{pmatrix}, \quad ba = \begin{pmatrix} \cos\theta & -\sin\theta \\ \sin\theta & \cos\theta \end{pmatrix}. \quad (2.27)$$

Recall that the trace of elements in C_2 is equal to -1, whence $\cos\theta = -1/2$, that is, $\theta = 2\pi/3, 4\pi/3$. When we choose $\theta = 4\pi/3$, we obtain the matrix representation of S_3 as

$$e = \begin{pmatrix} 1 & 0 \\ 0 & 1 \end{pmatrix}, \quad a = \begin{pmatrix} 1 & 0 \\ 0 & -1 \end{pmatrix}, \quad b = \begin{pmatrix} -\frac{1}{2} & -\frac{\sqrt{3}}{2} \\ -\frac{\sqrt{3}}{2} & \frac{1}{2} \end{pmatrix},$$

$$ab = \begin{pmatrix} -\frac{1}{2} & -\frac{\sqrt{3}}{2} \\ \frac{\sqrt{3}}{2} & -\frac{1}{2} \end{pmatrix}, \quad ba = \begin{pmatrix} -\frac{1}{2} & \frac{\sqrt{3}}{2} \\ -\frac{\sqrt{3}}{2} & -\frac{1}{2} \end{pmatrix}, \quad bab = \begin{pmatrix} -\frac{1}{2} & \frac{\sqrt{3}}{2} \\ \frac{\sqrt{3}}{2} & \frac{1}{2} \end{pmatrix}. \quad (2.28)$$

We can construct a larger group from two or more groups G_i, by means of certain products. A rather simple one is the *direct product*. We consider, e.g., two groups G_1 and G_2. Their direct product is denoted $G_1 \times G_2$, and its multiplication rule is defined as

$$(a_1, a_2)(b_1, b_2) = (a_1 b_1, a_2 b_2), \quad (2.29)$$

for $a_1, b_1 \in G_1$ and $a_2, b_2 \in G_2$.

The *semi-direct product* is a less trivial product between two groups G_1 and G_2, and it is defined such that

$$(a_1, a_2)(b_1, b_2) = (a_1 f_{a_2}(b_1), a_2 b_2), \quad (2.30)$$

for $a_1, b_1 \in G_1$ and $a_2, b_2 \in G_2$, where $f_{a_2}(b_1)$ denotes a homomorphic map from G_2 to the automorphisms of G_1. This semi-direct product is denoted by $G_1 \rtimes_f G_2$.

We consider the group G with a subgroup H and a normal subgroup N, whose elements are denoted h_i and n_j, respectively. When $G = NH = HN$ and $N \cap H = \{e\}$, the semi-direct product $N \rtimes_f H$ is isomorphic to G, $G \simeq N \rtimes_f H$, where we use the map f defined by

$$f_{h_i}(n_j) = h_i n_j (h_i)^{-1}. \quad (2.31)$$

For the notation of the semi-direct product, we will often omit f and denote it simply by $N \rtimes H$.

Example Let us study the semi-direct product $Z_3 \rtimes Z_2$. Here we denote the Z_3 and Z_2 generators by c and h, i.e., $c^3 = e$ and $h^2 = e$. In this case, (2.31) can be written

$$hch^{-1} = c^m, \quad (2.32)$$

where $m \neq 0$, because all the elements of Z_3 can be written c^m (the case $m = 0$ being inconsistent). When $m = 1$, the above relation is trivial and leads simply to the direct product $Z_3 \times Z_2$. Thus, only the case with $m = 2$ is non-trivial, i.e.,

$$hch^{-1} = c^2. \quad (2.33)$$

Indeed, this algebra is isomorphic to S_3, and h and c are identified with a and ab, respectively. Similarly, we can consider $Z_n \rtimes Z_m$. When we denote the Z_n and Z_m generators by a and b, respectively, they satisfy

$$a^n = b^m = e, \qquad bab^{-1} = a^k, \qquad (2.34)$$

where $k \neq 0$, although the case with $k = 1$ leads to the direct product $Z_n \times Z_m$.

References

1. Ramond, P.: Group Theory: A Physicist's Survey. Cambridge University Press, Cambridge (2010)
2. Miller, G.A., Dickson, H.F., Blichfeldt, L.E.: Theory and Applications of Finite Groups. Wiley, New York (1916)
3. Hamermesh, M.: Group Theory and Its Application to Physical Problems. Addison-Wesley, Reading (1962)
4. Georgi, H.: Front. Phys. **54**, 1 (1982)
5. Ludl, P.O.: arXiv:0907.5587 [hep-ph]
6. Grimus, W., Ludl, P.O.: J. Phys. A **45**, 233001 (2012). arXiv:1110.6376 [hep-ph]

Chapter 3
S_N

As introduced in the previous section, the symmetric group S_N consists of all possible permutations among N objects x_i with $i = 1, \ldots, N$:

$$(x_1, \ldots, x_N) \to (x_{i_1}, \ldots, x_{i_N}). \tag{3.1}$$

The group S_2 consists of two permutations

$$(x_1, x_2) \to (x_1, x_2), \qquad (x_1, x_2) \to (x_2, x_1). \tag{3.2}$$

This group is nothing but the Abelian group Z_2. Therefore, we study simple examples for $N = 3$ and 4, that is, S_3 and S_4.

3.1 S_3

We begin with S_3. Since some aspects of S_3 have already been discussed in Chap. 2, we summarize them briefly and go on to study other aspects such as tensor products. This group consists of all permutations among three objects (x_1, x_2, x_3), and so has order $3! = 6$. All six elements can be written as products of elements a and b:

$$a : (x_1, x_2, x_3) \to (x_2, x_1, x_3),$$
$$b : (x_1, x_2, x_3) \to (x_3, x_2, x_1),$$

together with the identity e, that is,

$$\{e, a, b, ab, ba, bab\}. \tag{3.3}$$

3.1.1 Conjugacy Classes

As studied in Chap. 2, S_3 has the following three conjugacy classes:

$$C_1 : \{e\}, \qquad C_2 : \{ab, ba\}, \qquad C_3 : \{a, b, bab\}. \tag{3.4}$$

Their orders are found from

$$(ab)^3 = (ba)^3 = e, \qquad a^2 = b^2 = (bab)^2 = e. \tag{3.5}$$

The elements $\{e, ab, ba\}$ correspond to even permutations, and $\{a, b, bab\}$ to odd permutations.

3.1.2 Characters and Representations

The characters and representations are studied in Chap. 2. The group S_3 has two singlet representations **1** and **1'**, and a doublet **2**. Their characters are summarized in Table 2.1. The characters for the singlets correspond to the representations on the singlets and the doublet representations are obtained in (2.28).

3.1.3 Tensor Products

Here, we consider tensor products of irreducible representations. Let us start by discussing the tensor products of two doublets (x_1, x_2) and (y_1, y_2). For example, each element $x_i y_j$ transforms under b according to

$$
\begin{aligned}
x_1 y_1 &\to \frac{x_1 y_1 + 3 x_2 y_2 + \sqrt{3}(x_1 y_2 + x_2 y_1)}{4}, \\
x_1 y_2 &\to \frac{\sqrt{3} x_1 y_1 - \sqrt{3} x_2 y_2 - x_1 y_2 + 3 x_2 y_1}{4}, \\
x_2 y_1 &\to \frac{\sqrt{3} x_1 y_1 - \sqrt{3} x_2 y_2 - x_2 y_1 + 3 x_1 y_2}{4}, \\
x_2 y_2 &\to \frac{3 x_1 y_1 + x_2 y_2 - \sqrt{3}(x_1 y_2 + x_2 y_1)}{4}.
\end{aligned}
\tag{3.6}
$$

Thus, it is found that

$$b(x_1 y_1 + x_2 y_2) = (x_1 y_1 + x_2 y_2), \qquad b(x_1 y_2 - x_2 y_1) = -(x_1 y_2 - x_2 y_1). \tag{3.7}$$

Therefore, these linear combinations correspond to the singlets

$$\mathbf{1} : x_1 y_1 + x_2 y_2, \qquad \mathbf{1'} : x_1 y_2 - x_2 y_1. \tag{3.8}$$

Furthermore, it is found that

$$b \begin{pmatrix} x_2 y_2 - x_1 y_1 \\ x_1 y_2 + x_2 y_1 \end{pmatrix} = \begin{pmatrix} -\frac{1}{2} & -\frac{\sqrt{3}}{2} \\ -\frac{\sqrt{3}}{2} & \frac{1}{2} \end{pmatrix} \begin{pmatrix} x_2 y_2 - x_1 y_1 \\ x_1 y_2 + x_2 y_1 \end{pmatrix}. \tag{3.9}$$

3.1 S_3

Hence, $(x_2 y_2 - x_2 y_2, x_1 y_2 + x_2 y_1)$ corresponds to the doublet, i.e.,

$$2 = \begin{pmatrix} x_2 y_2 - x_1 y_1 \\ x_1 y_2 + x_2 y_1 \end{pmatrix}. \tag{3.10}$$

Similarly, we can study the tensor product of the doublet (x_1, x_2) and the singlet $1' : y'$. Their products $x_i y'$ transform under b according to

$$\begin{aligned} x_1 y' &\to \frac{1}{2} x_1 y' + \frac{\sqrt{3}}{2} x_2 y', \\ x_2 y' &\to \frac{\sqrt{3}}{2} x_1 y' - \frac{1}{2} x_2 y'. \end{aligned} \tag{3.11}$$

They thus form a doublet

$$2 : \begin{pmatrix} -x_2 y' \\ x_1 y' \end{pmatrix}. \tag{3.12}$$

These tensor products are summarized as follows:

$$\begin{pmatrix} x_1 \\ x_2 \end{pmatrix}_2 \otimes \begin{pmatrix} y_1 \\ y_2 \end{pmatrix}_2 = (x_1 y_1 + x_2 y_2)_1 + (x_1 y_2 - x_2 y_1)_{1'} + \begin{pmatrix} x_1 y_2 + x_2 y_1 \\ x_1 y_1 - x_2 y_2 \end{pmatrix}_2,$$

$$\begin{pmatrix} x_1 \\ x_2 \end{pmatrix}_2 \otimes (y')_1 = \begin{pmatrix} x_1 y' \\ x_2 y' \end{pmatrix}_2, \quad \begin{pmatrix} x_1 \\ x_2 \end{pmatrix}_2 \otimes (y')_{1'} = \begin{pmatrix} -x_2 y' \\ x_1 y' \end{pmatrix}_2,$$

$$(x)_1 \otimes (y)_{1'} = (xy)_{1'}, \quad (x)_{1'} \otimes (y)_{1'} = (xy)_1. \tag{3.13}$$

Obviously, the tensor product of two trivial singlets is a trivial singlet.

Tensor products are important for applications to particle phenomenology. Matter and Higgs fields may be constructed to carry certain representations of discrete symmetries. The Lagrangian must be invariant under discrete symmetries. This implies that n-point couplings corresponding to a trivial singlet can appear in the Lagrangian.

In addition to the above (real) representation of S_3, another representation, i.e., the complex representation, is often used in the literature. Let us consider changes of representation basis. The permutations in S_3 in (3.1) are represented on the reducible triplet (x_1, x_2, x_3) by

$$\begin{pmatrix} 1 & 0 & 0 \\ 0 & 1 & 0 \\ 0 & 0 & 1 \end{pmatrix}, \quad \begin{pmatrix} 1 & 0 & 0 \\ 0 & 0 & 1 \\ 0 & 1 & 0 \end{pmatrix}, \quad \begin{pmatrix} 0 & 1 & 0 \\ 1 & 0 & 0 \\ 0 & 0 & 1 \end{pmatrix},$$

$$\begin{pmatrix} 0 & 1 & 0 \\ 0 & 0 & 1 \\ 1 & 0 & 0 \end{pmatrix}, \quad \begin{pmatrix} 0 & 0 & 1 \\ 0 & 1 & 0 \\ 1 & 0 & 0 \end{pmatrix}, \quad \begin{pmatrix} 0 & 0 & 1 \\ 1 & 0 & 0 \\ 0 & 1 & 0 \end{pmatrix}. \tag{3.14}$$

We change the representation through the unitary transformation, $U^\dagger g U$, e.g., using the unitary matrix

$$U = \begin{pmatrix} 1/\sqrt{3} & \sqrt{2/3} & 0 \\ 1/\sqrt{3} & -1/\sqrt{6} & -1/\sqrt{2} \\ 1/\sqrt{3} & -1/\sqrt{6} & 1/\sqrt{2} \end{pmatrix}. \tag{3.15}$$

Then, the six elements of S_3 are written as

$$\begin{pmatrix} 1 & 0 & 0 \\ 0 & 1 & 0 \\ 0 & 0 & 1 \end{pmatrix}, \begin{pmatrix} 1 & 0 & 0 \\ 0 & 1 & 0 \\ 0 & 0 & -1 \end{pmatrix}, \begin{pmatrix} 1 & 0 & 0 \\ 0 & -\frac{1}{2} & -\frac{\sqrt{3}}{2} \\ 0 & -\frac{\sqrt{3}}{2} & \frac{1}{2} \end{pmatrix},$$

$$\begin{pmatrix} 1 & 0 & 0 \\ 0 & -\frac{1}{2} & -\frac{\sqrt{3}}{2} \\ 0 & \frac{\sqrt{3}}{2} & -\frac{1}{2} \end{pmatrix}, \begin{pmatrix} 1 & 0 & 0 \\ 0 & -\frac{1}{2} & \frac{\sqrt{3}}{2} \\ 0 & \frac{\sqrt{3}}{2} & \frac{1}{2} \end{pmatrix}, \begin{pmatrix} 1 & 0 & 0 \\ 0 & -\frac{1}{2} & \frac{\sqrt{3}}{2} \\ 0 & -\frac{\sqrt{3}}{2} & -\frac{1}{2} \end{pmatrix}. \tag{3.16}$$

Note that this form is completely reducible and that the (bottom right) (2 × 2) submatrices are exactly the same as those for the doublet representation (2.28).

We can use another unitary matrix U to obtain a completely reducible form from the reducible representation matrices (3.14). For example, we can use the unitary matrix

$$U_w = \frac{1}{\sqrt{3}} \begin{pmatrix} 1 & 1 & 1 \\ 1 & w & w^2 \\ 1 & w^2 & w \end{pmatrix}, \tag{3.17}$$

which is called the magic matrix. Then, the six elements of S_3 are

$$\begin{pmatrix} 1 & 0 & 0 \\ 0 & 1 & 0 \\ 0 & 0 & 1 \end{pmatrix}, \begin{pmatrix} 1 & 0 & 0 \\ 0 & 0 & 1 \\ 0 & 1 & 0 \end{pmatrix}, \begin{pmatrix} 1 & 0 & 0 \\ 0 & 0 & w^2 \\ 0 & w & 0 \end{pmatrix},$$

$$\begin{pmatrix} 1 & 0 & 0 \\ 0 & w & 0 \\ 0 & 0 & w^2 \end{pmatrix}, \begin{pmatrix} 1 & 0 & 0 \\ 0 & 0 & w \\ 0 & w^2 & 0 \end{pmatrix}, \begin{pmatrix} 1 & 0 & 0 \\ 0 & w^2 & 0 \\ 0 & 0 & w \end{pmatrix}. \tag{3.18}$$

The (bottom right) (2 × 2) submatrices correspond to the doublet representation in the different basis, that is, the complex representation. In different bases, the multiplication rule does not change. For example, we obtain $2 \times 2 = 1 + 1' + 2$ in both the real and complex bases. However, elements of doublets on the left-hand side are written in a different way.

3.2 S_4

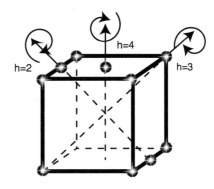

Fig. 3.1 The S_4 symmetry of a cube. The figure shows the transformations corresponding to the S_4 elements with $h = 2, 3$, and 4. Note that the group can also be considered as the symmetries of a regular octahedron in a way similar to the cube

3.2 S_4

We now discuss the group S_4, which consists of all permutations among four objects (x_1, x_2, x_3, x_4):

$$(x_1, x_2, x_3, x_4) \rightarrow (x_i, x_j, x_k, x_l). \tag{3.19}$$

The order of S_4 is $4! = 24$. We denote all the elements of S_4 as follows:

$$
\begin{aligned}
&a_1 : (x_1, x_2, x_3, x_4), & &a_2 : (x_2, x_1, x_4, x_3), \\
&a_3 : (x_3, x_4, x_1, x_2), & &a_4 : (x_4, x_3, x_2, x_1), \\
&b_1 : (x_1, x_4, x_2, x_3), & &b_2 : (x_4, x_1, x_3, x_2), \\
&b_3 : (x_2, x_3, x_1, x_4), & &b_4 : (x_3, x_2, x_4, x_1), \\
&c_1 : (x_1, x_3, x_4, x_2), & &c_2 : (x_3, x_1, x_2, x_4), \\
&c_3 : (x_4, x_2, x_1, x_3), & &c_4 : (x_2, x_4, x_3, x_1), \\
&d_1 : (x_1, x_2, x_4, x_3), & &d_2 : (x_2, x_1, x_3, x_4), \\
&d_3 : (x_4, x_3, x_1, x_2), & &d_4 : (x_3, x_4, x_2, x_1), \\
&e_1 : (x_1, x_3, x_2, x_4), & &e_2 : (x_3, x_1, x_4, x_2), \\
&e_3 : (x_2, x_4, x_1, x_3), & &e_4 : (x_4, x_2, x_3, x_1), \\
&f_1 : (x_1, x_4, x_3, x_2), & &f_2 : (x_4, x_1, x_2, x_3), \\
&f_3 : (x_3, x_2, x_1, x_4), & &f_4 : (x_2, x_3, x_4, x_1),
\end{aligned}
\tag{3.20}
$$

where we have shown the ordering of four objects after permutations. S_4 describes the symmetries of a cube, as shown in Fig. 3.1.

It is obvious that $x_1 + x_2 + x_3 + x_4$ is invariant under any permutation of S_4, so it is a trivial singlet. Thus, we use the vector space which is orthogonal to this singlet direction, viz.,

$$3 : \begin{pmatrix} A_x \\ A_y \\ A_z \end{pmatrix} = \begin{pmatrix} x_1 + x_2 - x_3 - x_4 \\ x_1 - x_2 + x_3 - x_4 \\ x_1 - x_2 - x_3 + x_4 \end{pmatrix}, \tag{3.21}$$

to construct matrix representations of S_4, that is, a triplet representation. In this triplet vector space, the elements of S_4 are represented by the following matrices:

$$a_1 = \begin{pmatrix} 1 & 0 & 0 \\ 0 & 1 & 0 \\ 0 & 0 & 1 \end{pmatrix}, \quad a_2 = \begin{pmatrix} 1 & 0 & 0 \\ 0 & -1 & 0 \\ 0 & 0 & -1 \end{pmatrix},$$

$$a_3 = \begin{pmatrix} -1 & 0 & 0 \\ 0 & 1 & 0 \\ 0 & 0 & -1 \end{pmatrix}, \quad a_4 = \begin{pmatrix} -1 & 0 & 0 \\ 0 & -1 & 0 \\ 0 & 0 & 1 \end{pmatrix},$$

$$b_1 = \begin{pmatrix} 0 & 0 & 1 \\ 1 & 0 & 0 \\ 0 & 1 & 0 \end{pmatrix}, \quad b_2 = \begin{pmatrix} 0 & 0 & 1 \\ -1 & 0 & 0 \\ 0 & -1 & 0 \end{pmatrix},$$

$$b_3 = \begin{pmatrix} 0 & 0 & -1 \\ 1 & 0 & 0 \\ 0 & -1 & 0 \end{pmatrix}, \quad b_4 = \begin{pmatrix} 0 & 0 & -1 \\ -1 & 0 & 0 \\ 0 & 1 & 0 \end{pmatrix},$$

$$c_1 = \begin{pmatrix} 0 & 1 & 0 \\ 0 & 0 & 1 \\ 1 & 0 & 0 \end{pmatrix}, \quad c_2 = \begin{pmatrix} 0 & 1 & 0 \\ 0 & 0 & -1 \\ -1 & 0 & 0 \end{pmatrix},$$

$$c_3 = \begin{pmatrix} 0 & -1 & 0 \\ 0 & 0 & 1 \\ -1 & 0 & 0 \end{pmatrix}, \quad c_4 = \begin{pmatrix} 0 & -1 & 0 \\ 0 & 0 & -1 \\ 1 & 0 & 0 \end{pmatrix},$$

$$d_1 = \begin{pmatrix} 1 & 0 & 0 \\ 0 & 0 & 1 \\ 0 & 1 & 0 \end{pmatrix}, \quad d_2 = \begin{pmatrix} 1 & 0 & 0 \\ 0 & 0 & -1 \\ 0 & -1 & 0 \end{pmatrix},$$

$$d_3 = \begin{pmatrix} -1 & 0 & 0 \\ 0 & 0 & 1 \\ 0 & -1 & 0 \end{pmatrix}, \quad d_4 = \begin{pmatrix} -1 & 0 & 0 \\ 0 & 0 & -1 \\ 0 & 1 & 0 \end{pmatrix},$$

$$e_1 = \begin{pmatrix} 0 & 1 & 0 \\ 1 & 0 & 0 \\ 0 & 0 & 1 \end{pmatrix}, \quad e_2 = \begin{pmatrix} 0 & 1 & 0 \\ -1 & 0 & 0 \\ 0 & 0 & -1 \end{pmatrix},$$

$$e_3 = \begin{pmatrix} 0 & -1 & 0 \\ 1 & 0 & 0 \\ 0 & 0 & -1 \end{pmatrix}, \quad e_4 = \begin{pmatrix} 0 & -1 & 0 \\ -1 & 0 & 0 \\ 0 & 0 & 1 \end{pmatrix},$$

(3.22)

3.2 S_4

$$f_1 = \begin{pmatrix} 0 & 0 & 1 \\ 0 & 1 & 0 \\ 1 & 0 & 0 \end{pmatrix}, \quad f_2 = \begin{pmatrix} 0 & 0 & 1 \\ 0 & -1 & 0 \\ -1 & 0 & 0 \end{pmatrix},$$

$$f_3 = \begin{pmatrix} 0 & 0 & -1 \\ 0 & 1 & 0 \\ -1 & 0 & 0 \end{pmatrix}, \quad f_4 = \begin{pmatrix} 0 & 0 & -1 \\ 0 & -1 & 0 \\ 1 & 0 & 0 \end{pmatrix}.$$

3.2.1 Conjugacy Classes

The elements of S_4 can be classified by their order h, i.e., the smallest positive integer such that $a^h = e$:

$$\begin{aligned} h &= 1: \quad \{a_1\}, \\ h &= 2: \quad \{a_2, a_3, a_4, d_1, d_2, e_1, e_4, f_1, f_3\}, \\ h &= 3: \quad \{b_1, b_2, b_3, b_4, c_1, c_2, c_3, c_4\}, \\ h &= 4: \quad \{d_3, d_4, e_2, e_3, f_2, f_4\}. \end{aligned} \tag{3.23}$$

Moreover, they are classified by the conjugacy classes according to:

$$\begin{aligned} C_1: & \quad \{a_1\}, & h &= 1, \\ C_3: & \quad \{a_2, a_3, a_4\}, & h &= 2, \\ C_6: & \quad \{d_1, d_2, e_1, e_4, f_1, f_3\}, & h &= 2, \\ C_8: & \quad \{b_1, b_2, b_3, b_4, c_1, c_2, c_3, c_4\}, & h &= 3, \\ C'_6: & \quad \{d_3, d_4, e_2, e_3, f_2, f_4\}, & h &= 4. \end{aligned} \tag{3.24}$$

3.2.2 Characters and Representations

The group S_4 includes five conjugacy classes, so there are five irreducible representations. For example, all its elements can be written as products of b_1 in C_8 and d_4 in $C_{6'}$, which satisfy

$$(b_1)^3 = e, \quad (d_4)^4 = e, \quad d_4(b_1)^2 d_4 = b_1,$$
$$d_4 b_1 d_4 = b_1 (d_4)^2 b_1. \tag{3.25}$$

The orthogonality relation (2.18) requires

$$\sum_\alpha [\chi_\alpha(C_1)]^2 = \sum_n m_n n^2 = m_1 + 4m_2 + 9m_3 + \cdots = 24, \tag{3.26}$$

Table 3.1 Characters of S_4 representations

	h	χ_1	$\chi_{1'}$	χ_2	χ_3	$\chi_{3'}$
C_1	1	1	1	2	3	3
C_3	2	1	1	2	-1	-1
C_6	2	1	-1	0	1	-1
C_6'	4	1	-1	0	-1	1
C_8	3	1	1	-1	0	0

like (2.20), and m_n also satisfy $m_1 + m_2 + m_3 + \cdots = 5$, because there must be five irreducible representations. Then, we can easily find the unique solution as $(m_1, m_2, m_3) = (2, 1, 2)$. Therefore, the irreducible representations of S_4 include two singlets **1** and **1'**, one doublet **2**, and two triplets **3** and **3'**, where **1** corresponds to a trivial singlet and **3** corresponds to (3.21) and (3.22). We can compute the character for each representation by a similar analysis to the one adopted for S_3. The characters are shown in Table 3.1.

For **2**, the representation matrices are, for example,

$$a_2(\mathbf{2}) = \begin{pmatrix} 1 & 0 \\ 0 & 1 \end{pmatrix}, \quad b_1(\mathbf{2}) = \begin{pmatrix} \omega & 0 \\ 0 & \omega^2 \end{pmatrix},$$

$$d_1(\mathbf{2}) = d_3(\mathbf{2}) = d_4(\mathbf{2}) = \begin{pmatrix} 0 & 1 \\ 1 & 0 \end{pmatrix}.$$

(3.27)

For **3'**, the representation matrices are, for example,

$$a_2(\mathbf{3'}) = \begin{pmatrix} 1 & 0 & 0 \\ 0 & -1 & 0 \\ 0 & 0 & -1 \end{pmatrix}, \quad b_1(\mathbf{3'}) = \begin{pmatrix} 0 & 0 & 1 \\ 1 & 0 & 0 \\ 0 & 1 & 0 \end{pmatrix},$$

$$d_1(\mathbf{3'}) = \begin{pmatrix} -1 & 0 & 0 \\ 0 & 0 & -1 \\ 0 & -1 & 0 \end{pmatrix}, \quad d_3(\mathbf{3'}) = \begin{pmatrix} 1 & 0 & 0 \\ 0 & 0 & -1 \\ 0 & 1 & 0 \end{pmatrix}, \quad (3.28)$$

$$d_4(\mathbf{3'}) = \begin{pmatrix} 1 & 0 & 0 \\ 0 & 0 & 1 \\ 0 & -1 & 0 \end{pmatrix}.$$

Note that $a_2(\mathbf{3'}) = a_2(\mathbf{3})$ and $b_1(\mathbf{3'}) = b_1(\mathbf{3})$, but $d_1(\mathbf{3'}) = -d_1(\mathbf{3})$, $d_3(\mathbf{3'}) = -d_3(\mathbf{3})$, and $d_4(\mathbf{3'}) = -d_4(\mathbf{3})$. This aspect should be obvious from the above character table.

3.2.3 Tensor Products

Finally, we present the tensor products. The tensor products of $\mathbf{3} \times \mathbf{3}$ can be decomposed as

$$(\mathbf{A})_3 \times (\mathbf{B})_3 = (\mathbf{A} \cdot \mathbf{B})_1 + \begin{pmatrix} \mathbf{A} \cdot \Sigma \cdot \mathbf{B} \\ \mathbf{A} \cdot \Sigma^* \cdot \mathbf{B} \end{pmatrix}_2$$

$$+ \begin{pmatrix} \{A_y B_z\} \\ \{A_z B_x\} \\ \{A_x B_y\} \end{pmatrix}_3 + \begin{pmatrix} [A_y B_z] \\ [A_z B_x] \\ [A_x B_y] \end{pmatrix}_{3'}, \qquad (3.29)$$

where

$$\mathbf{A} \cdot \mathbf{B} = A_x B_x + A_y B_y + A_z B_z,$$

$$\{A_i B_j\} = A_i B_j + A_j B_i,$$

$$[A_i B_j] = A_i B_j - A_j B_i, \qquad (3.30)$$

$$\mathbf{A} \cdot \Sigma \cdot \mathbf{B} = A_x B_x + \omega A_y B_y + \omega^2 A_z B_z,$$

$$\mathbf{A} \cdot \Sigma^* \cdot \mathbf{B} = A_x B_x + \omega^2 A_y B_y + \omega A_z B_z.$$

The tensor products of other representations can be decomposed in a similar way, e.g.,

$$(\mathbf{A})_{3'} \times (\mathbf{B})_{3'} = (\mathbf{A} \cdot \mathbf{B})_1 + \begin{pmatrix} \mathbf{A} \cdot \Sigma \cdot \mathbf{B} \\ \mathbf{A} \cdot \Sigma^* \cdot \mathbf{B} \end{pmatrix}_2 + \begin{pmatrix} \{A_y B_z\} \\ \{A_z B_x\} \\ \{A_x B_y\} \end{pmatrix}_3 + \begin{pmatrix} [A_y B_z] \\ [A_z B_x] \\ [A_x B_y] \end{pmatrix}_{3'},$$

$$(3.31)$$

$$(\mathbf{A})_3 \times (\mathbf{B})_{3'} = (\mathbf{A} \cdot \mathbf{B})_{1'} + \begin{pmatrix} \mathbf{A} \cdot \Sigma \cdot \mathbf{B} \\ -\mathbf{A} \cdot \Sigma^* \cdot \mathbf{B} \end{pmatrix}_2 + \begin{pmatrix} \{A_y B_z\} \\ \{A_z B_x\} \\ \{A_x B_y\} \end{pmatrix}_{3'} + \begin{pmatrix} [A_y B_z] \\ [A_z B_x] \\ [A_x B_y] \end{pmatrix}_3,$$

$$(3.32)$$

and

$$(\mathbf{A})_2 \times (\mathbf{B})_2 = \{A_x B_y\}_1 + [A_x B_y]_{1'} + \begin{pmatrix} A_y B_y \\ A_x B_x \end{pmatrix}_2, \qquad (3.33)$$

$$\begin{pmatrix} A_x \\ A_y \end{pmatrix}_2 \times \begin{pmatrix} B_x \\ B_y \\ B_z \end{pmatrix}_3 = \begin{pmatrix} (A_x + A_y) B_x \\ (\omega^2 A_x + \omega A_y) B_y \\ (\omega A_x + \omega^2 A_y) B_z \end{pmatrix}_3 + \begin{pmatrix} (A_x - A_y) B_x \\ (\omega^2 A_x - \omega A_y) B_y \\ (\omega A_x - \omega^2 A_y) B_z \end{pmatrix}_{3'}.$$

$$(3.34)$$

$$\begin{pmatrix} A_x \\ A_y \end{pmatrix}_2 \times \begin{pmatrix} B_x \\ B_y \\ B_z \end{pmatrix}_{3'} = \begin{pmatrix} (A_x + A_y)B_x \\ (\omega^2 A_x + \omega A_y)B_y \\ (\omega A_x + \omega^2 A_y)B_z \end{pmatrix}_{3'} + \begin{pmatrix} (A_x - A_y)B_x \\ (\omega^2 A_x - \omega A_y)B_y \\ (\omega A_x - \omega^2 A_y)B_z \end{pmatrix}_3.$$
(3.35)

In addition, we have decompositions $\mathbf{3} \times \mathbf{1'} = \mathbf{3'}$, $\mathbf{3'} \times \mathbf{1'} = \mathbf{3}$, and $\mathbf{2} \times \mathbf{1'} = \mathbf{2}$.

In the literature, several bases are used for S_4. The decomposition of tensor products, $\mathbf{r} \times \mathbf{r'} = \sum_m \mathbf{r}_m$, does not depend on the basis. For example, we obtain $\mathbf{3} \times \mathbf{3'} = \mathbf{1'} + \mathbf{2} + \mathbf{3} + \mathbf{3'}$ in any basis. However, the multiplication rules written in terms of components do depend on the basis we use. Here we have used the basis (3.27). In Appendix B, we show the relations between several bases and give the multiplication rules explicitly in terms of components.

Similarly, we can study the group S_N with $N > 4$. Here we give a brief comment on such groups. The group S_N with $N > 4$ has only one invariant subgroup, that is, the alternating group A_N. S_N has two one-dimensional representations: one is the trivial singlet, that is, invariant under all elements (symmetric representation), the other is a pseudo-singlet, that is, symmetric under the even permutation elements, but antisymmetric under the odd permutation elements. Group-theoretical aspects of S_5 are derived from those of S_4 by applying a theorem due to Frobenius (Frobenius formula), a graphical method (Young tableaux), and recursion formulas for characters (branching laws). The details can be found in the textbook [1], for example. Such analysis can be extended recursively from S_N to S_{N+1}.

References

1. Hamermesh, M.: Group Theory and Its Application to Physical Problems. Addison-Wesley, Reading (1962)

Chapter 4
A_N

In this chapter, we study the group A_N, consisting of all even permutations in S_N. These do indeed form a group, also called the alternating group. The order of this group is clearly $(N!)/2$. Let us consider a simple example. As discussed in Sect. 3.1, the even permutations in S_3 are

$$e : (x_1, x_2, x_3) \to (x_1, x_2, x_3),$$
$$a_4 : (x_1, x_2, x_3) \to (x_3, x_1, x_2), \quad (4.1)$$
$$a_5 : (x_1, x_2, x_3) \to (x_2, x_3, x_1),$$

while the odd permutations are

$$a_1 : (x_1, x_2, x_3) \to (x_2, x_1, x_3),$$
$$a_2 : (x_1, x_2, x_3) \to (x_3, x_2, x_1), \quad (4.2)$$
$$a_3 : (x_1, x_2, x_3) \to (x_1, x_3, x_2).$$

The three even permutations $\{e, a_4, a_5\}$ form the group A_3. Since $(a_4)^2 = a_5$ and $(a_4)^3 = e$, the group A_3 is nothing but Z_3. Therefore, we start by studying A_4, the smallest non-Abelian group.

4.1 A_4

The group A_4 consists of all even permutations in S_4 and thus has order $(4!)/2 = 12$. A_4 is the symmetry group of a tetrahedron as shown in Fig. 4.1. Indeed, the group A_4 is often denoted by T. Using the notation in Sect. 3.2, the 12 elements are:

Fig. 4.1 The A_4 symmetry of the tetrahedron

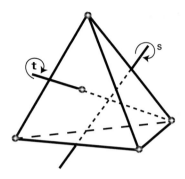

$$a_1 = \begin{pmatrix} 1 & 0 & 0 \\ 0 & 1 & 0 \\ 0 & 0 & 1 \end{pmatrix}, \quad a_2 = \begin{pmatrix} 1 & 0 & 0 \\ 0 & -1 & 0 \\ 0 & 0 & -1 \end{pmatrix},$$

$$a_3 = \begin{pmatrix} -1 & 0 & 0 \\ 0 & 1 & 0 \\ 0 & 0 & -1 \end{pmatrix}, \quad a_4 = \begin{pmatrix} -1 & 0 & 0 \\ 0 & -1 & 0 \\ 0 & 0 & 1 \end{pmatrix},$$

$$b_1 = \begin{pmatrix} 0 & 0 & 1 \\ 1 & 0 & 0 \\ 0 & 1 & 0 \end{pmatrix}, \quad b_2 = \begin{pmatrix} 0 & 0 & 1 \\ -1 & 0 & 0 \\ 0 & -1 & 0 \end{pmatrix},$$

$$b_3 = \begin{pmatrix} 0 & 0 & -1 \\ 1 & 0 & 0 \\ 0 & -1 & 0 \end{pmatrix}, \quad b_4 = \begin{pmatrix} 0 & 0 & -1 \\ -1 & 0 & 0 \\ 0 & 1 & 0 \end{pmatrix}, \qquad (4.3)$$

$$c_1 = \begin{pmatrix} 0 & 1 & 0 \\ 0 & 0 & 1 \\ 1 & 0 & 0 \end{pmatrix}, \quad c_2 = \begin{pmatrix} 0 & 1 & 0 \\ 0 & 0 & -1 \\ -1 & 0 & 0 \end{pmatrix},$$

$$c_3 = \begin{pmatrix} 0 & -1 & 0 \\ 0 & 0 & 1 \\ -1 & 0 & 0 \end{pmatrix}, \quad c_4 = \begin{pmatrix} 0 & -1 & 0 \\ 0 & 0 & -1 \\ 1 & 0 & 0 \end{pmatrix}.$$

As can be seen from this, A_4 is obviously isomorphic to $\Delta(12) \simeq (Z_2 \times Z_2) \rtimes Z_3$ (see Chap. 10 for explanation).

They are classified into conjugacy classes as follows:

$$\begin{aligned} C_1 &: \{a_1\}, & h &= 1, \\ C_3 &: \{a_2, a_3, a_4\}, & h &= 2, \\ C_4 &: \{b_1, b_2, b_3, b_4\}, & h &= 3, \\ C_4' &: \{c_1, c_2, c_3, c_4\}, & h &= 3, \end{aligned} \qquad (4.4)$$

where we have also shown the order h of each element in the conjugacy class. There are four conjugacy classes and there must therefore be four irreducible representations, i.e., $m_1 + m_2 + m_3 + \cdots = 4$.

4.1 A_4

The orthogonality relation (2.17) requires

$$\sum_\alpha [\chi_\alpha(C_1)]^2 = \sum_n m_n n^2 = m_1 + 4m_2 + 9m_3 + \cdots = 12, \qquad (4.5)$$

for the m_i, which must also satisfy $m_1 + m_2 + m_3 + \cdots = 4$. The only solution is $(m_1, m_2, m_3) = (3, 0, 1)$. That is, the A_4 group has three singlets, **1**, **1′**, and **1″**, and a single triplet **3**, where the triplet corresponds to (4.3).

Another algebraic definition of A_4 is often used in the literature. Let $a_1 = e$, $a_2 = s$, and $b_1 = t$. They satisfy the algebraic relations

$$s^2 = t^3 = (st)^3 = e. \qquad (4.6)$$

The closed algebra of these elements s and t is defined as A_4. It is straightforward to write all the a_i, b_i, and c_i in terms of s and t. Then, the conjugacy classes can be reexpressed as

$$\begin{aligned}
C_1 &: \{e\}, & h &= 1, \\
C_3 &: \{s, tst^2, t^2st\}, & h &= 2, \\
C_4 &: \{t, ts, st, sts\}, & h &= 3, \\
C_4' &: \{t^2, st^2, t^2s, tst\}, & h &= 3.
\end{aligned} \qquad (4.7)$$

We now use these to study the characters. First, we consider the characters of the three singlets. Because $s^2 = e$, the characters of C_3 have two possibilities, namely $\chi_\alpha(C_3) = \pm 1$. However, the two elements t and ts belong to the same conjugacy class C_4. This means that $\chi_\alpha(C_3)$ should have the value $\chi_\alpha(C_3) = 1$. Similarly, because $t^3 = e$, the characters $\chi_\alpha(t)$ can correspond to three possible values, i.e., $\chi_\alpha(t) = \omega^n$, $n = 0, 1, 2$, and all three values are consistent with the above conjugacy class structure. Thus, the three singlets **1**, **1′**, and **1″** are classified by the three values $\chi_\alpha(t) = 1, \omega,$ and ω^2, respectively. Clearly, $\chi_\alpha(C_4') = [\chi_\alpha(C_4)]^2$. Thus, generators such as $s = a_2$, $t = b_1$, and $t^2 = c_1$ are represented on the non-trivial singlets **1′** and **1″** by

$$\begin{aligned}
s(\mathbf{1'}) = a_2(\mathbf{1'}) = 1, &\quad t(\mathbf{1'}) = b_1(\mathbf{1'}) = \omega, &\quad t^2(\mathbf{1'}) = c_1(\mathbf{1'}) = \omega^2, \\
s(\mathbf{1''}) = a_2(\mathbf{1''}) = 1, &\quad t(\mathbf{1''}) = b_1(\mathbf{1''}) = \omega^2, &\quad t^2(\mathbf{1''}) = c_1(\mathbf{1''}) = \omega.
\end{aligned} \qquad (4.8)$$

These characters are shown in Table 4.1.

Next, we consider the characters for the triplet representation. Obviously, the matrices in (4.3) correspond to the triplet representation. We thus obtain their characters and the results are also shown in Table 4.1.

The tensor product of $\mathbf{3} \times \mathbf{3}$ can be decomposed as

$$(\mathbf{A})_3 \times (\mathbf{B})_3 = (\mathbf{A} \cdot \mathbf{B})_1 + (\mathbf{A} \cdot \Sigma \cdot \mathbf{B})_{1'} + (\mathbf{A} \cdot \Sigma^* \cdot \mathbf{B})_{1''}$$

$$+ \begin{pmatrix} \{A_y B_z\} \\ \{A_z B_x\} \\ \{A_x B_y\} \end{pmatrix}_3 + \begin{pmatrix} [A_y B_z] \\ [A_z B_x] \\ [A_x B_y] \end{pmatrix}_3. \qquad (4.9)$$

Table 4.1 Characters of A_4 representations

	h	χ_1	$\chi_{1'}$	$\chi_{1''}$	χ_3
C_1	1	1	1	1	3
C_3	2	1	1	1	-1
C_4	3	1	ω	ω^2	0
C_4'	3	1	ω^2	ω	0

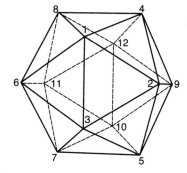

Fig. 4.2 The regular icosahedron

4.2 A_5

Next we study the group A_5. This group is isomorphic to the symmetry group of a regular icosahedron. Thus, it is of pedagogical interest to explain the group-theoretical aspects of A_5 in terms of the symmetries of a regular icosahedron [1]. As shown in Fig. 4.2, a regular icosahedron consists of 20 identical equilateral triangular faces, 30 edges and 12 vertices. The icosahedron is dual to a dodecahedron, whose symmetry group is also isomorphic to A_5.

The elements of A_5 correspond to all the proper rotations of the icosahedron. Such rotations are classified into five types, that is, the 0 rotation (identity), rotations by π about the midpoint of each edge, rotations by $2\pi/3$ about axes through the center of each face, and rotations by $2\pi/5$ and $4\pi/5$ about an axis through each vertex.

Following [1], we label the vertices by $n = 1, \ldots, 12$ in Fig. 4.2. Here, we define two elements a and b such that a corresponds to the rotation by π about the midpoint of the edge between vertices 1 and 2 while b corresponds to the rotation by $2\pi/3$ about the axis through the center of the triangular face 10-11-12. That is, these two elements correspond to the transformations acting on the 12 vertices as follows:

$$a : (1, 2, 3, 4, 5, 6, 7, 8, 9, 10, 11, 12) \to (2, 1, 4, 3, 8, 9, 12, 5, 6, 11, 10, 7),$$
$$b : (1, 2, 3, 4, 5, 6, 7, 8, 9, 10, 11, 12) \to (2, 3, 1, 5, 6, 4, 8, 9, 7, 11, 12, 10).$$

Then the product ab is given by the transformation

$$ab : (1, 2, 3, 4, 5, 6, 7, 8, 9, 10, 11, 12) \to (3, 2, 5, 1, 9, 7, 10, 6, 4, 12, 11, 8),$$

4.2 A_5

which is the rotation by $2\pi/5$ about the axis through vertex 2. All the elements of A_5 can be written in terms of products of these elements, which satisfy

$$a^2 = b^3 = (ab)^5 = e. \tag{4.10}$$

4.2.1 Conjugacy Classes

The order of A_5 is $(5!)/2 = 60$. The elements of A_5, i.e., all the rotations of the icosahedron, are classified into five conjugacy classes as follows:

$$\begin{aligned}
C_1 &: \{e\}, \\
C_{15} &: \{a(12), a(13), a(14), a(16), a(18), a(23), a(24), a(25), a(29), \\
&\quad a(35), a(36), a(37), a(48), a(49), a(59)\}, \\
C_{20} &: \{b(123), b(124), b(126), b(136), b(168), b(235), b(249), \\
&\quad b(259), b(357), b(367), \text{ and their inverse elements}\}, \\
C_{12} &: \{c(1), c(2), c(3), c(4), c(5), c(6), \text{ and their inverse elements}\}, \\
C'_{12} &: \{c^2(1), c^2(2), c^2(3), c^2(4), c^2(5), c^2(6), \text{ and their inverse elements}\}, \tag{4.11}
\end{aligned}$$

where $a(km)$, $b(kmn)$ and $c(k)$ denote respectively the rotation by π about the midpoint of the edge km, the rotation by $2\pi/3$ about the axis through the center of the face kmn, and the rotation by $2\pi/5$ about the axis through the vertex k. The conjugacy classes $C_1, C_{15}, C_{20}, C_{12}$, and C'_{12} contain 1, 15, 20, 12, and 12 elements, respectively. Since obviously $[a(km)]^2 = [b(kmn)]^3 = [c(k)]^5 = e$, we find $h = 2$ in C_{15}, $h = 3$ in C_{20}, $h = 5$ in C_{12}, and $h = 5$ in C'_{12}, where h denotes the order of each element in the conjugacy class, i.e., $g^h = e$.

4.2.2 Characters and Representations

The orthogonality relations (2.18) and (2.19) for A_5 are

$$m_1 + 4m_2 + 9m_3 + 16m_4 + 25m_5 + \cdots = 60, \tag{4.12}$$

$$m_1 + m_2 + m_3 + m_4 + m_5 + \cdots = 5. \tag{4.13}$$

Solving these equations, we obtain $(m_1, m_2, m_3, m_4, m_5) = (1, 0, 2, 1, 1)$. Therefore, the A_5 group has one trivial singlet **1**, two triplets **3** and **3'**, one quartet **4**, and one quintet **5**. The characters are shown in Table 4.2.

There are several ways to construct these representations, e.g., the Cummins–Patera basis [2], the Shirai basis [1], or the Feruglio–Paris basis [3]. Each of the relations is also summarized in [3]. Instead of a and b, which is called the Cummins–Patera basis [2], we use the generators of Shirai's basis [1], viz., $s = a$ and $t = bab$, which satisfy

$$s^2 = t^5 = \left(t^2 s t^3 s t^{-1} s t s t^{-1}\right)^3 = e. \tag{4.14}$$

Table 4.2 Characters of A_5 representations, where $\phi = (1+\sqrt{5})/2$

	h	χ_1	χ_3	$\chi_{3'}$	χ_4	χ_5
C_1	1	1	3	3	4	5
C_{15}	2	1	-1	-1	0	1
C_{20}	3	1	0	0	1	-1
C_{12}	5	1	ϕ	$1-\phi$	-1	0
C'_{12}	5	1	$1-\phi$	ϕ	-1	0

The generators s and t are represented by [1]

$$s = \frac{1}{2}\begin{pmatrix} -1 & \phi & \frac{1}{\phi} \\ \phi & \frac{1}{\phi} & 1 \\ \frac{1}{\phi} & 1 & -\phi \end{pmatrix}, \quad t = \frac{1}{2}\begin{pmatrix} 1 & \phi & \frac{1}{\phi} \\ -\phi & \frac{1}{\phi} & 1 \\ \frac{1}{\phi} & -1 & \phi \end{pmatrix}, \quad \text{on } \mathbf{3}, \quad (4.15)$$

$$s = \frac{1}{2}\begin{pmatrix} -\phi & \frac{1}{\phi} & 1 \\ \frac{1}{\phi} & -1 & \phi \\ 1 & \phi & \frac{1}{\phi} \end{pmatrix}, \quad t = \frac{1}{2}\begin{pmatrix} -\phi & -\frac{1}{\phi} & 1 \\ \frac{1}{\phi} & 1 & \phi \\ -1 & \phi & -\frac{1}{\phi} \end{pmatrix}, \quad \text{on } \mathbf{3'}, \quad (4.16)$$

$$s = \frac{1}{4}\begin{pmatrix} -1 & -1 & -3 & -\sqrt{5} \\ -1 & 3 & 1 & -\sqrt{5} \\ -3 & 1 & -1 & \sqrt{5} \\ -\sqrt{5} & -\sqrt{5} & \sqrt{5} & -1 \end{pmatrix},$$

$$t = \frac{1}{4}\begin{pmatrix} -1 & 1 & -3 & \sqrt{5} \\ -1 & -3 & 1 & \sqrt{5} \\ 3 & 1 & 1 & \sqrt{5} \\ \sqrt{5} & -\sqrt{5} & -\sqrt{5} & -1 \end{pmatrix}, \quad \text{on } \mathbf{4}, \quad (4.17)$$

$$s = \frac{1}{2}\begin{pmatrix} \frac{1-3\phi}{4} & \frac{\phi^2}{2} & -\frac{1}{2\phi^2} & \frac{\sqrt{5}}{2} & \frac{\sqrt{3}}{4\phi} \\ \frac{\phi^2}{2} & 1 & 1 & 0 & \frac{\sqrt{3}}{2\phi} \\ -\frac{1}{2\phi^2} & 1 & 0 & -1 & -\frac{\sqrt{3}\phi}{2} \\ \frac{\sqrt{5}}{2} & 0 & -1 & 1 & -\frac{\sqrt{3}}{2} \\ \frac{\sqrt{3}}{4\phi} & \frac{\sqrt{3}}{2\phi} & -\frac{\sqrt{3}\phi}{2} & -\frac{\sqrt{3}}{2} & \frac{3\phi-1}{4} \end{pmatrix},$$

$$t = \frac{1}{2}\begin{pmatrix} \frac{1-3\phi}{4} & -\frac{\phi^2}{2} & -\frac{1}{2\phi^2} & -\frac{\sqrt{5}}{2} & \frac{\sqrt{3}}{4\phi} \\ \frac{\phi^2}{2} & -1 & 1 & 0 & \frac{\sqrt{3}}{2\phi} \\ \frac{1}{2\phi^2} & 1 & 0 & -1 & \frac{\sqrt{3}\phi}{2} \\ -\frac{\sqrt{5}}{2} & 0 & 1 & 1 & \frac{\sqrt{3}}{2} \\ \frac{\sqrt{3}}{4\phi} & -\frac{\sqrt{3}}{2\phi} & -\frac{\sqrt{3}\phi}{2} & \frac{\sqrt{3}}{2} & \frac{3\phi-1}{4} \end{pmatrix}, \quad \text{on } \mathbf{5}, \quad (4.18)$$

where $\phi = (1+\sqrt{5})/2$. The multiplication rules are shown in Table 4.3.

4.2 A_5

Table 4.3 Multiplication rules for the group A_5

$3 \otimes 3 = 1 \oplus 3 \oplus 5$
$3' \otimes 3' = 1 \oplus 3' \oplus 5$
$3 \otimes 3' = 4 \oplus 5$
$3 \otimes 4 = 3' \oplus 4 \oplus 5$
$3' \otimes 4 = 3 \oplus 4 \oplus 5$
$3 \otimes 5 = 3 \oplus 3' \oplus 4 \oplus 5$
$3' \otimes 5 = 3 \oplus 3' \oplus 4 \oplus 5$
$4 \otimes 4 = 1 \oplus 3 \oplus 3' \oplus 4 \oplus 5$
$4 \otimes 5 = 3 \oplus 3' \oplus 4 \oplus 5 \oplus 5$
$5 \otimes 5 = 1 \oplus 3 \oplus 3' \oplus 4 \oplus 4 \oplus 5 \oplus 5$

4.2.3 Tensor Products

Concrete tensor products are partially available in [4], using the Shirai basis. Here we display the full set of tensor products:

$$\begin{pmatrix} x_1 \\ x_2 \\ x_3 \end{pmatrix}_3 \otimes \begin{pmatrix} y_1 \\ y_2 \\ y_3 \end{pmatrix}_3 = (x_1y_1 + x_2y_2 + x_3y_3)_1 \oplus \begin{pmatrix} x_3y_2 - x_2y_3 \\ x_1y_3 - x_3y_1 \\ x_2y_1 - x_1y_2 \end{pmatrix}_3$$

$$\oplus \begin{pmatrix} x_2y_2 - x_1y_1 \\ x_2y_1 + x_1y_2 \\ x_3y_2 + x_2y_3 \\ x_1y_3 + x_3y_1 \\ -\frac{1}{\sqrt{3}}(x_1y_1 + x_2y_2 - 2x_3y_3) \end{pmatrix}_5, \quad (4.19)$$

$$\begin{pmatrix} x_1 \\ x_2 \\ x_3 \end{pmatrix}_{3'} \otimes \begin{pmatrix} y_1 \\ y_2 \\ y_3 \end{pmatrix}_{3'} = (x_1y_1 + x_2y_2 + x_3y_3)_1 \oplus \begin{pmatrix} x_3y_2 - x_2y_3 \\ x_1y_3 - x_3y_1 \\ x_2y_1 - x_1y_2 \end{pmatrix}_{3'}$$

$$\oplus \begin{pmatrix} \frac{1}{2}(-\frac{1}{\phi}x_1y_1 - \phi x_2y_2 + \sqrt{5}x_3y_3) \\ x_2y_1 + x_1y_2 \\ -(x_3y_1 + x_1y_3) \\ x_2y_3 + x_3y_2 \\ \frac{1}{2\sqrt{3}}[(1 - 3\phi)x_1y_1 + (3\phi - 2)x_2y_2 + x_3y_3] \end{pmatrix}_5,$$

$$\begin{pmatrix} x_1 \\ x_2 \\ x_3 \end{pmatrix}_3 \otimes \begin{pmatrix} y_1 \\ y_2 \\ y_3 \end{pmatrix}_{3'}$$

$$= \begin{pmatrix} \frac{1}{\phi}x_3y_2 - \phi x_1y_3 \\ \phi x_3y_1 + \frac{1}{\phi}x_2y_3 \\ -\frac{1}{\phi}x_1y_1 + \phi x_2y_2 \\ x_2y_1 - x_1y_2 + x_3y_3 \end{pmatrix}_4 \oplus \begin{pmatrix} \frac{1}{2}(\phi^2 x_2y_1 + \frac{1}{\phi^2}x_1y_2 - \sqrt{5}x_3y_3) \\ -(\phi x_1y_1 + \frac{1}{\phi}x_2y_2) \\ \frac{1}{\phi}x_3y_1 - \phi x_2y_3 \\ \phi x_3y_2 + \frac{1}{\phi}x_1y_3 \\ \frac{\sqrt{3}}{2}(\frac{1}{\phi}x_2y_1 + \phi x_1y_2 + x_3y_3) \end{pmatrix}_5, \quad (4.20)$$

$$\begin{pmatrix} x_1 \\ x_2 \\ x_3 \end{pmatrix}_3 \otimes \begin{pmatrix} y_1 \\ y_2 \\ y_3 \\ y_4 \end{pmatrix}_4$$

$$= \begin{pmatrix} -\frac{1}{\phi^2}x_1y_3 + \frac{1}{\phi}x_2y_4 + x_3y_2 \\ -\frac{1}{\phi}x_1y_4 + x_2y_3 + \frac{1}{\phi^2}x_3y_1 \\ -x_1y_1 + \frac{1}{\phi^2}x_2y_2 + \frac{1}{\phi}x_3y_4 \end{pmatrix}_{3'} \oplus \begin{pmatrix} -x_1y_3 + x_2y_4 - x_3y_2 \\ -x_1y_4 - x_2y_3 + x_3y_1 \\ x_1y_1 + x_2y_2 + x_3y_4 \\ x_1y_2 - x_2y_1 - x_3y_3 \end{pmatrix}_4$$

$$\oplus \begin{pmatrix} \frac{1}{2}[(6\phi+5)x_1y_2 + (3\phi+4)x_2y_1 + (3\phi+1)x_3y_3] \\ -x_1y_1 + (3\phi+2)x_2y_2 - (3\phi+1)x_3y_4 \\ -(3\phi+1)x_1y_4 - x_2y_3 - (3\phi+2)x_3y_1 \\ -(3\phi+2)x_1y_3 - (3\phi+1)x_2y_4 + x_3y_2 \\ \frac{\sqrt{3}}{2}[x_1y_2 - (3\phi+2)x_2y_1 + 3(\phi+1)x_3y_3] \end{pmatrix}_5, \quad (4.21)$$

$$\begin{pmatrix} x_1 \\ x_2 \\ x_3 \end{pmatrix}_{3'} \otimes \begin{pmatrix} y_1 \\ y_2 \\ y_3 \\ y_4 \end{pmatrix}_4$$

$$= \begin{pmatrix} x_1y_3 + \phi x_2y_4 + \phi^2 x_3y_1 \\ -\phi x_1y_4 - \phi^2 x_2y_3 - x_3y_2 \\ -\phi^2 x_1y_2 - x_2y_1 - \phi x_3y_4 \end{pmatrix}_3 \oplus \begin{pmatrix} x_1y_4 - x_2y_3 + x_3y_2 \\ x_1y_3 + x_2y_4 - x_3y_1 \\ -x_1y_2 + x_2y_1 + x_3y_4 \\ -(x_1y_1 + x_2y_2 + x_3y_3) \end{pmatrix}_4$$

$$\oplus \begin{pmatrix} x_1y_1 - \phi^4 x_2y_2 + \phi^2(2\phi-1)x_3y_3 \\ x_1y_2 - \phi^4 x_2y_1 + \phi^2(2\phi-1)x_3y_4 \\ \phi^4 x_1y_3 - \phi^2(2\phi-1)x_2y_4 + x_3y_1 \\ \phi^2(2\phi-1)x_1y_4 - x_2y_3 - \phi^4 x_3y_2 \\ -\sqrt{3}\phi(\phi^2 x_1y_1 - x_2y_2 - \phi x_3y_3) \end{pmatrix}_5, \quad (4.22)$$

$$\begin{pmatrix} x_1 \\ x_2 \\ x_3 \end{pmatrix}_3 \otimes \begin{pmatrix} y_1 \\ y_2 \\ y_3 \\ y_4 \\ y_5 \end{pmatrix}_5 = \begin{pmatrix} x_1(y_1 + \frac{1}{\sqrt{3}}y_5) - x_2y_2 - x_3y_4 \\ -x_1y_2 - x_2(y_1 - \frac{1}{\sqrt{3}}y_5) - x_3y_3 \\ -x_1y_4 - x_2y_3 - \frac{2}{\sqrt{3}}x_3y_5 \end{pmatrix}_3$$

$$\oplus \begin{pmatrix} x_1y_2 - \frac{\phi}{2}x_2y_1 - \frac{\sqrt{3}}{2\phi^2}x_2y_5 - \frac{1}{\phi^2}x_3y_3 \\ -\frac{\sqrt{3}}{2}x_1y_5 - \frac{1}{2\phi^3}x_1y_1 + \frac{1}{\phi^2}x_2y_2 - x_3y_4 \\ -\frac{1}{\phi^2}x_1y_4 + x_2y_3 + \frac{\sqrt{5}}{2\phi}x_3y_1 - \frac{\sqrt{3}}{2\phi}x_3y_5 \end{pmatrix}_{3'}$$

$$\oplus \begin{pmatrix} \frac{1}{\phi^2}x_1y_2 + \frac{\phi^2-6}{2}x_2y_1 + \frac{\sqrt{3}}{2}\phi^2 x_2y_5 + \phi^2 x_3y_3 \\ -\frac{\phi+4}{2}x_1y_1 - \frac{\sqrt{3}}{2\phi^2}x_1y_5 - \phi^2 x_2y_2 - \frac{1}{\phi^2}x_3y_4 \\ \phi^2 x_1y_4 + \frac{1}{\phi^2}x_2y_3 - \frac{\sqrt{5}}{2}x_3y_1 - \frac{3\sqrt{3}}{2}x_3y_5 \\ \sqrt{5}(x_1y_3 + x_2y_4 + x_3y_2) \end{pmatrix}_4$$

4.2 A_5

$$\oplus \begin{pmatrix} x_1y_3 + x_2y_4 - 2x_3y_2 \\ x_1y_4 - x_2y_3 + 2x_3y_1 \\ -x_1y_1 + x_2y_2 - x_3y_4 + \sqrt{3}x_1y_5 \\ -x_1y_2 - x_2y_1 + x_3y_3 - \sqrt{3}x_2y_5 \\ -\sqrt{3}(x_1y_3 - x_2y_4) \end{pmatrix}_5, \quad (4.23)$$

$$\begin{pmatrix} x_1 \\ x_2 \\ x_3 \end{pmatrix}_{3'} \otimes \begin{pmatrix} y_1 \\ y_2 \\ y_3 \\ y_4 \\ y_5 \end{pmatrix}_5 = \begin{pmatrix} -\phi^2 x_1 y_2 + \frac{1}{2\phi} x_2 y_1 + \frac{\sqrt{3}}{2} \phi^2 x_2 y_5 + x_3 y_4 \\ \frac{2\phi+1}{2} x_1 y_1 + \frac{\sqrt{3}}{2} x_1 y_5 - x_2 y_2 - \phi^2 x_3 y_3 \\ x_1 y_3 + \phi^2 x_2 y_4 - \frac{\sqrt{5}}{2} \phi x_3 y_1 + \frac{\sqrt{3}}{2} \phi x_3 y_5 \end{pmatrix}_3$$

$$\oplus \begin{pmatrix} \frac{1}{2\phi} x_1 y_1 - x_2 y_2 + x_3 y_3 + \frac{3\phi-1}{2\sqrt{3}} x_1 y_5 \\ -x_1 y_2 + \frac{\phi}{2} x_2 y_1 - x_3 y_4 - \frac{3\phi-2}{2\sqrt{3}} x_2 y_5 \\ x_1 y_3 - x_2 y_4 - \frac{\sqrt{5}}{2} x_3 y_1 - \frac{1}{2\sqrt{3}} x_3 y_5 \end{pmatrix}_{3'}$$

$$\oplus \begin{pmatrix} \frac{1}{\sqrt{5}}(\frac{1}{\phi^2} x_1 y_1 + \phi^2 x_2 y_2 + \frac{1}{\phi^2} x_3 y_3 - \sqrt{3} \phi x_1 y_5) \\ \frac{1}{\sqrt{5}}(-\frac{1}{\phi^2} x_1 y_2 - \phi^2 x_2 y_1 - \frac{\sqrt{3}(\phi-3)}{\sqrt{5}} x_2 y_5 - \phi^2 x_3 y_4) \\ \frac{1}{\sqrt{5}}(\phi^2 x_1 y_3 - \frac{1}{\phi^2} x_2 y_4 + \sqrt{5} x_3 y_1 + \sqrt{3} x_3 y_5) \\ x_1 y_4 - x_2 y_3 + x_3 y_2 \end{pmatrix}_4$$

$$\oplus \begin{pmatrix} -(3\phi-1)x_1 y_4 + (2-3\phi)x_2 y_3 + x_3 y_2 \\ -2x_1 y_3 - 2x_2 y_4 - x_3 y_1 + \sqrt{15} x_3 y_5 \\ 2x_1 y_2 - (2-3\phi)x_2 y_1 - 2x_3 y_4 + \sqrt{3}\phi x_2 y_5 \\ (3\phi-1)x_1 y_1 + 2x_2 y_2 + 2x_3 y_3 - \frac{\sqrt{3}}{\phi} x_1 y_5 \\ \frac{\sqrt{3}}{\phi} x_1 y_4 - \phi\sqrt{3} x_2 y_3 - \sqrt{15} x_3 y_2 \end{pmatrix}_5, \quad (4.24)$$

$$\begin{pmatrix} x_1 \\ x_2 \\ x_3 \\ x_4 \end{pmatrix}_4 \otimes \begin{pmatrix} y_1 \\ y_2 \\ y_3 \\ y_4 \end{pmatrix}_4$$

$$= (x_1 y_1 + x_2 y_2 + x_3 y_3 + x_4 y_4)_1 \oplus \begin{pmatrix} x_1 y_3 + x_2 y_4 - x_3 y_1 - x_4 y_2 \\ -x_1 y_4 + x_2 y_3 - x_3 y_2 + x_4 y_1 \\ x_1 y_2 - x_2 y_1 - x_3 y_4 + x_4 y_3 \end{pmatrix}_3$$

$$\oplus \begin{pmatrix} x_1 y_4 + x_2 y_3 - x_3 y_2 - x_4 y_1 \\ -x_1 y_3 + x_2 y_4 + x_3 y_1 - x_4 y_2 \\ x_1 y_2 - x_2 y_1 + x_3 y_4 - x_4 y_3 \end{pmatrix}_{3'}$$

$$\oplus \begin{pmatrix} x_1 y_4 - \sqrt{5} x_2 y_3 - \sqrt{5} x_3 y_2 + x_4 y_1 \\ -\sqrt{5} x_1 y_3 + x_2 y_4 - \sqrt{5} x_3 y_1 + x_4 y_2 \\ -\sqrt{5} x_1 y_2 - \sqrt{5} x_2 y_1 + x_3 y_4 + x_4 y_3 \\ x_1 y_1 + x_2 y_2 + x_3 y_3 - 3 x_4 y_4 \end{pmatrix}_4$$

$$\oplus \begin{pmatrix} -\frac{\phi^2}{\sqrt{5}}x_1y_1 + \frac{1}{\sqrt{5}\phi^2}x_2y_2 + x_3y_3 \\ -\frac{1}{\sqrt{5}}x_1y_2 - \frac{1}{\sqrt{5}}x_2y_1 - x_3y_4 - x_4y_3 \\ \frac{1}{\sqrt{5}}x_1y_3 + x_2y_4 + \frac{1}{\sqrt{5}}x_3y_1 + x_4y_2 \\ -x_1y_4 - \frac{1}{\sqrt{5}}x_2y_3 - \frac{1}{\sqrt{5}}x_3y_2 - x_4y_1 \\ -\sqrt{\frac{3}{5}}(\frac{1}{\phi}x_1y_1 - \phi x_2y_2 + x_3y_3) \end{pmatrix}_5, \quad (4.25)$$

$$\begin{pmatrix} x_1 \\ x_2 \\ x_3 \\ x_4 \end{pmatrix}_4 \otimes \begin{pmatrix} y_1 \\ y_2 \\ y_3 \\ y_4 \\ y_5 \end{pmatrix}_5$$

$$= \begin{pmatrix} \frac{2}{\phi^2}x_1y_2 - (\phi+4)x_2y_1 + 2\phi^2 x_3y_4 + 2\sqrt{5}x_4y_3 - \frac{\sqrt{3}}{\phi^2}x_2y_5 \\ (\phi-5)x_1y_1 + \sqrt{3}\phi^2 x_1y_5 - 2\phi^2 x_2y_2 + \frac{2}{\phi^2}x_3y_3 + 2\sqrt{5}x_4y_4 \\ 2\phi^2 x_1y_3 - \frac{2}{\phi^2}x_2y_4 - \sqrt{5}x_3y_1 - 3\sqrt{3}x_3y_5 + 2\sqrt{5}x_4y_2 \end{pmatrix}_3$$

$$\oplus \begin{pmatrix} \frac{1}{\phi^2}x_1y_1 - \sqrt{3}\phi x_1y_5 - \frac{1}{\phi^2}x_2y_2 + \phi^2 x_3y_3 + \sqrt{5}x_4y_4 \\ -\phi^2 x_1y_2 - \phi^2 x_2y_1 + \frac{\sqrt{3}}{\phi}x_2y_5 - \frac{1}{\phi^2}x_3y_4 - \sqrt{5}x_4y_3 \\ \frac{1}{\phi^2}x_1y_3 - \phi^2 x_2y_4 + \sqrt{5}x_3y_1 + \sqrt{3}x_3y_5 + \sqrt{5}x_4y_2 \end{pmatrix}_{3'}$$

$$\oplus \begin{pmatrix} -\frac{\phi^2}{\sqrt{5}}x_1y_1 - \frac{1}{\sqrt{5}}x_2y_2 + \frac{1}{\sqrt{5}}x_3y_3 - x_4y_4 - \frac{\sqrt{3}}{\sqrt{5}\phi}x_1y_5 \\ -\frac{1}{\sqrt{5}}x_1y_2 + \frac{1}{\sqrt{5}\phi^2}x_2y_1 + \sqrt{\frac{3}{5}}\phi x_2y_5 - \frac{1}{\sqrt{5}}x_3y_4 + x_4y_3 \\ \frac{1}{\sqrt{5}}x_1y_3 - \frac{1}{\sqrt{5}}x_2y_4 + x_3y_1 - \sqrt{\frac{3}{5}}x_3y_5 - x_4y_2 \\ -x_1y_4 + x_2y_3 - x_3y_2 \end{pmatrix}_4$$

$$\oplus \begin{pmatrix} \frac{1}{2}(\phi^2 x_1y_4 - \frac{1}{\phi^2}x_2y_3 - 3x_3y_2 + 3x_4y_1 + \sqrt{\frac{5}{3}}x_4y_5) \\ \phi x_1y_3 - \frac{1}{\phi}x_2y_4 - x_3y_1 - x_4y_2 + \sqrt{\frac{5}{3}}x_3y_5 \\ \frac{1}{\phi}x_1y_2 + \frac{1}{\phi}x_2y_1 + \frac{1}{\sqrt{3}}\phi^2 x_2y_5 + \phi x_3y_4 - x_4y_3 \\ \phi x_1y_1 + \frac{1}{\sqrt{3}\phi^2}x_1y_5 - \phi x_2y_2 + \frac{1}{\phi}x_3y_3 - x_4y_4 \\ \frac{1}{2\sqrt{3}}(-(\phi-5)x_1y_4 + (\phi+4)x_2y_3 + \sqrt{5}x_3y_2 - \sqrt{5}x_4y_1) + \frac{3}{2}x_4y_5 \end{pmatrix}_5,$$

$$\oplus \begin{pmatrix} x_1y_4 - x_2y_3 - 2x_3y_2 + \frac{3}{2}x_4y_1 + \frac{\sqrt{15}}{2}x_4y_5 \\ \phi^2 x_1y_3 + \frac{1}{\phi^2}x_2y_4 - \frac{1}{2}x_3y_1 + \frac{\sqrt{15}}{2}x_3y_5 - x_4y_2 \\ -\frac{1}{\phi^2}x_1y_2 + \frac{3\phi-2}{2}x_2y_1 + \frac{\sqrt{3}}{2}\phi x_2y_5 + \phi^2 x_3y_4 - x_4y_3 \\ \frac{3\phi-1}{2}x_1y_1 - \phi^2 x_2y_2 - \frac{1}{\phi^2}x_3y_3 - x_4y_4 - \frac{\sqrt{3}}{2\phi}x_1y_5 \\ \sqrt{3}x_1y_4 + \sqrt{3}x_2y_3 + \frac{3}{2}x_4y_5 - \frac{\sqrt{15}}{2}x_4y_1 \end{pmatrix}_5 \quad (4.26)$$

$$\begin{pmatrix} x_1 \\ x_2 \\ x_3 \\ x_4 \\ x_5 \end{pmatrix}_5 \otimes \begin{pmatrix} y_1 \\ y_2 \\ y_3 \\ y_4 \\ y_5 \end{pmatrix}_5$$
$$= (x_1 y_1 + x_2 y_2 + x_3 y_3 + x_4 y_4 + x_5 y_5)_1$$
$$\oplus \begin{pmatrix} x_1 y_3 - x_3 y_1 + x_2 y_4 - x_4 y_2 + \sqrt{3}(x_3 y_5 - x_5 y_3) \\ x_1 y_4 - x_4 y_1 - (x_2 y_3 - x_3 y_2) - \sqrt{3}(x_4 y_5 - x_5 y_4) \\ -2(x_1 y_2 - x_2 y_1) - (x_3 y_4 - x_4 y_3) \end{pmatrix}_3$$
$$\oplus \begin{pmatrix} (2\phi+3)(x_1 y_4 - x_4 y_1) + 2\phi(x_2 y_3 - x_3 y_2) + \sqrt{3}(x_4 y_5 - x_5 y_4) \\ (\phi+3)(x_1 y_3 - x_3 y_1) + 2\phi(x_2 y_4 - x_4 y_2) - \sqrt{3}\phi^2(x_3 y_5 - x_5 y_3) \\ -\phi(x_1 y_2 - x_2 y_1) - \sqrt{15}\phi(x_2 y_5 - x_5 y_2) + 2\phi(x_3 y_4 - x_4 y_3) \end{pmatrix}_{3'}$$
$$\oplus \begin{pmatrix} \sqrt{5}(\frac{\phi^2}{2}(x_1 y_4 + x_4 y_1) + x_2 y_3 + x_3 y_2 + \frac{\sqrt{3}}{2\phi}(x_4 y_5 + x_5 y_4)) \\ \sqrt{5}(\frac{1}{2\phi^2}(x_1 y_3 + x_3 y_1) - (x_2 y_4 + x_4 y_2) + \frac{\sqrt{3}}{2}\phi(x_3 y_5 + x_5 y_3)) \\ \frac{\sqrt{5}}{2}(-\sqrt{5}(x_1 y_2 + x_2 y_1) + \sqrt{3}(x_2 y_5 + x_5 y_2) + 2(x_3 y_4 + x_4 y_3)) \\ 3x_1 y_1 - 2(x_2 y_2 + x_3 y_3 + x_4 y_4) + 3x_5 y_5 \end{pmatrix}_4$$
$$\oplus \begin{pmatrix} \frac{1}{\sqrt{5}}(\frac{1}{2\phi}(x_1 y_4 - x_4 y_1) - (x_2 y_3 - x_3 y_2) + \frac{\phi^2}{2\sqrt{3}}(x_4 y_5 - x_5 y_4)) \\ \frac{1}{\sqrt{5}}(\frac{\phi}{2}(x_1 y_3 - x_3 y_1) - (x_2 y_4 - x_4 y_2) + \frac{1}{2\sqrt{3}\phi^2}(x_3 y_5 - x_5 y_3)) \\ \frac{1}{2\sqrt{5}}(x_1 y_2 - x_2 y_1) - \frac{1}{2\sqrt{3}}(x_2 y_5 - x_5 y_2) - \frac{1}{\sqrt{5}}(x_3 y_4 - x_4 y_3) \\ -\frac{1}{\sqrt{3}}(x_1 y_5 - x_5 y_1) \end{pmatrix}_4$$
$$\oplus \begin{pmatrix} -x_1 y_1 - \frac{11}{3\sqrt{15}}(x_1 y_5 + x_5 y_1) + \frac{4}{3}x_2 y_2 - \frac{4\sqrt{5}}{15}(\phi x_3 y_3 + \frac{1}{\phi}x_4 y_4) + x_5 y_5 \\ \frac{4}{3}(x_1 y_2 + x_2 y_1) + \frac{1}{\sqrt{15}}(x_2 y_5 + x_5 y_2) + \frac{2}{\sqrt{5}}(x_3 y_4 + x_4 y_3)) \\ \frac{4}{3\sqrt{5}}(-\phi(x_1 y_3 + x_3 y_1) + 2(x_2 y_4 + x_4 y_2) - \frac{2-3\phi}{\sqrt{3}}(x_3 y_5 + x_5 y_3)) \\ \frac{4\sqrt{5}}{15}(-\frac{1}{\phi}(x_1 y_4 + x_4 y_1) + 2(x_2 y_3 + x_3 y_2) - \frac{\sqrt{3}}{3}(3\phi-1)(x_4 y_5 + x_5 y_4)) \\ x_1 y_5 + x_5 y_1 + \frac{1}{3\sqrt{15}}(-11 x_1 y_1 + 4 x_2 y_2 + 11 x_5 y_5) - \frac{4\sqrt{15}}{45}((2-3\phi)x_3 y_3 + (3\phi-1)x_4 y_4) \end{pmatrix}_5$$
$$\oplus \begin{pmatrix} -\frac{3\sqrt{5}}{4}(x_1 y_1 - x_5 y_5) - \frac{\sqrt{3}}{4}(x_1 y_5 + x_5 y_1) + \sqrt{5}x_2 y_2 - \phi^2 x_3 y_3 + \frac{1}{\phi^2}x_4 y_4 \\ \sqrt{5}(x_1 y_2 + x_2 y_1) + \sqrt{3}(x_2 y_5 + x_5 y_2) + x_3 y_4 + x_4 y_3 \\ -\phi^2(x_1 y_3 + x_3 y_1) + x_2 y_4 + x_4 y_2 + \frac{\sqrt{3}}{\phi}(x_3 y_5 + x_5 y_3) \\ \frac{1}{\phi^2}(x_1 y_4 + x_4 y_1) + (x_2 y_3 + x_3 y_2) - \sqrt{3}\phi(x_4 y_5 + x_5 y_4) \\ -\frac{\sqrt{3}}{4}(x_1 y_1 - 4 x_2 y_2 - x_5 y_5) + \frac{3\sqrt{5}}{4}(x_1 y_5 + x_5 y_1) + \frac{\sqrt{3}}{\phi}x_3 y_3 - \sqrt{3}\phi x_4 y_4 \end{pmatrix}_5$$
(4.27)

References

1. Shirai, K.: J. Phys. Soc. Jpn. **61**, 2735 (1992)
2. Cummins, C.J., Patera, J.: J. Math. Phys. **29**, 1736 (1988)
3. Feruglio, F., Paris, A.: J. High Energy Phys. **1103**, 101 (2011). arXiv:1101.0393 [hep-ph]
4. Everett, L.L., Stuart, A.J.: Phys. Rev. D **79**, 085005 (2009). arXiv:0812.1057 [hep-ph]

Chapter 5
T'

In this section, we study the group T', which is the double covering group of $A_4 = T$. Instead of (4.6) for the case of A_4, we consider the following algebraic relations:

$$s^2 = r, \qquad r^2 = t^3 = (st)^3 = e, \qquad rt = tr. \qquad (5.1)$$

The closed algebra generated by r, s, and t forms the group T', which contains 24 elements.

5.1 Conjugacy Classes

The 24 elements of T' are classified by their orders according to:

$$\begin{aligned}
h &= 1: \quad \{e\}, \\
h &= 2: \quad \{r\}, \\
h &= 3: \quad \{t, t^2, ts, st, rst^2, rt^2s, rtst, rsts\}, \\
h &= 4: \quad \{s, rs, tst^2, t^2st, rtst^2, rt^2st\}, \\
h &= 6: \quad \{rt, rst, rts, rt^2, sts, st^2, t^2s, tst\}.
\end{aligned} \qquad (5.2)$$

Moreover, these elements fall into seven conjugacy classes:

$$\begin{aligned}
C_1 &: \quad \{e\}, & h &= 1, \\
C_1' &: \quad \{r\}, & h &= 2, \\
C_4 &: \quad \{t, rsts, st, ts\}, & h &= 3, \\
C_4' &: \quad \{t^2, rtst, rt^2s, rst^2\}, & h &= 3, \\
C_6 &: \quad \{s, rs, tst^2, t^2st, rtst^2, rt^2st\}, & h &= 4, \\
C_4'' &: \quad \{rt, sts, rst, rts\}, & h &= 6, \\
C_4''' &: \quad \{rt^2, tst, t^2s, st^2\}, & h &= 6.
\end{aligned} \qquad (5.3)$$

Table 5.1 Characters of T' representations

	h	χ_1	$\chi_{1'}$	$\chi_{1''}$	χ_2	$\chi_{2'}$	$\chi_{2''}$	χ_3
C_1	1	1	1	1	2	2	2	3
C'_1	2	1	1	1	-2	-2	-2	3
C_4	3	1	ω	ω^2	-1	$-\omega$	$-\omega^2$	0
C'_4	3	1	ω^2	ω	-1	$-\omega^2$	$-\omega$	0
C''_4	6	1	ω	ω^2	1	ω	ω^2	0
C'''_4	6	1	ω^2	ω	1	ω^2	ω	0
C_6	4	1	1	1	0	0	0	-1

5.2 Characters and Representations

The orthogonality relations (2.18) and (2.19) for T' lead to

$$m_1 + 2^2 m_2 + 3^2 m_3 + \cdots = 24, \tag{5.4}$$

$$m_1 + m_2 + m_3 + \cdots = 7. \tag{5.5}$$

The solution is $(m_1, m_2, m_3) = (3, 3, 1)$. Therefore, we find three singlets, three doublets, and a triplet in T'.

Now consider the characters, which are obtained by a similar analysis to $A_4 = T$. We begin with the singlets. Because $s^4 = r^2 = e$, there are four possibilities for $\chi_\alpha(s) = i^n$ ($n = 0, 1, 2, 3$). However, since t and ts belong to the same conjugacy class, namely C_4, the only value of the character that is consistent with the conjugacy class structure is $\chi_\alpha(s) = 1$ for the singlets. This also implies that $\chi_\alpha(r) = 1$. Then, as for $A_4 = T$, the three singlets are classified by three possible values of $\chi_\alpha(t) = \omega^n$. That is, the three singlets **1**, **1'**, and **1''** are classified by the three values $\chi_\alpha(t) = 1$, ω, and ω^2, respectively. These are shown in Table 5.1.

Next consider the three doublet representations **2**, **2'**, and **2''**, and the triplet representation **3** for r. The element r commutes with all elements. This implies by Schur's lemma that r can be represented by

$$\lambda_{2,2',2''} \begin{pmatrix} 1 & 0 \\ 0 & 1 \end{pmatrix}, \tag{5.6}$$

on **2**, **2'**, and **2''**, and

$$\lambda_3 \begin{pmatrix} 1 & 0 & 0 \\ 0 & 1 & 0 \\ 0 & 0 & 1 \end{pmatrix}, \tag{5.7}$$

on **3**. In addition, the possible values of $\lambda_{2,2',2''}$ and λ_3 must be equal to $\lambda_{2,2',2''} = \pm 1$ and $\lambda_3 = \pm 1$ because $r^2 = e$. Thus, we obtain the possible values of the characters as $\chi_2(r), \chi_{2'}(r), \chi_{2''}(r) = \pm 2$ and $\chi_3(r) = \pm 3$. On the other hand, the second orthogonality relation between e and r gives

$$\sum_\alpha \chi_{D_\alpha}(e)^* \chi_{D_\alpha}(r) = 3 + 2\chi_2(r) + 2\chi_{2'}(r) + 2\chi_{2''}(r) + 3\chi_3(r) = 0, \tag{5.8}$$

5.2 Characters and Representations

where $\chi_{1,1',1''}(r) = 1$ has been used. We obtain the solution

$$\chi_2(r) = \chi_{2'}(r) = \chi_{2''}(r) = -2, \qquad \chi_3(r) = 3.$$

These results are summarized in Table 5.1.

The element r is therefore represented by

$$r = -\begin{pmatrix} 1 & 0 \\ 0 & 1 \end{pmatrix}, \qquad (5.9)$$

on **2**, **2'**, and **2''**, and

$$r = \begin{pmatrix} 1 & 0 & 0 \\ 0 & 1 & 0 \\ 0 & 0 & 1 \end{pmatrix}, \qquad (5.10)$$

on **3**.

We now consider the doublet representation of t. We use the basis diagonalizing t. Because $t^3 = e$, the element t can be written in the form

$$\begin{pmatrix} \omega^k & 0 \\ 0 & \omega^\ell \end{pmatrix}, \qquad (5.11)$$

with $k, \ell = 0, 1, 2$. However, if $k = \ell$, the above matrix would become proportional to the (2×2) identity matrix, that is, the element t would also commute with all the elements. In fact, it is nothing but a singlet representation. Hence, we should have the condition $k \neq \ell$ and it follows that there are three possible values for the trace of the above values as $\omega^k + \omega^\ell = -\omega^n$ with $k, \ell, n = 0, 1, 2$ and $k \neq \ell, \ell \neq n$, $n \neq k$. That is, the characters of t for the three doublets **2**, **2'**, and **2''** are classified by $\chi_2(t) = -1$, $\chi_{2'}(t) = -\omega$, and $\chi_{2''}(t) = -\omega^2$. These are shown in Table 5.1. The element t is thus represented by

$$t = \begin{pmatrix} \omega^2 & 0 \\ 0 & \omega \end{pmatrix}, \quad \text{on } \mathbf{2}, \qquad (5.12)$$

$$t = \begin{pmatrix} 1 & 0 \\ 0 & \omega^2 \end{pmatrix}, \quad \text{on } \mathbf{2'}, \qquad (5.13)$$

and

$$t = \begin{pmatrix} \omega & 0 \\ 0 & 1 \end{pmatrix}, \quad \text{on } \mathbf{2''}. \qquad (5.14)$$

Since we have found explicit (2×2) matrices for r and t on all three doublets, it is straightforward to calculate the explicit forms of rt and rt^2, which belong to the conjugacy classes C'_4 and C''_4, respectively. Then, it is also straightforward to compute the characters of C'_4 and C''_4 for the doublets using the explicit forms of the (2×2) matrices for rt and rt^2. These are shown in Table 5.1.

In order to determine the character of t for the triplet $\chi_3(t)$, we exploit the second orthogonality relation between e and t:

$$\sum_\alpha \chi_{D_\alpha}(e)^* \chi_{D_\alpha}(t) = 0. \qquad (5.15)$$

Since all the characters $\chi_\alpha(t)$ except $\chi_3(t)$ have been derived in the above, the orthogonality relation (5.15) requires $\chi_3(t) = 0$, that is, $\chi_3(C_4) = 0$. Similarly, it is found that $\chi_3(C_4') = \chi_3(C_4'') = \chi_3(C_4''') = 0$, as shown in Table 5.1.

We now consider the explicit form of the (3×3) matrix for t on the triplet. We take the basis to diagonalize t. Since $t^3 = e$ and $\chi_3(t) = 0$, we obtain

$$t = \begin{pmatrix} 1 & 0 & 0 \\ 0 & \omega & 0 \\ 0 & 0 & \omega^2 \end{pmatrix}, \quad \text{on } \mathbf{3}. \tag{5.16}$$

Finally, we study the characters of C_6, which contains s, for the doublets and the triplet. Here, we use the first orthogonality relation between the trivial singlet representation and the doublet representation **2**:

$$\sum_{g \in G} \chi_1(g)^* \chi_2(g) = 0. \tag{5.17}$$

Since all the characters except $\chi_2(C_6)$ have already been identified, this orthogonality relation (5.17) requires $\chi_2(C_6) = 0$. Similarly, we find $\chi_{2'}(C_6) = \chi_{2''}(C_6) = 0$. In addition, the character of C_6 for the triplet $\chi_3(C_6)$ is also determined using the orthogonality relation $\sum_{g \in G} \chi_1(g)^* \chi_2(g) = 0$ and the known values of the other characters. As a result, we obtain $\chi_3(C_6) = -1$. We have thus found all characters of the T' group. These are summarized in Table 5.1.

Let us now investigate the explicit form of s on the doublets and triplet. On the doublets, this element must act as a (2×2) unitary matrix which satisfies $\text{tr}(s) = 0$ and $s^2 = r$. Recall that the doublet representation for r has already been obtained in (5.9). Thus, the element s could be represented by

$$s = -\frac{1}{\sqrt{3}} \begin{pmatrix} i & \sqrt{2}p \\ \sqrt{2}\bar{p} & -i \end{pmatrix}, \quad p = e^{i\phi}, \tag{5.18}$$

on the doublet representations. For example, for **2**, this representation of s satisfies

$$\text{tr}(st) = -\frac{i}{\sqrt{3}}(\omega^2 - \omega) = -1, \tag{5.19}$$

so the ambiguity of p cannot be removed. Similarly, we can study the explicit form of s on the triplet.

Here, we summarize the doublet and triplet representations:

$$t = \begin{pmatrix} \omega^2 & 0 \\ 0 & \omega \end{pmatrix}, \quad r = \begin{pmatrix} -1 & 0 \\ 0 & -1 \end{pmatrix}, \quad s = -\frac{1}{\sqrt{3}} \begin{pmatrix} i & \sqrt{2}p \\ -\sqrt{2}\bar{p} & -i \end{pmatrix} \quad \text{on } \mathbf{2}, \tag{5.20}$$

$$t = \begin{pmatrix} 1 & 0 \\ 0 & \omega^2 \end{pmatrix}, \quad r = \begin{pmatrix} -1 & 0 \\ 0 & -1 \end{pmatrix}, \quad s = -\frac{1}{\sqrt{3}} \begin{pmatrix} i & \sqrt{2}p \\ -\sqrt{2}\bar{p} & -i \end{pmatrix} \quad \text{on } \mathbf{2'}, \tag{5.21}$$

$$t = \begin{pmatrix} \omega & 0 \\ 0 & 1 \end{pmatrix}, \quad r = \begin{pmatrix} -1 & 0 \\ 0 & -1 \end{pmatrix}, \quad s = -\frac{1}{\sqrt{3}} \begin{pmatrix} i & \sqrt{2}p \\ -\sqrt{2}\bar{p} & -i \end{pmatrix} \quad \text{on } \mathbf{2''}, \tag{5.22}$$

5.3 Tensor Products

$$t = \begin{pmatrix} 1 & 0 & 0 \\ 0 & \omega & 0 \\ 0 & 0 & \omega^2 \end{pmatrix}, \quad r = \begin{pmatrix} 1 & 0 & 0 \\ 0 & 1 & 0 \\ 0 & 0 & 1 \end{pmatrix},$$

$$s = \frac{1}{3} \begin{pmatrix} -1 & 2p_1 & 2p_1 p_2 \\ 2\bar{p}_1 & -1 & 2p_2 \\ 2\bar{p}_1 \bar{p}_2 & 2\bar{p}_2 & -1 \end{pmatrix} \quad \text{on } \mathbf{3},$$

(5.23)

where $p_1 = e^{i\phi_1}$ and $p_2 = e^{i\phi_2}$.

5.3 Tensor Products

First, we study the tensor product of **2** and **2**, i.e.,

$$\begin{pmatrix} x_1 \\ x_2 \end{pmatrix}_{\mathbf{2}} \otimes \begin{pmatrix} y_1 \\ y_2 \end{pmatrix}_{\mathbf{2}}.$$

(5.24)

Let us investigate the transformation properties of elements $x_i y_j$, for $i, j = 1, 2$, under t, r, and s. It is easily found that

$$\begin{pmatrix} x_1 \\ x_2 \end{pmatrix}_{\mathbf{2}(\mathbf{2}')} \otimes \begin{pmatrix} y_1 \\ y_2 \end{pmatrix}_{\mathbf{2}(\mathbf{2}'')} = \left(\frac{x_1 y_2 - x_2 y_1}{\sqrt{2}} \right)_{\mathbf{1}} \oplus \begin{pmatrix} \frac{i}{\sqrt{2}} p_1 p_2 \bar{p}(x_1 y_2 + x_2 y_1) \\ p_2 \bar{p}^2 x_1 y_1 \\ x_2 y_2 \end{pmatrix}_{\mathbf{3}}.$$

(5.25)

Similarly, we obtain

$$\begin{pmatrix} x_1 \\ x_2 \end{pmatrix}_{\mathbf{2}'(\mathbf{2})} \otimes \begin{pmatrix} y_1 \\ y_2 \end{pmatrix}_{\mathbf{2}'(\mathbf{2}'')} = \left(\frac{x_1 y_2 - x_2 y_1}{\sqrt{2}} \right)_{\mathbf{1}''} \oplus \begin{pmatrix} p_1 \bar{p}^2 x_1 y_1 \\ x_2 y_2 \\ \frac{i}{\sqrt{2}} \bar{p} \bar{p}_2 (x_1 y_2 + x_2 y_1) \end{pmatrix}_{\mathbf{3}},$$

(5.26)

$$\begin{pmatrix} x_1 \\ x_2 \end{pmatrix}_{\mathbf{2}''(\mathbf{2})} \otimes \begin{pmatrix} y_1 \\ y_2 \end{pmatrix}_{\mathbf{2}''(\mathbf{2}')} = \left(\frac{x_1 y_2 - x_2 y_1}{\sqrt{2}} \right)_{\mathbf{1}'} \oplus \begin{pmatrix} x_2 y_2 \\ \frac{i}{\sqrt{2}} \bar{p} \bar{p}_1 (x_1 y_2 + x_2 y_1) \\ \bar{p}^2 \bar{p}_1 \bar{p}_2 x_1 y_1 \end{pmatrix}_{\mathbf{3}}.$$

(5.27)

We can also compute other products such as $\mathbf{2} \times \mathbf{2}'$, $\mathbf{2} \times \mathbf{2}''$, and $\mathbf{2}' \times \mathbf{2}''$. It is found that

$$\mathbf{2} \times \mathbf{2}' = \mathbf{2}'' \times \mathbf{2}'', \qquad \mathbf{2} \times \mathbf{2}'' = \mathbf{2}' \times \mathbf{2}', \qquad \mathbf{2}' \times \mathbf{2}'' = \mathbf{2} \times \mathbf{2}.$$

(5.28)

Moreover, a similar analysis leads to

$$\begin{pmatrix} x_1 \\ x_2 \\ x_3 \end{pmatrix}_{\mathbf{3}} \otimes \begin{pmatrix} y_1 \\ y_2 \\ y_3 \end{pmatrix}_{\mathbf{3}} = \left[x_1 y_1 + p_1^2 p_2 (x_2 y_3 + x_3 y_2) \right]_{\mathbf{1}}$$

$$\oplus \left[x_3 y_3 + \bar{p}_1 \bar{p}_2^2 (x_1 y_2 + x_2 y_1) \right]_{\mathbf{1}'}$$

$$\oplus \left[(x_2 y_2 + \bar{p}_1 p_2 (x_1 y_3 + x_3 y_1) \right]_{1''}$$

$$\oplus \frac{1}{3} \begin{pmatrix} 2x_1 y_1 - p_1^2 p_2 (x_2 y_3 + x_3 y_3) \\ 2 p_1 p_2^2 x_3 y_3 - x_1 y_2 - x_2 y_1 \\ 2 p_1 \bar{p}_2 x_2 y_2 - x_1 y_3 - x_3 y_1 \end{pmatrix}_3$$

$$\oplus \frac{1}{2} \begin{pmatrix} x_2 y_3 - x_3 y_2 \\ \bar{p}_1^2 \bar{p}_2 (x_1 y_2 - x_2 y_1) \\ \bar{p}_1^2 \bar{p}_2 (x_3 y_1 - x_1 y_3) \end{pmatrix}_3, \tag{5.29}$$

$$\begin{pmatrix} x_1 \\ x_2 \end{pmatrix}_{2,2',2''} \otimes \begin{pmatrix} y_1 \\ y_2 \\ y_3 \end{pmatrix}_3 = \begin{pmatrix} -i\sqrt{2} p p_1 x_2 y_2 + x_1 y_1 \\ i\sqrt{2} \bar{p} p_1 p_2 x_1 y_3 - x_2 y_1 \end{pmatrix}_{2,2',2''}$$

$$\oplus \begin{pmatrix} -i\sqrt{2} p p_2 x_2 y_3 + x_1 y_2 \\ i\sqrt{2} \bar{p} \bar{p}_1 x_1 y_1 - x_2 y_2 \end{pmatrix}_{2',2'',2}$$

$$\oplus \begin{pmatrix} -i\sqrt{2} p \bar{p}_1 \bar{p}_2 x_2 y_1 + x_1 y_3 \\ i\sqrt{2} \bar{p} \bar{p}_2 x_1 y_2 - x_2 y_3 \end{pmatrix}_{2'',2,2'}, \tag{5.30}$$

$$(x)_{1'(1'')} \otimes \begin{pmatrix} y_1 \\ y_2 \end{pmatrix}_{2,2',2''} = \begin{pmatrix} xy_1 \\ xy_2 \end{pmatrix}_{2'(2''),2''(2),2(2')}, \tag{5.31}$$

$$(x)_{1'} \otimes \begin{pmatrix} y_1 \\ y_2 \\ y_3 \end{pmatrix}_3 = \begin{pmatrix} xy_3 \\ \bar{p}_1^2 \bar{p}_2 x y_1 \\ \bar{p}_1 \bar{p}_2^2 x y_2 \end{pmatrix}_3, \quad (x)_{1''} \otimes \begin{pmatrix} y_1 \\ y_2 \\ y_3 \end{pmatrix}_3 = \begin{pmatrix} xy_2 \\ \bar{p}_1 p_2 x y_3 \\ \bar{p}_1^2 \bar{p}_2 x y_1 \end{pmatrix}_3. \tag{5.32}$$

The representations for p' can be obtained in general by transforming p as follows:

$$\Phi_2(p') = \begin{pmatrix} 1 & 0 \\ 0 & e^{-i\gamma} \end{pmatrix} \Phi_2(p), \quad p' = p e^{i\gamma}, \tag{5.33}$$

$$\Phi_3(p') = \begin{pmatrix} 1 & 0 & 0 \\ 0 & e^{-i\gamma} & 0 \\ 0 & 0 & e^{-i(\alpha+\beta)} \end{pmatrix} \Phi_3(p), \quad p'_1 = p_1 e^{i\alpha}, \ p'_2 = p_2 e^{-i\beta}. \tag{5.34}$$

If one takes the parameters $p = i$ and $p_1 = p_2 = 1$, then the generator s simplifies to

$$s = -\frac{i}{\sqrt{3}} \begin{pmatrix} 1 & \sqrt{2} \\ \sqrt{2} & -1 \end{pmatrix}, \quad \text{on } 2, 2', 2'', \tag{5.35}$$

$$s = \frac{1}{3} \begin{pmatrix} -1 & 2 & 2 \\ 2 & -1 & 2 \\ 2 & 2 & -1 \end{pmatrix}, \quad \text{on } 3. \tag{5.36}$$

These tensor products can also be simplified to:

$$\begin{pmatrix} x_1 \\ x_2 \end{pmatrix}_{2(2')} \otimes \begin{pmatrix} y_1 \\ y_2 \end{pmatrix}_{2(2'')} = \begin{pmatrix} \frac{x_1 y_2 - x_2 y_1}{\sqrt{2}} \end{pmatrix}_1 \oplus \begin{pmatrix} \frac{x_1 y_2 + x_2 y_1}{\sqrt{2}} \\ -x_1 y_1 \\ x_2 y_2 \end{pmatrix}_3, \tag{5.37}$$

5.3 Tensor Products

$$\begin{pmatrix} x_1 \\ x_2 \end{pmatrix}_{2'(2)} \otimes \begin{pmatrix} y_1 \\ y_2 \end{pmatrix}_{2'(2'')} = \left(\frac{x_1 y_2 - x_2 y_1}{\sqrt{2}} \right)_{1''} \oplus \begin{pmatrix} -x_1 y_1 \\ x_2 y_2 \\ \frac{x_1 y_2 + x_2 y_1}{\sqrt{2}} \end{pmatrix}_{3}, \quad (5.38)$$

$$\begin{pmatrix} x_1 \\ x_2 \end{pmatrix}_{2''(2)} \otimes \begin{pmatrix} y_1 \\ y_2 \end{pmatrix}_{2''(2')} = \left(\frac{x_1 y_2 - x_2 y_1}{\sqrt{2}} \right)_{1'} \oplus \begin{pmatrix} x_2 y_2 \\ \frac{x_1 y_2 + x_2 y_1}{\sqrt{2}} \\ -x_1 y_1 \end{pmatrix}_{3}, \quad (5.39)$$

$$\begin{pmatrix} x_1 \\ x_2 \\ x_3 \end{pmatrix}_3 \otimes \begin{pmatrix} y_1 \\ y_2 \\ y_3 \end{pmatrix}_3 = [x_1 y_1 + x_2 y_3 + x_3 y_2]_1$$

$$\oplus [x_3 y_3 + x_1 y_2 + x_2 y_1]_{1'} \oplus [x_2 y_2 + x_1 y_3 + x_3 y_1]_{1''}$$

$$\oplus \frac{1}{3} \begin{pmatrix} 2x_1 y_1 - x_2 y_3 - x_3 y_2 \\ 2x_3 y_3 - x_1 y_2 - x_2 y_1 \\ 2x_2 y_2 - x_1 y_3 - x_3 y_1 \end{pmatrix}_3$$

$$\oplus \frac{1}{2} \begin{pmatrix} x_2 y_3 - x_3 y_2 \\ x_1 y_2 - x_2 y_1 \\ x_3 y_1 - x_1 y_3 \end{pmatrix}_3, \quad (5.40)$$

$$\begin{pmatrix} x_1 \\ x_2 \end{pmatrix}_{2, 2', 2''} \otimes \begin{pmatrix} y_1 \\ y_2 \\ y_3 \end{pmatrix}_3 = \begin{pmatrix} \sqrt{2} x_2 y_2 + x_1 y_1 \\ \sqrt{2} x_1 y_3 - x_2 y_1 \end{pmatrix}_{2, 2', 2''} \oplus \begin{pmatrix} \sqrt{2} x_2 y_3 + x_1 y_2 \\ \sqrt{2} x_1 y_1 - x_2 y_2 \end{pmatrix}_{2', 2'', 2}$$

$$\oplus \begin{pmatrix} \sqrt{2} x_2 y_1 + x_1 y_3 \\ \sqrt{2} x_1 y_2 - x_2 y_3 \end{pmatrix}_{2'', 2, 2'}, \quad (5.41)$$

$$(x)_{1'(1'')} \otimes \begin{pmatrix} y_1 \\ y_2 \end{pmatrix}_{2, 2', 2''} = \begin{pmatrix} x y_1 \\ x y_2 \end{pmatrix}_{2'(2''), 2''(2), 2(2')}, \quad (5.42)$$

$$(x)_{1'} \otimes \begin{pmatrix} y_1 \\ y_2 \\ y_3 \end{pmatrix}_3 = \begin{pmatrix} x y_3 \\ x y_1 \\ x y_2 \end{pmatrix}_3, \quad (x)_{1''} \otimes \begin{pmatrix} y_1 \\ y_2 \\ y_3 \end{pmatrix}_3 = \begin{pmatrix} x y_2 \\ x y_3 \\ x y_1 \end{pmatrix}_3. \quad (5.43)$$

When $p = e^{i\pi/12}$, and $p_1 = p_2 = \omega$, the representations and their tensor products are as given in Appendix E.2.

Chapter 6
D_N

In this chapter, we discuss the dihedral group, which is denoted by D_N. It is the symmetry group of the regular polygon with N sides. This group is isomorphic to $Z_N \rtimes Z_2$ and is also denoted by $\Delta(2N)$. It consists of cyclic rotations Z_N and reflections. That is, it is generated by two generators a and b, which act on the N edges x_i ($i = 1, \ldots, N$) of the N-polygon according to

$$a : (x_1, x_2, \ldots, x_N) \to (x_N, x_1, \ldots, x_{N-1}), \tag{6.1}$$

$$b : (x_1, x_2, \ldots, x_N) \to (x_1, x_N, \ldots, x_2). \tag{6.2}$$

These two generators satisfy

$$a^N = e, \qquad b^2 = e, \qquad bab = a^{-1}, \tag{6.3}$$

where the third equation is equivalent to $aba = b$. The order of D_N is $2N$, and all of the $2N$ elements can be written in the form $a^m b^k$, with $m = 0, \ldots, N-1$ and $k = 0, 1$. The third equation in (6.3) implies that the Z_N subgroup including a^m is a normal subgroup of D_N. Thus, D_N corresponds to a semi-direct product between Z_N including a^m and Z_2 including b^k, i.e., $Z_N \rtimes Z_2$. Equation (6.1) corresponds to the (reducible) N-dimensional representation. The simple doublet representation is

$$a = \begin{pmatrix} \cos 2\pi/N & -\sin 2\pi/N \\ \sin 2\pi/N & \cos 2\pi/N \end{pmatrix}, \qquad b = \begin{pmatrix} 1 & 0 \\ 0 & -1 \end{pmatrix}. \tag{6.4}$$

6.1 D_N with N Even

The groups D_N have different features for N even and odd. We begin by studying D_N when N is even.

6.1.1 Conjugacy Classes

The algebraic relations (6.3) tell us that a^m and a^{N-m} belong to the same conjugacy class and also that b and $a^{2m}b$ belong to the same conjugacy class. When N is even, D_N has the following $3 + N/2$ conjugacy classes:

$$
\begin{array}{lll}
C_1: & \{e\}, & h = 1, \\
C_2^{(1)}: & \{a, a^{N-1}\}, & h = N, \\
\vdots & \vdots & \vdots \\
C_2^{(N/2-1)}: & \{a^{N/2-1}, a^{N/2+1}\}, & h = N/\gcd(N, N/2 - 1), \\
C_1': & \{a^{N/2}\}, & h = 2, \\
C_{N/2}: & \{b, a^2 b, \ldots, a^{N-2} b\}, & h = 2, \\
C_{N/2}': & \{ab, a^3 b, \ldots, a^{N-1} b\}, & h = 2,
\end{array}
\quad (6.5)
$$

where we have also shown the order h of each element in the conjugacy class and gcd stands for greatest common divisor. This implies that there are $3 + N/2$ irreducible representations. Furthermore, the orthogonality relation (2.18) requires

$$\sum_\alpha [\chi_\alpha(C_1)]^2 = \sum_n m_n n^2 = m_1 + 4m_2 + 9m_3 + \cdots = 2N, \quad (6.6)$$

for the m_i, which also satisfy $m_1 + m_2 + m_3 + \cdots = 3 + N/2$. The solution is found to be $(m_1, m_2) = (4, N/2 - 1)$. Therefore, it is found that there are four singlets and $(N/2 - 1)$ doublets.

6.1.2 Characters and Representations

We start by studying singlets when N is even, in which case there are four singlets. Since the generators satisfy $b^2 = e$ in $C_{N/2}$ and $(ab)^2 = e$ in $C_{N/2}'$, the characters $\chi_\alpha(g)$ for the four singlets should be $\chi_\alpha(C_{N/2}) = \pm 1$ and $\chi_\alpha(C_{N/2}') = \pm 1$. Therefore, there are four possible combinations of $\chi_\alpha(C_{N/2}) = \pm 1$ and $\chi_\alpha(C_{N/2}') = \pm 1$ and they correspond to four singlets, $\mathbf{1}_{\pm\pm}$, which are presented in Table 6.1.

Now consider the doublet representations, that is, (2×2) matrix representations. As can be seen from (6.4), these (2×2) matrices correspond to one of the doublet representations. The (2×2) matrix representations for the generic doublet $\mathbf{2}_k$ are obtained by replacing

$$a \to a^k. \quad (6.7)$$

6.1 D_N with N Even

Table 6.1 Characters of $D_{N\,\text{even}}$ representations

	h	$\chi_{1_{++}}$	$\chi_{1_{+-}}$	$\chi_{1_{-+}}$	$\chi_{1_{--}}$	χ_{2_k}
C_1	1	1	1	1	1	2
$C_2^{(1)}$	N	1	-1	-1	1	$2\cos(2\pi k/N)$
$\vdots$						
$C_2^{(N/2-1)}$	$N/\gcd(N, N/2-1)$	1	$(-1)^{(N/2-1)}$	$(-1)^{(N/2-1)}$	1	$2\cos[2\pi k(N/2-1)/N]$
C_1'	2	1	$(-1)^{N/2}$	$(-1)^{N/2}$	1	-2
$C_{N/2}$	2	1	1	-1	-1	0
$C'_{N/2}$	2	1	-1	1	-1	0

Thus, a and b are represented for the doublet $\mathbf{2}_k$ by

$$a = \begin{pmatrix} \cos 2\pi k/N & -\sin 2\pi k/N \\ \sin 2\pi k/N & \cos 2\pi k/N \end{pmatrix}, \quad b = \begin{pmatrix} 1 & 0 \\ 0 & -1 \end{pmatrix}, \tag{6.8}$$

where $k = 1, \ldots, N/2 - 1$ for N even and $k = 1, \ldots, (N-1)/2$ for $N = $ odd. In the expression for the doublet $\mathbf{2}_k$ by

$$\mathbf{2}_k = \begin{pmatrix} x_k \\ y_k \end{pmatrix}, \tag{6.9}$$

the generator a is the Z_N rotation on the two-dimensional real coordinates (x_k, y_k), and the generator b is the reflection along y_k, i.e., $y_k \to -y_k$. These transformations can be represented on the complex coordinate z_k and its conjugate $\bar{z}_{-k}$. The bases transform according to

$$\begin{pmatrix} z_k \\ \bar{z}_{-k} \end{pmatrix} = U \begin{pmatrix} x_k \\ y_k \end{pmatrix}, \quad U = \frac{1}{\sqrt{2}} \begin{pmatrix} 1 & i \\ 1 & -i \end{pmatrix}. \tag{6.10}$$

In the complex basis, the generators a and b can be expressed by $\tilde{a} = UaU^{-1}$ and $\tilde{b} = UbU^{-1}$:

$$\tilde{a} = \begin{pmatrix} \exp(2\pi i k/N) & 0 \\ 0 & \exp(-2\pi i k/N) \end{pmatrix}, \quad \tilde{b} = \begin{pmatrix} 0 & 1 \\ 1 & 0 \end{pmatrix}. \tag{6.11}$$

This complex basis may be useful. Actually, the generator $\tilde{a}$ is the diagonal matrix. This implies that, in the doublet $\mathbf{2}_k$, which is denoted by

$$\mathbf{2}_k = \begin{pmatrix} z_k \\ \bar{z}_{-k} \end{pmatrix}, \tag{6.12}$$

each of the up and down components z_k and $\bar{z}_{-k}$, respectively, has definite Z_N charge. Indeed, the Z_N charges of z_k and $\bar{z}_{-k}$ are equal to k and $-k$, respectively.

The characters of these matrices for the doublets $\mathbf{2}_k$ are presented in Table 6.1. It is easy to show that these characters satisfy the orthogonality relations (2.16) and (2.17).

6.1.3 Tensor Products

In the next step, we discuss the tensor products of the group D_N with N even. Let us start with $\mathbf{2}_k \times \mathbf{2}_{k'}$, i.e.,

$$\begin{pmatrix} z_k \\ \bar{z}_{-k} \end{pmatrix}_{\mathbf{2}_k} \otimes \begin{pmatrix} z_{k'} \\ \bar{z}_{-k'} \end{pmatrix}_{\mathbf{2}_{k'}}, \tag{6.13}$$

where $k, k' = 1, \ldots, N/2 - 1$. Note that $z_k z_{k'}$, $z_k \bar{z}_{-k'}$, $\bar{z}_{-k} z_{k'}$, and $\bar{z}_{-k} \bar{z}_{-k'}$ have definite Z_N charges, i.e., $k + k'$, $k - k'$, $-k + k'$, and $-k - k'$, respectively. For the case with $k + k' \neq N/2$ and $k - k' \neq 0$, they are decomposed into two doublets as

$$\begin{pmatrix} z_k \\ \bar{z}_{-k} \end{pmatrix}_{\mathbf{2}_k} \otimes \begin{pmatrix} z_{k'} \\ \bar{z}_{-k'} \end{pmatrix}_{\mathbf{2}_{k'}} = \begin{pmatrix} z_k z_{k'} \\ \bar{z}_{-k} \bar{z}_{-k'} \end{pmatrix}_{\mathbf{2}_{k+k'}} \oplus \begin{pmatrix} z_k \bar{z}_{-k'} \\ \bar{z}_{-k} z_{k'} \end{pmatrix}_{\mathbf{2}_{k-k'}}. \tag{6.14}$$

In the case of $k + k' = N/2$, the matrix a is represented on the above (reducible) doublet $(z_k z_{k'}, \bar{z}_{-k} \bar{z}_{-k'})$ by

$$a \begin{pmatrix} z_k z_{k'} \\ \bar{z}_{-k} \bar{z}_{-k'} \end{pmatrix} = \begin{pmatrix} -1 & 0 \\ 0 & -1 \end{pmatrix} \begin{pmatrix} z_k z_{k'} \\ \bar{z}_{-k} \bar{z}_{-k'} \end{pmatrix}. \tag{6.15}$$

Since a is proportional to the (2×2) identity matrix for $(z_k z_{k'}, \bar{z}_{-k} \bar{z}_{-k'})$ with $k + k' = N/2$, we can diagonalize another matrix b in this vector space $(z_k z_{k'}, \bar{z}_{-k} \bar{z}_{-k'})$. Such a basis is $(z_k z_{k'} + \bar{z}_{-k} \bar{z}_{-k'}, z_k z_{k'} - \bar{z}_{-k} \bar{z}_{-k'})$, and the eigenvalues of b are

$$b \begin{pmatrix} z_k z_{k'} + \bar{z}_{-k} \bar{z}_{-k'} \\ z_k z_{k'} - \bar{z}_{-k} \bar{z}_{-k'} \end{pmatrix} = \begin{pmatrix} 1 & 0 \\ 0 & -1 \end{pmatrix} \begin{pmatrix} z_k z_{k'} + \bar{z}_{-k} \bar{z}_{-k'} \\ z_k z_{k'} - \bar{z}_{-k} \bar{z}_{-k'} \end{pmatrix}. \tag{6.16}$$

Thus, $z_k z_{k'} + \bar{z}_{-k} \bar{z}_{-k'}$ and $z_k z_{k'} - \bar{z}_{-k} \bar{z}_{-k'}$ correspond to $\mathbf{1}_{+-}$ and $\mathbf{1}_{-+}$, respectively.

When $k - k' = 0$, a similar decomposition is obtained for the (reducible) doublet $(z_k \bar{z}_{-k'}, \bar{z}_{-k} z_{k'})$. The generator a is the (2×2) identity matrix on the vector space $(z_k \bar{z}_{-k'}, \bar{z}_{-k} z_{k'})$ with $k - k' = 0$. Therefore, we can take the basis

$$(z_k \bar{z}_{-k'} + \bar{z}_{-k} z_{k'}, z_k \bar{z}_{-k'} - \bar{z}_{-k} z_{k'}),$$

where b is diagonalized. That is, $z_k \bar{z}_{-k'} + \bar{z}_{-k} z_{k'}$ and $z_k \bar{z}_{-k'} - \bar{z}_{-k} z_{k'}$ correspond to $\mathbf{1}_{++}$ and $\mathbf{1}_{--}$, respectively.

Now, we study the tensor products of the doublets $\mathbf{2}_k$ and singlets, for example, $\mathbf{1}_{--} \times \mathbf{2}_k$. Here we denote the vector space for the singlet $\mathbf{1}_{--}$ by w, where $aw = w$ and $bw = -w$. It is easily found that $(w z_k, -w \bar{z}_k)$ is nothing but the doublet $\mathbf{2}_k$, that

6.1 D_N with N Even

is, $\mathbf{1}_{--} \times \mathbf{2}_k = \mathbf{2}_k$. Similar results are obtained for other singlets. Furthermore, it is straightforward to study the tensor products among singlets.

Hence, the tensor products of D_N irreducible representations with N even can be summarized as follows:

$$\begin{pmatrix} z_k \\ \bar{z}_{-k} \end{pmatrix}_{\mathbf{2}_k} \otimes \begin{pmatrix} z_{k'} \\ \bar{z}_{-k'} \end{pmatrix}_{\mathbf{2}_{k'}} = \begin{pmatrix} z_k z_{k'} \\ \bar{z}_{-k}\bar{z}_{-k'} \end{pmatrix}_{\mathbf{2}_{k+k'}} \oplus \begin{pmatrix} z_k \bar{z}_{-k'} \\ \bar{z}_{-k} z_{k'} \end{pmatrix}_{\mathbf{2}_{k-k'}}, \qquad (6.17)$$

for $k + k' \neq N/2$ and $k - k' \neq 0$,

$$\begin{pmatrix} z_k \\ \bar{z}_{-k} \end{pmatrix}_{\mathbf{2}_k} \otimes \begin{pmatrix} z_{k'} \\ \bar{z}_{-k'} \end{pmatrix}_{\mathbf{2}_{k'}} = (z_k z_{k'} + \bar{z}_{-k}\bar{z}_{-k'})_{\mathbf{1}_{+-}} \oplus (z_k z_{k'} - \bar{z}_{-k}\bar{z}_{-k'})_{\mathbf{1}_{-+}}$$

$$\oplus \begin{pmatrix} z_k \bar{z}_{-k'} \\ \bar{z}_{-k} z_{k'} \end{pmatrix}_{\mathbf{2}_{k-k'}}, \qquad (6.18)$$

for $k + k' = N/2$ and $k - k' \neq 0$,

$$\begin{pmatrix} z_k \\ \bar{z}_{-k} \end{pmatrix}_{\mathbf{2}_k} \otimes \begin{pmatrix} z_{k'} \\ \bar{z}_{-k'} \end{pmatrix}_{\mathbf{2}_{k'}} = (z_k \bar{z}_{-k'} + \bar{z}_{-k} z_{k'})_{\mathbf{1}_{++}} \oplus (z_k \bar{z}_{-k'} - \bar{z}_{-k} z_{k'})_{\mathbf{1}_{--}}$$

$$\oplus \begin{pmatrix} z_k z_{k'} \\ \bar{z}_{-k}\bar{z}_{-k'} \end{pmatrix}_{\mathbf{2}_{k+k'}}, \qquad (6.19)$$

for $k + k' \neq N/2$ and $k - k' = 0$,

$$\begin{pmatrix} z_k \\ \bar{z}_{-k} \end{pmatrix}_{\mathbf{2}_k} \otimes \begin{pmatrix} z_{k'} \\ \bar{z}_{-k'} \end{pmatrix}_{\mathbf{2}_{k'}} = (z_k \bar{z}_{-k'} + \bar{z}_{-k} z_{k'})_{\mathbf{1}_{++}} \oplus (z_k \bar{z}_{-k'} - \bar{z}_{-k} z_{k'})_{\mathbf{1}_{--}}$$

$$\oplus (z_k z_{k'} + \bar{z}_{-k}\bar{z}_{-k'})_{\mathbf{1}_{+-}} \oplus (z_k z_{k'} - \bar{z}_{-k}\bar{z}_{-k'})_{\mathbf{1}_{-+}}, \quad (6.20)$$

for $k + k' = N/2$ and $k - k' = 0$, and

$$(w)_{\mathbf{1}_{++}} \otimes \begin{pmatrix} z_k \\ \bar{z}_{-k} \end{pmatrix}_{\mathbf{2}_k} = \begin{pmatrix} w z_k \\ w \bar{z}_{-k} \end{pmatrix}_{\mathbf{2}_k},$$

$$(w)_{\mathbf{1}_{--}} \otimes \begin{pmatrix} z_k \\ \bar{z}_{-k} \end{pmatrix}_{\mathbf{2}_k} = \begin{pmatrix} w z_k \\ -w \bar{z}_{-k} \end{pmatrix}_{\mathbf{2}_k},$$

$$(w)_{\mathbf{1}_{+-}} \otimes \begin{pmatrix} z_k \\ \bar{z}_{-k} \end{pmatrix}_{\mathbf{2}_k} = \begin{pmatrix} w \bar{z}_{-k} \\ w z_k \end{pmatrix}_{\mathbf{2}_k}, \qquad (6.21)$$

$$(w)_{\mathbf{1}_{-+}} \otimes \begin{pmatrix} z_k \\ \bar{z}_{-k} \end{pmatrix}_{\mathbf{2}_k} = \begin{pmatrix} w \bar{z}_{-k} \\ -w z_k \end{pmatrix}_{\mathbf{2}_k},$$

$$\mathbf{1}_{s_1 s_2} \otimes \mathbf{1}_{s'_1 s'_2} = \mathbf{1}_{s''_1 s''_2}, \qquad (6.22)$$

with $s_i, s'_i, s''_i = \pm$ ($i = 1, 2$), where $s''_i = +$ for $(s_i, s'_i) = (+, +)$ and $(-, -)$, and $s''_i = -$ for $(s_i, s'_i) = (+, -)$ and $(-, +)$. Hereafter, this sign rule for s''_i will be denoted by $s''_i = s_i s'_i$ ($i = 1, 2$) for simplicity.

Note that the above multiplication rules are the same for the complex basis and the real basis. For example, in both bases we get

$$2_k \otimes 2_{k'} = 2'_{k+k} + 2_{k-k'},$$

for $k + k' \neq N/2$ and $k - k' \neq 0$. On the other hand, the elements of the doublets are written in a different way, although they transform according to (6.10).

6.2 D_N with N Odd

We now consider D_N with N odd, carrying out a similar study of the conjugacy classes, characters, representations, and tensor products.

6.2.1 Conjugacy Classes

The group D_N with N odd has the following $2 + (N-1)/2$ conjugacy classes:

$$\begin{aligned}
C_1: &\quad \{e\}, &&h = 1, \\
C_2^{(1)}: &\quad \{a, a^{N-1}\}, &&h = N, \\
&\quad \vdots \quad\quad\quad \vdots &&\vdots \quad\quad\quad\quad\quad\quad\quad (6.23)\\
C_2^{((N-1)/2)}: &\quad \{a^{(N-1)/2}, a^{(N+1)/2}\}, &&h = N/\gcd(N, (N-1)/2), \\
C_N: &\quad \{b, ab, \ldots, a^{N-1}b\}, &&h = 2.
\end{aligned}$$

That is, there are $2 + (N-1)/2$ irreducible representations. Furthermore, the orthogonality relation (2.18) requires the same equation as (6.6) for the m_i, which also satisfy $m_1 + m_2 + m_3 + \cdots = 2 + (N-1)/2$. The solution is found to be $(m_1, m_2) = (2, (N-1)/2)$. Thus, there are two singlets and $(N-1)/2$ doublets.

6.2.2 Characters and Representations

We study the two singlets of D_N with N odd. Since $b^2 = e$ holds in C_N, the characters $\chi_\alpha(g)$ for the two singlets should be $\chi_\alpha(C_N) = \pm 1$. Since both b and ab belong to the same conjugacy class C_N, the characters $\chi_\alpha(a)$ for the two singlets

6.2 D_N with N Odd

Table 6.2 Characters of $D_{N=\text{odd}}$ representations

	h	χ_{1_+}	χ_{1_-}	χ_{2k}
C_1	1	1	1	2
$C_2^{(1)}$	N	1	1	$2\cos(2\pi k/N)$
$\vdots$				
$C_2^{((N-1)/2)}$	$N/\gcd(N,(N-1)/2)$	1	1	$2\cos[2\pi k(N-1)/2N]$
C_N	2	1	-1	0

must always satisfy $\chi_\alpha(a) = 1$. That is, there are two singlets $\mathbf{1}_+$ and $\mathbf{1}_-$. Their characters are determined by whether the conjugacy class includes b or not, as shown in Table 6.2.

The doublet representations of D_N with N odd are the same as those in D_N with N even. Their characters are also shown in Table 6.2.

6.2.3 Tensor Products

Let us discuss the tensor products of the irreducible representations of D_N with N odd. We can analyze them in a similar way to those of D_N with N even. The results are summarized as follows:

$$\begin{pmatrix} z_k \\ \bar{z}_{-k} \end{pmatrix}_{2_k} \otimes \begin{pmatrix} z_{k'} \\ \bar{z}_{-k'} \end{pmatrix}_{2_{k'}} = \begin{pmatrix} z_k z_{k'} \\ \bar{z}_{-k}\bar{z}_{-k'} \end{pmatrix}_{2_{k+k'}} \oplus \begin{pmatrix} z_k \bar{z}_{-k'} \\ \bar{z}_{-k} z_{k'} \end{pmatrix}_{2_{k-k'}}, \quad (6.24)$$

for $k - k' \neq 0$, where $k, k' = 1, \ldots, N/2 - 1$,

$$\begin{pmatrix} z_k \\ \bar{z}_{-k} \end{pmatrix}_{2_k} \otimes \begin{pmatrix} z_{k'} \\ \bar{z}_{-k'} \end{pmatrix}_{2_{k'}} = (z_k \bar{z}_{-k'} + \bar{z}_{-k} z_{k'})\mathbf{1}_+ \oplus (z_k \bar{z}_{-k'} - \bar{z}_{-k} z_{k'})\mathbf{1}_-$$

$$\oplus \begin{pmatrix} z_k z_{k'} \\ \bar{z}_{-k}\bar{z}_{-k'} \end{pmatrix}_{2_{k+k'}}, \quad (6.25)$$

for $k - k' = 0$, and

$$(w)_{\mathbf{1}_{++}} \otimes \begin{pmatrix} z_k \\ \bar{z}_{-k} \end{pmatrix}_{2_k} = \begin{pmatrix} wz_k \\ w\bar{z}_{-k} \end{pmatrix}_{2_k},$$

$$(w)_{\mathbf{1}_-} \otimes \begin{pmatrix} z_k \\ \bar{z}_{-k} \end{pmatrix}_{2_k} = \begin{pmatrix} wz_k \\ -w\bar{z}_{-k} \end{pmatrix}_{2_k}, \quad (6.26)$$

$$\mathbf{1}_s \otimes \mathbf{1}_{s'} = \mathbf{1}_{s''}, \quad (6.27)$$

where $s'' = ss'$.

Fig. 6.1 The D_4 symmetry group of a square

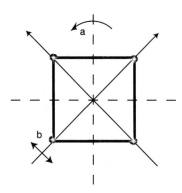

6.3 D_4

In this and the next section, we present simple examples of D_N. The smallest non-Abelian group D_N is D_3. However, D_3 corresponds to the group of all possible permutations of three objects, that is, it is just S_3. We thus examine D_4 and D_5 as simple examples.

The group D_4 is the symmetry group of a square, which is generated by the $\pi/2$ rotation a and the reflection b. These satisfy $a^4 = e$, $b^2 = e$, and $bab = a^{-1}$ (see Fig. 6.1). D_4 thus consists of the eight elements $a^m b^k$ with $m = 0, 1, 2, 3$ and $k = 0, 1$. D_4 has the following five conjugacy classes,

$$\begin{aligned} C_1: & \quad \{e\}, & h &= 1, \\ C_2: & \quad \{a, a^3\}, & h &= 4, \\ C_1': & \quad \{a^2\}, & h &= 2, \\ C_2': & \quad \{b, a^2 b\}, & h &= 2, \\ C_2'': & \quad \{ab, a^3 b\}, & h &= 2, \end{aligned} \quad (6.28)$$

where h is the order of each element in the conjugacy class.

D_4 has four singlets $\mathbf{1}_{++}$, $\mathbf{1}_{+-}$, $\mathbf{1}_{-+}$, and $\mathbf{1}_{--}$, and one doublet $\mathbf{2}$. The characters are shown in Table 6.3. The tensor products are:

$$\begin{pmatrix} z \\ \bar{z} \end{pmatrix}_2 \otimes \begin{pmatrix} z' \\ \bar{z}' \end{pmatrix}_2 = (z\bar{z}' + \bar{z}z')_{\mathbf{1}_{++}} \oplus (z\bar{z}' - \bar{z}z')_{\mathbf{1}_{--}}$$
$$\oplus (zz' + \bar{z}\bar{z}')_{\mathbf{1}_{+-}} \oplus (zz' - \bar{z}\bar{z}')_{\mathbf{1}_{-+}}, \quad (6.29)$$

$$(w)_{\mathbf{1}_{++}} \otimes \begin{pmatrix} z \\ \bar{z} \end{pmatrix}_2 = \begin{pmatrix} wz \\ w\bar{z} \end{pmatrix}_2, \quad (w)_{\mathbf{1}_{--}} \otimes \begin{pmatrix} z \\ \bar{z} \end{pmatrix}_2 = \begin{pmatrix} wz \\ -w\bar{z} \end{pmatrix}_2,$$

$$(w)_{\mathbf{1}_{+-}} \otimes \begin{pmatrix} z \\ \bar{z} \end{pmatrix}_2 = \begin{pmatrix} w\bar{z} \\ wz \end{pmatrix}_2, \quad (w)_{\mathbf{1}_{-+}} \otimes \begin{pmatrix} z \\ \bar{z} \end{pmatrix}_2 = \begin{pmatrix} w\bar{z} \\ -wz \end{pmatrix}_2, \quad (6.30)$$

6.4 D_5

Table 6.3 Characters of D_4 representations

	h	$\chi_{1_{++}}$	$\chi_{1_{+-}}$	$\chi_{1_{-+}}$	$\chi_{1_{--}}$	χ_2
C_1	1	1	1	1	1	2
C_2	4	1	−1	−1	1	0
C'_1	2	1	1	1	1	−2
C'_2	2	1	1	−1	−1	0
C''_2	2	1	−1	1	−1	0

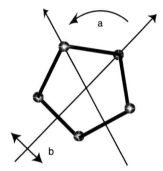

Fig. 6.2 The D_5 symmetry group of a regular pentagon

$$\mathbf{1}_{s_1 s_2} \otimes \mathbf{1}_{s'_1 s'_2} = \mathbf{1}_{s''_1 s''_2}, \tag{6.31}$$

where $s''_1 = s_1 s'_1$ and $s''_2 = s_2 s'_2$.

6.4 D_5

The group D_5 is the symmetry group of a regular pentagon. This is generated by the $2\pi/5$ rotation a and the reflection b (see Fig. 6.2). The generators satisfy $a^5 = e$, $b^2 = e$, and $bab = a^{-1}$. D_5 thus contains the 10 elements $a^m b^k$ with $m = 0, 1, 2, 3, 4$ and $k = 0, 1$. They are classified into the following four conjugacy classes:

$$\begin{aligned} C_1: &\quad \{e\}, & h &= 1, \\ C_2^{(1)}: &\quad \{a, a^4\}, & h &= 5, \\ C_2^{(2)}: &\quad \{a^2, a^3\}, & h &= 5, \\ C_5: &\quad \{b, ab, a^2 b, a^3 b, a^4 b\}, & h &= 2. \end{aligned} \tag{6.32}$$

D_5 has two singlets $\mathbf{1}_+$ and $\mathbf{1}_-$, and two doublets $\mathbf{2}_1$ and $\mathbf{2}_2$. Their characters are shown in Table 6.4.

Table 6.4 Characters of D_5 representations

	h	χ_{1+}	χ_{1-}	χ_{2_1}	χ_{2_2}
C_1	1	1	1	2	2
$C_2^{(1)}$	5	1	1	$2\cos(2\pi/5)$	$2\cos(4\pi/5)$
$C_2^{(2)}$	5	1	1	$2\cos(4\pi/5)$	$2\cos(8\pi/5)$
C_5	2	1	-1	0	0

The tensor products are

$$\begin{pmatrix} z \\ \bar{z} \end{pmatrix}_{2_2} \otimes \begin{pmatrix} z' \\ \bar{z}' \end{pmatrix}_{2_1} = \begin{pmatrix} zz' \\ \bar{z}\bar{z}' \end{pmatrix}_{2_2} \oplus \begin{pmatrix} z\bar{z}' \\ \bar{z}z' \end{pmatrix}_{2_1}, \quad (6.33)$$

$$\begin{pmatrix} z_k \\ \bar{z}_{-k} \end{pmatrix}_{2_k} \otimes \begin{pmatrix} z'_k \\ \bar{z}'_{-k} \end{pmatrix}_{2_k} = (z_k \bar{z}'_{-k} + \bar{z}_{-k} z'_k)_{1_+}$$

$$\oplus (z_k \bar{z}'_{-k} - \bar{z}_{-k} z'_k)_{1_-} \oplus \begin{pmatrix} z_k z'_k \\ \bar{z}_{-k} \bar{z}'_{-k} \end{pmatrix}_{2_{2k}}, \quad (6.34)$$

$$(w)_{1_+} \otimes \begin{pmatrix} z_k \\ \bar{z}_{-k} \end{pmatrix}_{2_k} = \begin{pmatrix} w z_k \\ w \bar{z}_{-k} \end{pmatrix}_{2_k},$$

$$(w)_{1_-} \otimes \begin{pmatrix} z_k \\ \bar{z}_{-k} \end{pmatrix}_{2_k} = \begin{pmatrix} w z_k \\ -w \bar{z}_{-k} \end{pmatrix}_{2_k}, \quad (6.35)$$

$$\mathbf{1}_s \otimes \mathbf{1}_{s'} = \mathbf{1}_{s''}, \quad (6.36)$$

where $s'' = ss'$.

Chapter 7
Q_N

The binary dihedral group is denoted by Q_N, where N is even. It consists of the elements $a^m b^k$ with $m = 0, \ldots, N-1$ and $k = 0, 1$, where the generators a and b satisfy

$$a^N = e, \qquad b^2 = a^{N/2}, \qquad b^{-1}ab = a^{-1}. \tag{7.1}$$

The order of Q_N is $2N$. The generator a can be represented by the same (2×2) matrix as for D_N, i.e.,

$$a = \begin{pmatrix} \exp(2\pi i k/N) & 0 \\ 0 & \exp(-2\pi i k/N) \end{pmatrix}. \tag{7.2}$$

Note that $a^{N/2} = e$ for k even and $a^{N/2} = -e$ for k odd. This leads to $b^2 = e$ for k even and $b^2 = -e$ for k odd. Thus, the generators a and b are represented by (2×2) matrices, e.g.,

$$a = \begin{pmatrix} \exp(2\pi i k/N) & 0 \\ 0 & \exp(-2\pi i k/N) \end{pmatrix}, \qquad b = \begin{pmatrix} 0 & i \\ i & 0 \end{pmatrix}, \tag{7.3}$$

for $k =$ odd,

$$a = \begin{pmatrix} \exp(2\pi i k/N) & 0 \\ 0 & \exp(-2\pi i k/N) \end{pmatrix}, \qquad b = \begin{pmatrix} 0 & 1 \\ 1 & 0 \end{pmatrix}, \tag{7.4}$$

for $k =$ even.

7.1 Q_N with $N = 4n$

The Q_N groups have different features between $N = 4n$ and $4n + 2$. We begin by investigating Q_N with $N = 4n$.

7.1.1 Conjugacy Classes

The conjugacy classes are given by the algebraic relations (7.1). The elements are classified into the following $(3 + N/2)$ conjugacy classes:

$$
\begin{aligned}
&C_1: &&\{e\}, &&h = 1, \\
&C_2^{(1)}: &&\{a, a^{N-1}\}, &&h = N, \\
&\quad\vdots &&\quad\vdots &&\quad\vdots \\
&C_2^{(N/2-1)}: &&\{a^{N/2-1}, a^{N/2+1}\}, &&h = N/\gcd(N, N/2 - 1), \\
&C_1': &&\{a^{N/2}\}, &&h = 2, \\
&C_{N/2}: &&\{b, a^2 b, \ldots, a^{N-2} B\}, &&h = 4, \\
&C_{N/2}': &&\{ab, a^3 b, \ldots, a^{N-1} B\}, &&h = 4,
\end{aligned}
\qquad (7.5)
$$

where h is the order of each element in the given conjugacy class. These are almost the same as the conjugacy classes of D_N with N even. There must be $(3 + N/2)$ irreducible representations, and similarly to $D_{N\,\text{even}}$, there are four singlets and $(N/2 - 1)$ doublets.

7.1.2 Characters and Representations

The characters of Q_N for doublets are the same as those of $D_{N\,\text{even}}$, and are shown in Table 7.1. We study the characters of the singlets of Q_N with $N = 4n$. We then have the relation

$$b^2 = a^{2n}. \qquad (7.6)$$

Since $b^4 = e$ holds in $C_{N/2}$, the characters $\chi_\alpha(b)$ for the four singlets must be $\chi_\alpha(b) = e^{\pi i n/2}$ with $n = 0, 1, 2, 3$. Furthermore, note that the element ba^2 belongs to the same conjugacy class as b. That implies $\chi_\alpha(a^2) = 1$ for the four singlets. Using (7.6), we have $\chi_\alpha(b^2) = 1$, that is, $\chi_\alpha(b) = \pm 1$. Thus, the characters for the singlets of Q_N with $N = 4n$ are the same as those of $D_{N\,\text{even}}$, as shown in Table 7.1.

7.1.3 Tensor Products

The tensor products of the irreducible representations of Q_N can be analyzed in a similar way to those of D_N with N even. The results for Q_N with $N = 4n$ are:

7.1 Q_N with $N = 4n$

Table 7.1 Characters of Q_N representations for $N = 4n$

	h	χ_{1++}	χ_{1+-}	χ_{1-+}	χ_{1--}	χ_{2k}
C_1	1	1	1	1	1	2
$C_2^{(1)}$	N	1	-1	-1	1	$2\cos(2\pi k/N)$
$\vdots$						
$C_2^{(N/2-1)}$	$N/\gcd(N, N/2-1)$	1	$(-1)^{(N/2-1)}$	$(-1)^{(N/2-1)}$	1	$2\cos[2\pi k(N/2-1)/N]$
C_1'	2	1	$(-1)^{N/2}$	$(-1)^{N/2}$	1	-2
$C_{N/2}$	4	1	1	-1	-1	0
$C_{N/2}'$	4	1	-1	1	-1	0

$$\begin{pmatrix} z_k \\ \bar{z}_{-k} \end{pmatrix}_{2_k} \otimes \begin{pmatrix} z_{k'} \\ \bar{z}_{-k'} \end{pmatrix}_{2_{k'}} = \begin{pmatrix} z_k z_{k'} \\ (-1)^{kk'} \bar{z}_{-k} \bar{z}_{-k'} \end{pmatrix}_{2_{k+k'}}$$
$$\oplus \begin{pmatrix} z_k \bar{z}_{-k'} \\ (-1)^{kk'} \bar{z}_{-k} z_{k'} \end{pmatrix}_{2_{k-k'}}, \quad (7.7)$$

for $k + k' \neq N/2$ and $k - k' \neq 0$,

$$\begin{pmatrix} z_k \\ \bar{z}_{-k} \end{pmatrix}_{2_k} \otimes \begin{pmatrix} z_{k'} \\ \bar{z}_{-k'} \end{pmatrix}_{2_{k'}} = \left(z_k z_{k'} + (-1)^{kk'} \bar{z}_{-k} \bar{z}_{-k'}\right)_{1_{+-}}$$
$$\oplus \left(z_k z_{k'} - (-1)^{kk'} \bar{z}_{-k} \bar{z}_{-k'}\right)_{1_{-+}}$$
$$\oplus \begin{pmatrix} z_k \bar{z}_{-k'} \\ (-1)^{kk'} \bar{z}_{-k} z_{k'} \end{pmatrix}_{2_{k-k'}}, \quad (7.8)$$

for $k + k' = N/2$ and $k - k' \neq 0$,

$$\begin{pmatrix} z_k \\ \bar{z}_{-k} \end{pmatrix}_{2_k} \otimes \begin{pmatrix} z_{k'} \\ \bar{z}_{-k'} \end{pmatrix}_{2_{k'}} = \left(z_k \bar{z}_{-k'} + (-1)^{kk'} \bar{z}_{-k} z_{k'}\right)_{1_{++}}$$
$$\oplus \left(z_k \bar{z}_{-k'} - (-1)^{kk'} \bar{z}_{-k} z_{k'}\right)_{1_{--}}$$
$$\oplus \begin{pmatrix} z_k z_{k'} \\ (-1)^{kk'} \bar{z}_{-k} \bar{z}_{-k'} \end{pmatrix}_{2_{k+k'}}, \quad (7.9)$$

for $k + k' \neq N/2$ and $k - k' = 0$,

$$\begin{pmatrix} z_k \\ \bar{z}_{-k} \end{pmatrix}_{2_k} \otimes \begin{pmatrix} z_{k'} \\ \bar{z}_{-k'} \end{pmatrix}_{2_{k'}} = \left(z_k \bar{z}_{-k'} + (-1)^{kk'} \bar{z}_{-k} z_{k'}\right)_{1_{++}}$$
$$\oplus \left(z_k \bar{z}_{-k'} - (-1)^{kk'} \bar{z}_{-k} z_{k'}\right)_{1_{--}}$$

$$\oplus \left(z_k z_{k'} + (-1)^{kk'} \bar{z}_{-k} \bar{z}_{-k'} \right)_{1_{+-}}$$

$$\oplus \left(z_k z_{k'} - (-1)^{kk'} \bar{z}_{-k} \bar{z}_{-k'} \right)_{1_{-+}}, \quad (7.10)$$

for $k + k' = N/2$ and $k - k' = 0$,

$$(w)_{1_{++}} \otimes \begin{pmatrix} z_k \\ \bar{z}_{-k} \end{pmatrix}_{2_k} = \begin{pmatrix} w z_k \\ w \bar{z}_{-k} \end{pmatrix}_{2_k}, \quad (w)_{1_{--}} \otimes \begin{pmatrix} z_k \\ \bar{z}_{-k} \end{pmatrix}_{2_k} = \begin{pmatrix} w z_k \\ -w \bar{z}_{-k} \end{pmatrix}_{2_k},$$

$$(w)_{1_{+-}} \otimes \begin{pmatrix} z_k \\ \bar{z}_{-k} \end{pmatrix}_{2_k} = \begin{pmatrix} w \bar{z}_{-k} \\ w z_k \end{pmatrix}_{2_k}, \quad (w)_{1_{-+}} \otimes \begin{pmatrix} z_k \\ \bar{z}_{-k} \end{pmatrix}_{2_k} = \begin{pmatrix} w \bar{z}_{-k} \\ -w z_k \end{pmatrix}_{2_k},$$

$$(7.11)$$

$$\mathbf{1}_{s_1 s_2} \otimes \mathbf{1}_{s'_1 s'_2} = \mathbf{1}_{s''_1 s''_2}, \quad (7.12)$$

where $s''_1 = s_1 s'_1$ and $s''_2 = s_2 s'_2$. It should be noted that some minus signs differ from those occurring in the tensor products for D_N.

7.2 Q_N with $N = 4n + 2$

As for Q_N with $N = 4n$, we now investigate Q_N with $N = 4n + 2$.

7.2.1 Conjugacy Classes

The conjugacy classes of Q_N with $N = 4n + 2$ are exactly the same as those of Q_N with $N = 4n$.

7.2.2 Characters and Representations

The characters of Q_N for doublets are the same as those of D_N even, and are shown in Table 7.2.

Let us consider the four singlets of Q_N for $N = 4n + 2$. In this case we have the relation

$$b^2 = a^{2n+1}. \quad (7.13)$$

Since b and $a^2 b$ are in the same conjugacy class, the characters $\chi_\alpha(a^2)$ for the four singlets must be $\chi_\alpha(a^2) = 1$. Thus, we obtain $\chi_\alpha(a) = \pm 1$. When $\chi_\alpha(a) = 1$, the relation (7.13) leads to the two possibilities $\chi_\alpha(b) = \pm 1$. On the other hand, when $\chi_\alpha(a) = -1$, the relation (7.13) gives the alternative possibilities $\chi_\alpha(b) = \pm i$. Thus, there are four possibilities corresponding to the four singlets. Note also that $\chi_\alpha(a) = \chi_\alpha(b^2)$ is satisfied for all the singlets.

7.2 Q_N with $N = 4n + 2$

Table 7.2 Characters of Q_N representations for $N = 4n + 2$

h		$\chi_{1_{++}}$	$\chi_{1_{+-}}$	$\chi_{1_{-+}}$	$\chi_{1_{--}}$	χ_{2k}
C_1	1	1	1	1	1	2
$C_2^{(1)}$	N	1	-1	-1	1	$2\cos(2\pi k/N)$
$\vdots$						
$C_2^{(N/2-1)}$	$N/\gcd(N, N/2-1)$	1	$(-1)^{(N/2-1)}$	$(-1)^{(N/2-1)}$	1	$2\cos[2\pi k(N/2-1)/N]$
C_1'	2	1	$(-1)^{N/2}$	$(-1)^{N/2}$	1	-2
$C_{N/2}$	4	1	i	$-i$	-1	0
$C_{N/2}'$	4	1	$-i$	i	-1	0

7.2.3 Tensor Products

We can obtain the tensor products of representations of Q_N with $N = 4n + 2$ in a similar way to those of Q_N with $N = 4n$, as follows:

$$\begin{pmatrix} z_k \\ \bar{z}_{-k} \end{pmatrix}_{2_k} \otimes \begin{pmatrix} z_{k'} \\ \bar{z}_{-k'} \end{pmatrix}_{2_{k'}} = \begin{pmatrix} z_k z_{k'} \\ (-1)^{kk'} \bar{z}_{-k} \bar{z}_{-k'} \end{pmatrix}_{2_{k+k'}} \oplus \begin{pmatrix} z_k \bar{z}_{-k'} \\ (-1)^{kk'} \bar{z}_{-k} z_{k'} \end{pmatrix}_{2_{k-k'}}, \tag{7.14}$$

for $k + k' \neq N/2$ and $k - k' \neq 0$,

$$\begin{pmatrix} z_k \\ \bar{z}_{-k} \end{pmatrix}_{2_k} \otimes \begin{pmatrix} z_{k'} \\ \bar{z}_{-k'} \end{pmatrix}_{2_{k'}} = (z_k z_{k'} + \bar{z}_{-k} \bar{z}_{-k'}) \mathbf{1}_{+-}$$

$$\oplus (z_k z_{k'} - \bar{z}_{-k} \bar{z}_{-k'}) \mathbf{1}_{-+}$$

$$\oplus \begin{pmatrix} z_k \bar{z}_{-k'} \\ (-1)^{kk'} \bar{z}_{-k} z_{k'} \end{pmatrix}_{2_{k-k'}}, \tag{7.15}$$

for $k + k' = N/2$ and $k - k' \neq 0$,

$$\begin{pmatrix} z_k \\ \bar{z}_{-k} \end{pmatrix}_{2_k} \otimes \begin{pmatrix} z_{k'} \\ \bar{z}_{-k'} \end{pmatrix}_{2_{k'}} = (z_k \bar{z}_{-k'} + (-1)^{kk'} \bar{z}_{-k} z_{k'}) \mathbf{1}_{++}$$

$$\oplus (z_k \bar{z}_{-k'} - (-1)^{kk'} \bar{z}_{-k} z_{k'}) \mathbf{1}_{--}$$

$$\oplus \begin{pmatrix} z_k z_{k'} \\ (-1)^{kk'} \bar{z}_{-k} \bar{z}_{-k'} \end{pmatrix}_{2_{k+k'}}, \tag{7.16}$$

for $k + k' \neq N/2$ and $k - k' = 0$,

$$\begin{pmatrix} z_k \\ \bar{z}_{-k} \end{pmatrix}_{2_k} \otimes \begin{pmatrix} z_{k'} \\ \bar{z}_{-k'} \end{pmatrix}_{2_{k'}} = (z_k \bar{z}_{-k'} + (-1)^{kk'} \bar{z}_{-k} z_{k'}) \mathbf{1}_{++}$$

$$\oplus \left(z_k \bar{z}_{-k'} - (-1)^{kk'} \bar{z}_{-k} z_{k'}\right)_{1_{--}}$$
$$\oplus \left(z_k z_{k'} + \bar{z}_{-k} \bar{z}_{-k'}\right)_{1_{+-}}$$
$$\oplus \left(z_k z_{k'} - \bar{z}_{-k} \bar{z}_{-k'}\right)_{1_{-+}}, \qquad (7.17)$$

for $k + k' = N/2$ and $k - k' = 0$,

$$(w)_{1_{++}} \otimes \begin{pmatrix} z_k \\ \bar{z}_{-k} \end{pmatrix}_{2_k} = \begin{pmatrix} w z_k \\ w \bar{z}_{-k} \end{pmatrix}_{2_k}, \qquad (w)_{1_{--}} \otimes \begin{pmatrix} z_k \\ \bar{z}_{-k} \end{pmatrix}_{2_k} = \begin{pmatrix} w z_k \\ -w \bar{z}_{-k} \end{pmatrix}_{2_k},$$

$$(w)_{1_{+-}} \otimes \begin{pmatrix} z_k \\ \bar{z}_{-k} \end{pmatrix}_{2_k} = \begin{pmatrix} w \bar{z}_{-k} \\ w z_k \end{pmatrix}_{2_k}, \qquad (w)_{1_{-+}} \otimes \begin{pmatrix} z_k \\ \bar{z}_{-k} \end{pmatrix}_{2_k} = \begin{pmatrix} w \bar{z}_{-k} \\ -w z_k \end{pmatrix}_{2_k},$$
$$(7.18)$$

$$\mathbf{1}_{s_1 s_2} \otimes \mathbf{1}_{s'_1 s'_2} = \mathbf{1}_{s''_1 s''_2}, \qquad (7.19)$$

where $s''_1 = s_1 s'_1$ and $s''_2 = s_2 s'_2$.

7.3 Q_4

Simple examples of Q_N are useful for applications. Here we present the results for Q_4 and in the next section we discuss Q_6.

The group Q_4 contains the eight elements $a^m b^k$, for $m = 0, 1, 2, 3$ and $k = 0, 1$, where a and b satisfy $a^4 = e$, $b^2 = a^2$, and $b^{-1} a b = a^{-1}$. These elements are classified into five conjugacy classes:

$$\begin{aligned}
C_1: & \quad \{e\}, & h &= 1, \\
C_2: & \quad \{a, a^3\}, & h &= 4, \\
C'_1: & \quad \{a^2\}, & h &= 2, \\
C'_2: & \quad \{b, a^2 b\}, & h &= 4, \\
C''_2: & \quad \{ab, a^3 b\}, & h &= 4.
\end{aligned} \qquad (7.20)$$

Q_4 has four singlets $\mathbf{1}_{++}$, $\mathbf{1}_{+-}$, $\mathbf{1}_{-+}$, and $\mathbf{1}_{--}$, and one doublet $\mathbf{2}$. The characters are shown in Table 7.3. The tensor products are:

$$\begin{pmatrix} z \\ \bar{z} \end{pmatrix}_2 \otimes \begin{pmatrix} z' \\ \bar{z}' \end{pmatrix}_2 = (z\bar{z}' - \bar{z}z')_{1_{++}} \oplus (z\bar{z}' + \bar{z}z')_{1_{--}}$$
$$\oplus (zz' - \bar{z}\bar{z}')_{1_{+-}} \oplus (zz' + \bar{z}\bar{z}')_{1_{-+}}, \qquad (7.21)$$

7.4 Q_6

Table 7.3 Characters of Q_4 representations

	h	$\chi_{1_{++}}$	$\chi_{1_{+-}}$	$\chi_{1_{-+}}$	$\chi_{1_{--}}$	χ_2
C_1	1	1	1	1	1	2
C_2	4	1	-1	-1	1	$2\cos(\pi/2)$
C'_1	2	1	1	1	1	-2
C'_2	4	1	1	-1	-1	0
C''_2	4	1	-1	1	-1	0

$$(w)_{1_{++}} \otimes \begin{pmatrix} z \\ \bar{z} \end{pmatrix}_2 = \begin{pmatrix} wz \\ w\bar{z} \end{pmatrix}_2, \quad (w)_{1_{--}} \otimes \begin{pmatrix} z \\ \bar{z} \end{pmatrix}_2 = \begin{pmatrix} wz \\ -w\bar{z} \end{pmatrix}_2,$$

$$(w)_{1_{+-}} \otimes \begin{pmatrix} z \\ \bar{z} \end{pmatrix}_2 = \begin{pmatrix} w\bar{z} \\ wz \end{pmatrix}_2, \quad (w)_{1_{-+}} \otimes \begin{pmatrix} z \\ \bar{z} \end{pmatrix}_2 = \begin{pmatrix} w\bar{z} \\ -wz \end{pmatrix}_2, \tag{7.22}$$

$$\mathbf{1}_{s_1 s_2} \otimes \mathbf{1}_{s'_1 s'_2} = \mathbf{1}_{s''_1 s''_2}, \tag{7.23}$$

where $s''_1 = s_1 s'_1$ and $s''_2 = s_2 s'_2$. Some minus signs differ from those occurring in the tensor products for D_4.

7.4 Q_6

The group Q_6 has 12 elements $a^m b^k$, for $m = 0, 1, 2, 3, 4, 5$ and $k = 0, 1$, where a and b satisfy $a^6 = e$, $b^2 = a^3$, and $b^{-1}ab = a^{-1}$. These elements are classified into six conjugacy classes:

$$\begin{aligned}
C_1: &\quad \{e\}, & h &= 1, \\
C_2^{(1)}: &\quad \{a, a^5\}, & h &= 6, \\
C_2^{(2)}: &\quad \{a^2, a^4\}, & h &= 3, \\
C'_1: &\quad \{a^3\}, & h &= 2, \\
C_3: &\quad \{b, a^2 b, a^4 b\}, & h &= 4, \\
C'_3: &\quad \{ab, a^3 b, a^5 b\}, & h &= 4.
\end{aligned} \tag{7.24}$$

Q_6 has four singlets $\mathbf{1}_{++}$, $\mathbf{1}_{+-}$, $\mathbf{1}_{-+}$, and $\mathbf{1}_{--}$, and two doublets $\mathbf{2}_1$ and $\mathbf{2}_2$. The characters are shown in Table 7.4.

The tensor products are:

$$\begin{pmatrix} z \\ \bar{z} \end{pmatrix}_{2_2} \otimes \begin{pmatrix} z' \\ \bar{z}' \end{pmatrix}_{2_1} = (zz' - \bar{z}\bar{z}')_{\mathbf{1}_{+-}} \oplus (zz' + \bar{z}\bar{z}')_{\mathbf{1}_{-+}} \oplus \begin{pmatrix} z\bar{z}' \\ \bar{z}z' \end{pmatrix}_{2_1}, \tag{7.25}$$

Table 7.4 Characters of Q_6 representations

	h		$\chi_{1_{++}}$	$\chi_{1_{+-}}$	$\chi_{1_{-+}}$	$\chi_{1_{--}}$	χ_{2_1}	χ_{2_2}
C_1	1	1	1	1	1	1	2	2
$C_2^{(1)}$	6	1	−1	−1	1	1	$2\cos(2\pi/6)$	$2\cos(4\pi/6)$
$C_2^{(2)}$	3	1	1	1	1	1	$2\cos(4\pi/6)$	$2\cos(8\pi/6)$
C_1'	2	1	1	1	1	1	−2	2
C_3	4	1	i	−i	−1	0	0	
C_3'	4	1	−i	i	−1	0	0	

$$\begin{pmatrix} z \\ \bar{z} \end{pmatrix}_{2k} \otimes \begin{pmatrix} z' \\ \bar{z}' \end{pmatrix}_{2k} = (z\bar{z}' - \bar{z}z')_{1_{++}} \oplus (z\bar{z}' + \bar{z}z')_{1_{--}} \oplus \begin{pmatrix} zz' \\ -\bar{z}\bar{z}' \end{pmatrix}_{2_{k'}}, \quad (7.26)$$

for $k, k' = 1, 2$ and $k' \neq k$, and

$$(w)_{1_{++}} \otimes \begin{pmatrix} z_k \\ \bar{z}_{-k} \end{pmatrix}_{2_k} = \begin{pmatrix} wz_k \\ w\bar{z}_{-k} \end{pmatrix}_{2_k},$$

$$(w)_{1_{--}} \otimes \begin{pmatrix} z_k \\ \bar{z}_{-k} \end{pmatrix}_{2_k} = \begin{pmatrix} wz_k \\ -w\bar{z}_{-k} \end{pmatrix}_{2_k},$$

$$(w)_{1_{+-}} \otimes \begin{pmatrix} z_k \\ \bar{z}_{-k} \end{pmatrix}_{2_k} = \begin{pmatrix} w\bar{z}_{-k} \\ wz_k \end{pmatrix}_{2_k},$$

$$(w)_{1_{-+}} \otimes \begin{pmatrix} z_k \\ \bar{z}_{-k} \end{pmatrix}_{2_k} = \begin{pmatrix} w\bar{z}_{-k} \\ -wz_k \end{pmatrix}_{2_k},$$

(7.27)

$$\mathbf{1}_{s_1 s_2} \otimes \mathbf{1}_{s_1' s_2'} = \mathbf{1}_{s_1'' s_2''}, \quad (7.28)$$

where $s_1'' = s_1 s_1'$ and $s_2'' = s_2 s_2'$.

Chapter 8
QD_{2N}

8.1 Generic Aspects

Here, we briefly study generic aspects of QD_{2N}. Let us start with the semi-direct product $Z_{2^{N'-1}} \rtimes Z_2$, which has order $2^{N'}$. The Abelian groups $Z_{2^{N'-1}}$ and Z_2 are generated by two generators a and b, respectively, which satisfy

$$a^{2^{N'-1}} = 1, \qquad b^2 = 1. \tag{8.1}$$

Because of the semi-direct product, we require

$$bab^{-1} = a^m. \tag{8.2}$$

If $m = 1 \mod 2^{N'-1}$, this corresponds to the direct product. Thus, we require $m \neq 1 \mod 2^{N'-1}$. Furthermore, using $a = ba^m b$, it is found that

$$a^m = (ba^m b) \cdots (ba^m b) = ba^{m^2} b. \tag{8.3}$$

Note that $a^m = bab$, because $a = ba^m b$. Then consistency requires $a^{m^2} = a$. This implies

$$m^2 = 1 \mod 2^{N'-1}. \tag{8.4}$$

It is generally known that the solutions of the above equation are $m = \pm 1$ for $N' \leq 3$ and $m = \pm 1, 2^{N'-2} \pm 1$ for $N' \geq 4$.

For $m = 1$, the result is nothing but a direct product group. For $m = -1$, one finds that it is identified as a dihedral group. For the non-trivial solution $m = 2^{N'-2} - 1$, the group $Z_{2^{N'-1}} \rtimes Z_2$ is called the quasi-dihedral group.[1] Therefore, we define $QD_{2^{N'}}$ for $N' \geq 4$, e.g., QD_{16}, QD_{32}, etc.

Hereafter we use the notation $N \equiv 2^{N'-1}$ for convenience. Then, the group QD_{2N} is isomorphic to $Z_N \rtimes Z_2$ and the generators a and b of Z_N and Z_2, respectively,

---

[1] No name has been adopted for the non-trivial solution $m = 2^{N'-2} + 1$.

satisfy

$$a^N = 1, \qquad b^2 = 1, \qquad bab = a^{N/2-1}. \tag{8.5}$$

All elements of QD_{2N} can be written in the form $a^k b^\ell$ for $k = 0, \ldots, N-1$ and $\ell = 0, 1$. The generators a and b are represented, e.g., by

$$a = \begin{pmatrix} \rho & 0 \\ 0 & \rho^{N/2-1} \end{pmatrix}, \qquad b = \begin{pmatrix} 0 & 1 \\ 1 & 0 \end{pmatrix}, \tag{8.6}$$

where $\rho = e^{2\pi i/N}$.

8.1.1 Conjugacy Classes

The algebraic relations (8.5) tell us that a^k and $a^{k(N/2-1)}$ belong to the same conjugacy class and also that b and $a^{m(N/2-2)}b$ belong to the same conjugacy class. The group QD_{2N} has the following $(3 + N/2)$ conjugacy classes:

$$\begin{array}{lll}
C_1 : & \{e\}, & h = 1, \\
C_2^{[k]} : & \{a^k, a^{k(N/2-1)}\}, & h = N/\gcd(N, k), \\
C_1' : & \{a^{N/2}\}, & h = 2, \\
C_{N/2} : & \{b, a^2 b, \ldots, a^{N-2} b\}, & h = 2, \\
C_{N/2}' : & \{ab, a^3 b, \ldots, a^{N-1} b\}, & h = 4,
\end{array} \tag{8.7}$$

where $[k] = k$ or $k(N/2-1) \bmod N$ with $k = 1, \ldots, N-1$ except $N/2$. We have also shown the order h of each element in the given conjugacy class. This implies that there are $(3 + N/2)$ irreducible representations. Furthermore, the orthogonality relation (2.18) requires

$$\sum_\alpha [\chi_\alpha(C_1)]^2 = \sum_n m_n n^2 = m_1 + 4m_2 + 9m_3 + \cdots = 2N, \tag{8.8}$$

for the m_i, which also satisfy $m_1 + m_2 + m_3 + \cdots = 3 + N/2$. The solution is $(m_1, m_2) = (4, N/2 - 1)$, so there are four singlets and $(N/2 - 1)$ doublets.

8.1.2 Characters and Representations

We now turn to the characters and representations. The group QD_{2N} has four singlets and $(N/2 - 1)$ doublets. We denote the four singlets by $\mathbf{1}_{ss'}$, with $s, s' = \pm$. The characters of a and b are obtained as $\chi_{1_{\pm s'}}(a) = \pm 1$ for any s' and $\chi_{1_{s\pm}}(b) = \pm 1$ for any s.

Now consider the doublets. The two generators a and b are represented, e.g., by

$$a = \begin{pmatrix} \rho^k & 0 \\ 0 & \rho^{k(N/2-1)} \end{pmatrix}, \qquad b = \begin{pmatrix} 0 & 1 \\ 1 & 0 \end{pmatrix}, \tag{8.9}$$

8.1 Generic Aspects

Table 8.1 Characters of QD_{2N} representations, where $\rho = e^{2\pi i/N}$. Note that $\rho^{mk} + \rho^{mk(N/2-1)} = 2\cos(2\pi mk/N)$ when mk is even, and $\rho^{mk} + \rho^{mk(N/2-1)} = 2i\sin(2\pi mk/N)$ when mk is odd

	h	χ_{1++}	χ_{1-+}	χ_{1--}	χ_{1+-}	$\chi_{2_{[k]}}$
C_1	1	1	1	1	1	2
$C_2^{[m]}$	$N/\gcd(N,m)$	1	$(-1)^m$	$(-1)^m$	1	$\rho^{mk} + \rho^{mk(N/2-1)}$
C_1'	2	1	$(-1)^{N/2} = 1$	$(-1)^{N/2} = 1$	1	-2
$C_{N/2}$	2	1	1	-1	-1	0
$C_{N/2}'$	4	1	-1	1	-1	0

on the doublet $2_{[k]}$. We also denote the vector $2_{[k]}$ by

$$2_{[k]} = \begin{pmatrix} x_k \\ x_{(N/2-1)k} \end{pmatrix}, \tag{8.10}$$

where each of the up and down components x_k and $x_{(N/2-1)k}$, respectively, has definite Z_N charge. The characters are shown in Table 8.1.

8.1.3 Tensor Products

We now discuss the tensor products of representations of the QD_{2N} group. Let us start with $2_k \times 2_{k'}$, i.e.,

$$\begin{pmatrix} x_k \\ x_{(N/2-1)k} \end{pmatrix}_{2_k} \otimes \begin{pmatrix} y_{k'} \\ y_{(N/2-1)k'} \end{pmatrix}_{2_{k'}}, \tag{8.11}$$

where $k, k' = 1, \ldots, N-1$ except $N/2$. Hence, the tensor products of irreducible representations of QD_{2N} are generally given by

$$\begin{pmatrix} x_k \\ x_{(N/2-1)k} \end{pmatrix}_{2_{[k]}} \otimes \begin{pmatrix} y_{k'} \\ y_{(N/2-1)k'} \end{pmatrix}_{2_{[k']}}$$
$$= \begin{pmatrix} x_k y_{k'} \\ x_{(N/2-1)k} y_{(N/2-1)k'} \end{pmatrix}_{2_{[k+k']}} \oplus \begin{pmatrix} x_k y_{(N/2-1)k'} \\ x_{N/2-1}k y_{k'} \end{pmatrix}_{2_{[k+(N/2-1)k']}}, \tag{8.12}$$

for $k + k', k + (N/2 - 1)k' \neq 0, N/2$. In certain cases, the above representation becomes reducible. For example, if $k + k' = 0 \mod N$, the doublet $2_{[k+k']}$ can be reduced according to

$$\begin{pmatrix} x_k y_{k'} \\ x_{(N/2-1)k} y_{(N/2-1)k'} \end{pmatrix}_{2_{[k+k']}}$$
$$= (x_k y_{k'} + x_{(N/2-1)k} y_{(N/2-1)k'}) \mathbf{1}_{++} \oplus (x_k y_{k'} - x_{(N/2-1)k} y_{(N/2-1)k'}) \mathbf{1}_{+-}. \tag{8.13}$$

Similarly, when $k + k' = N/2 \mod N$, the doublet $2_{[k+k']}$ can be reduced according to

$$\begin{pmatrix} x_k y_{k'} \\ x_{(N/2-1)k} y_{(N/2-1)k'} \end{pmatrix}_{2_{[k+k']}}$$
$$= (x_k y_{k'} + x_{(N/2-1)k} y_{(N/2-1)k'}) \mathbf{1}_{-+} \oplus (x_k y_{k'} - x_{(N/2-1)k} y_{(N/2-1)k'}) \mathbf{1}_{--}. \tag{8.14}$$

The tensor products between singlets and doublets are:

$$(w)_{1_{++}} \otimes \begin{pmatrix} x_k \\ x_{(N/2-1)k} \end{pmatrix}_{2_{[k]}} = \begin{pmatrix} w x_k \\ w x_{(N/2-1)k} \end{pmatrix}_{2_{[k]}}, \tag{8.15}$$

$$(w)_{1_{--}} \otimes \begin{pmatrix} x_k \\ x_{(N/2-1)k} \end{pmatrix}_{2_{[k]}} = \begin{pmatrix} w x_k \\ -w x_{(N/2-1)k} \end{pmatrix}_{2_{[k+N/2]}}, \tag{8.16}$$

$$(w)_{1_{+-}} \otimes \begin{pmatrix} x_k \\ x_{(N/2-1)k} \end{pmatrix}_{2_{[k]}} = \begin{pmatrix} w x_k \\ -w x_{(N/2-1)k} \end{pmatrix}_{2_{[k]}}, \tag{8.17}$$

$$(w)_{1_{-+}} \otimes \begin{pmatrix} x_k \\ x_{(N/2-1)k} \end{pmatrix}_{2_{[k]}} = \begin{pmatrix} w x_k \\ w x_{(N/2-1)k} \end{pmatrix}_{2_{[k+N/2]}}. \tag{8.18}$$

Finally, the tensor products among singlets are

$$\mathbf{1}_{s_1 s_2} \otimes \mathbf{1}_{s'_1 s'_2} = \mathbf{1}_{s''_1 s''_2}, \tag{8.19}$$

where $s''_i = s_i s'_i$ ($i = 1, 2$).

8.2 QD_{16}

We discuss the simple example $N' = 4$ and $N = 8$, i.e., QD_{16}. This group is generated by the $2\pi/8$ rotation a and the reflection b. These generators satisfy

$$a^8 = 1, \qquad b^2 = 1, \qquad bab = a^3. \tag{8.20}$$

The group QD_{16} contains the 16 elements $a^m b^k$ with $m = 0, \ldots, 7$ and $k = 0, 1$. They are classified into the following seven conjugacy classes:

8.2 QD_{16}

Table 8.2 Characters of QD_{16} representations

	h	χ_{1++}	χ_{1-+}	χ_{1--}	χ_{1+-}	χ_{2_1}	χ_{2_2}	χ_{2_5}
C_1	1	1	1	1	1	2	2	2
$C_2^{[1]}$	8	1	-1	-1	1	$\sqrt{2}i$	0	$-\sqrt{2}i$
$C_2^{[2]}$	4	1	1	1	1	0	-2	0
$C_2^{[5]}$	8	1	-1	-1	1	$-\sqrt{2}i$	0	$\sqrt{2}i$
C_1'	2	1	1	1	1	-2	2	-2
C_4	2	1	1	-1	-1	0	0	0
C_4'	2	1	-1	1	-1	0	0	0

$$\begin{aligned}
C_1: &\quad \{e\}, &h&=1,\\
C_2^{[1]}: &\quad \{a,a^3\}, &h&=8,\\
C_2^{[2]}: &\quad \{a^2,a^6\}, &h&=4,\\
C_2^{[5]}: &\quad \{a^5,a^7\}, &h&=8,\\
C_1': &\quad \{a^4\}, &h&=2,\\
C_4: &\quad \{b,a^2b,\ldots,a^6b\}, &h&=2,\\
C_4': &\quad \{ab,a^3b,\ldots,a^7b\}, &h&=4.
\end{aligned} \quad (8.21)$$

QD_{16} has four singlets $\mathbf{1}_{\pm\pm}$, and three doublets $\mathbf{2}_1$, $\mathbf{2}_2$, and $\mathbf{2}_3$. Their characters are shown in Table 8.2.

We define the three doublets

$$\mathbf{2}_1 = \begin{pmatrix} x_1 \\ x_3 \end{pmatrix}, \quad \mathbf{2}_2 = \begin{pmatrix} x_2 \\ x_6 \end{pmatrix}, \quad \mathbf{2}_5 = \begin{pmatrix} x_5 \\ x_7 \end{pmatrix}. \tag{8.22}$$

Then the tensor products are

$$\begin{pmatrix} x_1 \\ x_3 \end{pmatrix}_{\mathbf{2}_1} \otimes \begin{pmatrix} y_1 \\ y_3 \end{pmatrix}_{\mathbf{2}_1} = \begin{pmatrix} x_1 y_1 \\ x_3 y_3 \end{pmatrix}_{\mathbf{2}_2} \oplus (x_1 y_3 \pm x_3 y_1)_{\mathbf{1}_{-\pm}}, \tag{8.23}$$

$$\begin{pmatrix} x_1 \\ x_3 \end{pmatrix}_{\mathbf{2}_1} \otimes \begin{pmatrix} y_2 \\ y_6 \end{pmatrix}_{\mathbf{2}_2} = \begin{pmatrix} x_3 y_2 \\ x_1 y_6 \end{pmatrix}_{\mathbf{2}_5} \oplus \begin{pmatrix} x_3 y_6 \\ x_1 y_2 \end{pmatrix}_{\mathbf{2}_1}, \tag{8.24}$$

$$\begin{pmatrix} x_1 \\ x_3 \end{pmatrix}_{\mathbf{2}_1} \otimes \begin{pmatrix} y_5 \\ y_7 \end{pmatrix}_{\mathbf{2}_5} = \begin{pmatrix} x_3 y_7 \\ x_1 y_5 \end{pmatrix}_{\mathbf{2}_2} \oplus (x_1 y_7 \pm x_3 y_5)_{\mathbf{1}_{+\pm}}, \tag{8.25}$$

$$\begin{pmatrix} x_2 \\ x_6 \end{pmatrix}_{\mathbf{2}_2} \otimes \begin{pmatrix} y_2 \\ y_6 \end{pmatrix}_{\mathbf{2}_2} = (x_2 y_2 \pm x_6 y_6)_{\mathbf{1}_{-\pm}} \oplus (x_2 y_6 \pm x_6 y_2)_{\mathbf{1}_{+\pm}}, \tag{8.26}$$

$$\begin{pmatrix} x_2 \\ x_6 \end{pmatrix}_{\mathbf{2}_2} \otimes \begin{pmatrix} y_5 \\ y_7 \end{pmatrix}_{\mathbf{2}_5} = \begin{pmatrix} x_6 y_7 \\ x_2 y_5 \end{pmatrix}_{\mathbf{2}_5} \oplus \begin{pmatrix} x_2 y_7 \\ x_6 y_5 \end{pmatrix}_{\mathbf{2}_1}, \tag{8.27}$$

$$\begin{pmatrix} x_5 \\ x_7 \end{pmatrix}_{\mathbf{2}_5} \otimes \begin{pmatrix} y_5 \\ y_7 \end{pmatrix}_{\mathbf{2}_5} = \begin{pmatrix} x_5 y_5 \\ x_7 y_7 \end{pmatrix}_{\mathbf{2}_2} \oplus (x_5 y_7 \pm x_7 y_5)_{\mathbf{1}_{-\pm}}, \tag{8.28}$$

$$(w)_{\mathbf{1}_{-\pm}} \otimes \begin{pmatrix} y_1 \\ y_3 \end{pmatrix}_{\mathbf{2}_1} = \begin{pmatrix} w y_1 \\ \pm w y_3 \end{pmatrix}_{\mathbf{2}_5}, \quad (w)_{\mathbf{1}_{-\pm}} \otimes \begin{pmatrix} y_2 \\ y_6 \end{pmatrix}_{\mathbf{2}_2} = \begin{pmatrix} w y_6 \\ \pm w y_2 \end{pmatrix}_{\mathbf{2}_2}, \tag{8.29}$$

$$(w)_{\mathbf{1}_{-\pm}} \otimes \begin{pmatrix} y_5 \\ y_7 \end{pmatrix}_{\mathbf{2}_5} = \begin{pmatrix} wy_5 \\ \pm wy_7 \end{pmatrix}_{\mathbf{2}_1}, \tag{8.30}$$

$$\mathbf{1}_{+\pm} \otimes \mathbf{1}_{+\pm} = \mathbf{1}_{++}, \qquad \mathbf{1}_{-\pm} \otimes \mathbf{1}_{+\pm} = \mathbf{1}_{-+}, \qquad \mathbf{1}_{-\pm} \otimes \mathbf{1}_{-\pm} = \mathbf{1}_{++}, \tag{8.31}$$

$$\mathbf{1}_{+\pm} \otimes \mathbf{1}_{+\mp} = \mathbf{1}_{+-}, \qquad \mathbf{1}_{-\pm} \otimes \mathbf{1}_{+\mp} = \mathbf{1}_{--}, \qquad \mathbf{1}_{-\pm} \otimes \mathbf{1}_{-\mp} = \mathbf{1}_{+-}. \tag{8.32}$$

Chapter 9
$\Sigma(2N^2)$

9.1 Generic Aspects

In this chapter, we investigate the discrete group $\Sigma(2N^2)$, which is isomorphic to $(Z_N \times Z'_N) \rtimes Z_2$. Let us denote the generators of Z_N and Z'_N by a and a', respectively, and the Z_2 generator by b. These generators satisfy

$$a^N = a'^N = b^2 = e, \qquad aa' = a'a, \qquad bab = a'. \tag{9.1}$$

Therefore, all elements of $\Sigma(2N^2)$ are given by

$$g = b^k a^m a'^n, \tag{9.2}$$

for $k = 0, 1$ and $m, n = 0, 1, \ldots, N-1$.

Since these generators, a, a', and b, can be represented, e.g., by

$$a = \begin{pmatrix} 1 & 0 \\ 0 & \rho \end{pmatrix}, \qquad a' = \begin{pmatrix} \rho & 0 \\ 0 & 1 \end{pmatrix}, \qquad b = \begin{pmatrix} 0 & 1 \\ 1 & 0 \end{pmatrix}, \tag{9.3}$$

where $\rho = e^{2\pi i/N}$, all elements of $\Sigma(2N^2)$ can be expressed by the 2×2 matrices

$$\begin{pmatrix} \rho^m & 0 \\ 0 & \rho^n \end{pmatrix}, \qquad \begin{pmatrix} 0 & \rho^m \\ \rho^n & 0 \end{pmatrix}. \tag{9.4}$$

9.1.1 Conjugacy Classes

The conjugacy classes of $\Sigma(2N^2)$ are easily found using the algebraic relations

$$\begin{aligned} b(a^l a'^m)b^{-1} &= a^m a'^l, & b(ba^l a'^m)b^{-1} &= ba^m a'^l, \\ a^k(ba^l a'^m)a^{-k} &= ba^{l-k} a'^{m+k}, & a'^k(ba^l a'^m)a'^{-k} &= ba^{l+k} a'^{m-k}, \end{aligned} \tag{9.5}$$

which are given by (9.3). Hence, we find that the group $\Sigma(2N^2)$ has the following conjugacy classes:

$$
\begin{array}{lll}
C_1: & \{e\}, & h=1, \\
C_1^{(1)}: & \{aa'\}, & h=N, \\
\quad\vdots & \quad\vdots & \quad\vdots \\
C_1^{(k)}: & \{a^k a'^k\}, & h=N/\gcd(N,k), \\
\quad\vdots & \quad\vdots & \quad\vdots \\
C_1^{(N-1)}: & \{a^{N-1} a'^{N-1}\}, & h=N/\gcd(N,N-1), \\
C_N'^{(k)}: & \{ba^k, ba^{k-1}a', \ldots, ba'^k, \ldots, ba^{k+1}a'^{N-1}\}, & h=2N/\gcd(N,k), \\
C_2^{(l,m)}: & \{a^l a'^m, a'^l a^m\}, & h=N/\gcd(N,l,m),
\end{array}
\tag{9.6}
$$

where $l > m$ for $l, m = 0, \ldots, N-1$. The number of conjugacy classes $C_2^{(l,m)}$ is given by $N(N-1)/2$ and the total number of conjugacy classes of $\Sigma(2N^2)$ is

$$N(N-1)/2 + N + N = (N^2 + 3N)/2.$$

9.1.2 Characters and Representations

The orthogonality relations (2.18) and (2.19) for $\Sigma(2N^2)$ give

$$m_1 + 2^2 m_2 + \cdots = 2N^2, \qquad m_1 + m_2 + \cdots = (N^2 + 3N)/2. \tag{9.7}$$

The solution is

$$(m_1, m_2) = (2N, N(N-1)/2),$$

so there are $2N$ singlets and $N(N-1)/2$ doublets.

Let us first discuss the singlets. Since a and a' belong to the same conjugacy class $C_2^{(1,0)}$, the characters $\chi_\alpha(g)$ for the singlets should satisfy $\chi_\alpha(a) = \chi_\alpha(a')$. Because $b^2 = e$ and $a^N = e$, possible values of $\chi_\alpha(g)$ for the singlets are $\chi_\alpha(a) = \rho^n$ and $\chi_\alpha(b) = \pm 1$. Then we have a total of $2N$ combinations, which correspond to the $2N$ singlets $\mathbf{1}_{\pm n}$ for $n = 0, 1, \ldots, N-1$. These characters are summarized in Table 9.1.

Now consider the doublet representations. In (9.3), the generators are represented in the doublet representation. Similarly, (2×2) matrix representations for generic doublets $\mathbf{2}_{p,q}$ are obtained by replacing

$$a \to a^p a'^q \quad \text{and} \quad a' \to a^q a'^p. \tag{9.8}$$

That is, for doublets $\mathbf{2}_{p,q}$, the generators a, a', and b are

$$a = \begin{pmatrix} \rho^q & 0 \\ 0 & \rho^p \end{pmatrix}, \quad a' = \begin{pmatrix} \rho^p & 0 \\ 0 & \rho^q \end{pmatrix}, \quad b = \begin{pmatrix} 0 & 1 \\ 1 & 0 \end{pmatrix}. \tag{9.9}$$

9.1 Generic Aspects

Table 9.1 Characters of $\Sigma(2N^2)$ representations

h		$\chi_{1_{+n}}$	$\chi_{1_{-n}}$	$\chi_{2_{p,q}}$
C_1	1	1	1	2
$C_1^{(1)}$	N	ρ^{2n}	ρ^{2n}	$2\rho^{p+q}$
$\vdots$				
$C_1^{(N-1)}$	$N/\gcd(N, N-1)$	$\rho^{2n(N-1)}$	$\rho^{2n(N-1)}$	$2\rho^{(N-1)(p+q)}$
$C_N'^{(k)}$	$2N/\gcd(N,k)$	ρ^{kn}	$-\rho^{kn}$	0
$C_2^{(l,m)}$	$N/\gcd(N,l,m)$	$\rho^{(l+m)n}$	$\rho^{(l+m)n}$	$\rho^{lq+mp} + \rho^{lp+mq}$

Let us denote the doublet $\mathbf{2}_{p,q}$ by

$$\mathbf{2}_{p,q} = \begin{pmatrix} x_q \\ x_p \end{pmatrix}, \tag{9.10}$$

where we take $p > q$ and $q = 0, 1, \ldots, N - 2$. Then, each of the up and down components x_q and x_p, respectively, has a definite $Z_N \times Z_N'$ charge. That is, x_q and x_p have $(q, 0)$ and $(0, p)$ $Z_N \times Z_N'$ charges, respectively. The characters for the doublets are also summarized in Table 9.1.

9.1.3 Tensor Products

We now consider the tensor products of doublets $\mathbf{2}_{p,q}$ in $\Sigma(2N^2)$. Taking into account the $Z_N \times Z_N'$ charges, their tensor products are given by

$$\begin{pmatrix} x_q \\ x_p \end{pmatrix}_{\mathbf{2}_{q,p}} \otimes \begin{pmatrix} y_{q'} \\ y_{p'} \end{pmatrix}_{\mathbf{2}_{q',p'}} = \begin{pmatrix} x_q y_{q'} \\ x_p y_{p'} \end{pmatrix}_{\mathbf{2}_{q+q',p+p'}} \oplus \begin{pmatrix} x_p y_{q'} \\ x_q y_{p'} \end{pmatrix}_{\mathbf{2}_{q'+p,q+p'}}, \tag{9.11}$$

for $q + q' \neq p + p' \mod(N)$ and $q + p' \neq p + q' \mod(N)$,

$$\begin{pmatrix} x_q \\ x_p \end{pmatrix}_{\mathbf{2}_{q,p}} \otimes \begin{pmatrix} y_{q'} \\ y_{p'} \end{pmatrix}_{\mathbf{2}_{q',p'}} = (x_q y_{q'} + x_p y_{p'})\mathbf{1}_{+,q+q'} \oplus (x_q y_{q'} - x_p y_{p'})\mathbf{1}_{-,q+q'}$$

$$\oplus \begin{pmatrix} x_p y_{q'} \\ x_q y_{p'} \end{pmatrix}_{\mathbf{2}_{q'+p,q+p'}}, \tag{9.12}$$

for $q + q' = p + p' \mod(N)$ and $q + p' \neq p + q' \mod(N)$,

$$\begin{pmatrix} x_q \\ x_p \end{pmatrix}_{\mathbf{2}_{q,p}} \otimes \begin{pmatrix} y_{q'} \\ y_{p'} \end{pmatrix}_{\mathbf{2}_{q',p'}} = (x_p y_{q'} + x_q y_{p'})\mathbf{1}_{+,q+p'} \oplus (x_p y_{q'} - x_q y_{p'})\mathbf{1}_{-,q+p'}$$

$$\oplus \begin{pmatrix} x_q y_{q'} \\ x_p y_{p'} \end{pmatrix}_{\mathbf{2}_{q+q',p+p'}}, \tag{9.13}$$

for $q + q' \neq p + p' \mod(N)$ and $q + p' = p + q' \mod(N)$, and

$$\begin{pmatrix} x_q \\ x_p \end{pmatrix}_{2_{q,p}} \otimes \begin{pmatrix} y_{q'} \\ y_{p'} \end{pmatrix}_{2_{q',p'}} = (x_q y_{q'} + x_p y_{p'}) 1_{+,q+q'} \oplus (x_q y_{q'} - x_p y_{p'}) 1_{-,q+q'}$$

$$\oplus (x_p y_{q'} + x_q y_{p'}) 1_{+,q+p'} \oplus (x_p y_{q'} - x_q y_{p'}) 1_{-,q+p'}, \quad (9.14)$$

for $q + q' = p + p' \mod(N)$ and $q + p' = p + q' \mod(N)$.

Furthermore, we obtain the tensor products between singlets and doublets as

$$(y)_{1_{s,n}} \otimes \begin{pmatrix} x_q \\ x_p \end{pmatrix}_{2_{q,p}} = \begin{pmatrix} y x_q \\ y x_p \end{pmatrix}_{2_{q+n,p+n}}. \quad (9.15)$$

These tensor products are independent of $s = \pm$. The tensor products of singlets are simply given by

$$1_{sn} \otimes 1_{s'n'} = 1_{ss',n+n'}. \quad (9.16)$$

9.2 $\Sigma(18)$

In this and the following two sections, we present some simple examples of $\Sigma(2N^2)$. The simplest group $\Sigma(2N^2)$ is $\Sigma(2)$, which is nothing but the Abelian group Z_2. The next is the group $\Sigma(8)$, which is isomorphic to D_4. Consequently, the simplest non-trivial example is $\Sigma(18)$.

In $\Sigma(18)$, there are eighteen elements $b^k a^m a'^n$ for $k = 0, 1$, and $m, n = 0, 1, 2$, where a, a', and b satisfy $b^2 = e$, $a^3 = a'^3 = e$, $aa' = a'a$, and $bab = a'$. These elements are classified into nine conjugacy classes:

$$\begin{array}{lll}
C_1: & \{e\}, & h = 1, \\
C_1^{(1)}: & \{aa'\}, & h = 3, \\
C_1^{(2)}: & \{a^2 a'^2\}, & h = 3, \\
C_3'^{(0)}: & \{b, ba'^2 a, ba' a^2\}, & h = 2, \\
C_3'^{(1)}: & \{ba', ba, ba'^2 a^2\}, & h = 6, \\
C_3'^{(2)}: & \{ba'^2, ba'a, ba^2\}, & h = 6, \\
C_2^{(1,0)}: & \{a, a'\}, & h = 3, \\
C_2^{(2,0)}: & \{a^2, a'^2\}, & h = 3, \\
C_2^{(2,1)}: & \{a^2 a', aa'^2\}, & h = 3,
\end{array} \quad (9.17)$$

where h is the order of each element in the given conjugacy class.

9.2 Σ(18)

Table 9.2 Characters of Σ(18) representations

	h	χ_{1+0}	χ_{1+1}	χ_{1+2}	χ_{1-0}	χ_{1-1}	χ_{1-2}	$\chi_{2_{1,0}}$	$\chi_{2_{2,0}}$	$\chi_{2_{2,1}}$
C_1	1	1	1	1	1	1	1	2	2	2
$C_1^{(1)}$	3	1	ρ^2	ρ	1	ρ^2	ρ	2ρ	$2\rho^2$	2
$C_1^{(2)}$	3	1	ρ	ρ^2	1	ρ	ρ^2	$2\rho^2$	2ρ	2
$C_3'^{(0)}$	2	1	1	1	-1	-1	-1	0	0	0
$C_3'^{(1)}$	6	1	ρ	ρ^2	-1	$-\rho$	$-\rho^2$	0	0	0
$C_3'^{(2)}$	6	1	ρ^2	ρ	-1	$-\rho^2$	$-\rho$	0	0	0
$C_2^{(1,0)}$	3	1	ρ	ρ^2	1	ρ	ρ^2	$-\rho^2$	$-\rho$	-1
$C_1^{(2,0)}$	3	1	ρ^2	ρ	1	ρ^2	ρ	$-\rho$	$-\rho^2$	-1
$C_1^{(3,0)}$	3	1	1	1	1	1	1	-1	-1	-1

Σ(18) has six singlets $\mathbf{1}_{\pm,n}$ with $n=0,1,2$, and three doublets $\mathbf{2}_{p,q}$ with $(p,q)=(1,0),(2,0),(2,1)$. The characters are shown in Table 9.2.

The tensor products between doublets are as follows:

$$\begin{pmatrix} x_2 \\ x_1 \end{pmatrix}_{\mathbf{2}_{2,1}} \otimes \begin{pmatrix} y_2 \\ y_1 \end{pmatrix}_{\mathbf{2}_{2,1}} = (x_1 y_2 + x_2 y_1)_{\mathbf{1}_{+,0}} \oplus (x_1 y_2 - x_2 y_1)_{\mathbf{1}_{-,0}} \oplus \begin{pmatrix} x_1 y_1 \\ x_2 y_2 \end{pmatrix}_{\mathbf{2}_{2,1}},$$

$$\begin{pmatrix} x_2 \\ x_0 \end{pmatrix}_{\mathbf{2}_{2,0}} \otimes \begin{pmatrix} y_2 \\ y_0 \end{pmatrix}_{\mathbf{2}_{2,0}} = (x_0 y_2 + x_2 y_0)_{\mathbf{1}_{+,2}} \oplus (x_0 y_2 - x_2 y_0)_{\mathbf{1}_{-,2}} \oplus \begin{pmatrix} x_2 y_2 \\ x_0 y_0 \end{pmatrix}_{\mathbf{2}_{1,0}},$$

$$\begin{pmatrix} x_1 \\ x_0 \end{pmatrix}_{\mathbf{2}_{1,0}} \otimes \begin{pmatrix} y_1 \\ y_0 \end{pmatrix}_{\mathbf{2}_{1,0}} = (x_0 y_1 + x_1 y_0)_{\mathbf{1}_{+,1}} \oplus (x_0 y_1 - x_1 y_0)_{\mathbf{1}_{-,1}} \oplus \begin{pmatrix} x_1 y_1 \\ x_0 y_0 \end{pmatrix}_{\mathbf{2}_{2,0}},$$

$$\begin{pmatrix} x_2 \\ x_1 \end{pmatrix}_{\mathbf{2}_{2,1}} \otimes \begin{pmatrix} y_2 \\ y_0 \end{pmatrix}_{\mathbf{2}_{2,0}} = (x_2 y_2 + x_1 y_0)_{\mathbf{1}_{+,1}} \oplus (x_2 y_2 - x_1 y_0)_{\mathbf{1}_{-,1}} \oplus \begin{pmatrix} x_2 y_0 \\ x_1 y_2 \end{pmatrix}_{\mathbf{2}_{2,0}},$$

$$\begin{pmatrix} x_2 \\ x_1 \end{pmatrix}_{\mathbf{2}_{2,1}} \otimes \begin{pmatrix} y_1 \\ y_0 \end{pmatrix}_{\mathbf{2}_{1,0}} = (x_1 y_1 + x_2 y_0)_{\mathbf{1}_{+,2}} \oplus (x_1 y_1 - x_2 y_0)_{\mathbf{1}_{-,2}} \oplus \begin{pmatrix} x_1 y_0 \\ x_2 y_1 \end{pmatrix}_{\mathbf{2}_{1,0}},$$

$$\begin{pmatrix} x_2 \\ x_0 \end{pmatrix}_{\mathbf{2}_{2,0}} \otimes \begin{pmatrix} y_1 \\ y_0 \end{pmatrix}_{\mathbf{2}_{1,0}} = (x_2 y_1 + x_0 y_0)_{\mathbf{1}_{+,0}} \oplus (x_2 y_1 - x_0 y_0)_{\mathbf{1}_{-,0}} \oplus \begin{pmatrix} x_2 y_0 \\ x_0 y_1 \end{pmatrix}_{\mathbf{2}_{2,1}}.$$

(9.18)

The tensor products between singlets are:

$$\begin{aligned}
&\mathbf{1}_{\pm,0} \otimes \mathbf{1}_{\pm,0} = \mathbf{1}_{+,0}, & &\mathbf{1}_{\pm,1} \otimes \mathbf{1}_{\pm,1} = \mathbf{1}_{+,2}, & &\mathbf{1}_{\pm,2} \otimes \mathbf{1}_{\pm,2} = \mathbf{1}_{+,1}, \\
&\mathbf{1}_{\pm,1} \otimes \mathbf{1}_{\pm,0} = \mathbf{1}_{+,1}, & &\mathbf{1}_{\pm,2} \otimes \mathbf{1}_{\pm,0} = \mathbf{1}_{+,2}, & &\mathbf{1}_{\pm,2} \otimes \mathbf{1}_{\pm,1} = \mathbf{1}_{+,0}, \\
&\mathbf{1}_{\pm,0} \otimes \mathbf{1}_{\mp,0} = \mathbf{1}_{-,0}, & &\mathbf{1}_{\pm,1} \otimes \mathbf{1}_{\mp,1} = \mathbf{1}_{-,2}, & &\mathbf{1}_{\pm,2} \otimes \mathbf{1}_{\mp,2} = \mathbf{1}_{-,1}, \\
&\mathbf{1}_{\pm,1} \otimes \mathbf{1}_{\mp,0} = \mathbf{1}_{-,1}, & &\mathbf{1}_{\pm,2} \otimes \mathbf{1}_{\pm,0} = \mathbf{1}_{-,2}, & &\mathbf{1}_{\pm,2} \otimes \mathbf{1}_{\mp,1} = \mathbf{1}_{-,0}.
\end{aligned}$$

(9.19)

And finally, the tensor products between singlets and doublets are:

$$(y)\mathbf{1}_{\pm,0} \otimes \begin{pmatrix} x_2 \\ x_1 \end{pmatrix}_{2_{2,1}} = \begin{pmatrix} yx_2 \\ yx_1 \end{pmatrix}_{2_{2,1}}, \quad (y)\mathbf{1}_{\pm,1} \otimes \begin{pmatrix} x_2 \\ x_1 \end{pmatrix}_{2_{2,1}} = \begin{pmatrix} yx_1 \\ yx_2 \end{pmatrix}_{2_{2,0}},$$

$$(y)\mathbf{1}_{\pm,2} \otimes \begin{pmatrix} x_2 \\ x_1 \end{pmatrix}_{2_{2,1}} = \begin{pmatrix} yx_2 \\ yx_1 \end{pmatrix}_{2_{1,0}}, \quad (y)\mathbf{1}_{\pm,0} \otimes \begin{pmatrix} x_2 \\ x_0 \end{pmatrix}_{2_{2,0}} = \begin{pmatrix} yx_2 \\ yx_0 \end{pmatrix}_{2_{2,0}},$$

$$(y)\mathbf{1}_{\pm,1} \otimes \begin{pmatrix} x_2 \\ x_0 \end{pmatrix}_{2_{2,0}} = \begin{pmatrix} yx_0 \\ yx_2 \end{pmatrix}_{2_{1,0}}, \quad (y)\mathbf{1}_{\pm,2} \otimes \begin{pmatrix} x_2 \\ x_0 \end{pmatrix}_{2_{2,0}} = \begin{pmatrix} yx_0 \\ yx_2 \end{pmatrix}_{2_{2,1}} \quad (9.20)$$

$$(y)\mathbf{1}_{\pm,0} \otimes \begin{pmatrix} x_1 \\ x_0 \end{pmatrix}_{2_{1,0}} = \begin{pmatrix} yx_1 \\ yx_0 \end{pmatrix}_{2_{1,0}}, \quad (y)\mathbf{1}_{\pm,1} \otimes \begin{pmatrix} x_1 \\ x_0 \end{pmatrix}_{2_{1,0}} = \begin{pmatrix} yx_1 \\ yx_0 \end{pmatrix}_{2_{2,1}},$$

$$(y)\mathbf{1}_{\pm,2} \otimes \begin{pmatrix} x_1 \\ x_0 \end{pmatrix}_{2_{1,0}} = \begin{pmatrix} yx_0 \\ yx_1 \end{pmatrix}_{2_{2,0}}.$$

9.3 $\Sigma(32)$

The next example is the group $\Sigma(32)$, which has thirty-two elements, $b^k a^m a'^n$ for $k = 0, 1$, and $m, n = 0, 1, 2, 3$, where a, a', and b satisfy $b^2 = e$, $a^4 = a'^4 = e$, $aa' = a'a$, and $bab = a'$. These elements are classified into fourteen conjugacy classes:

$$\begin{aligned}
&C_1: &&\{e\}, &&h=1, \\
&C_1^{(1)}: &&\{aa'\}, &&h=4, \\
&C_1^{(2)}: &&\{a^2 a'^2\}, &&h=2, \\
&C_1^{(3)}: &&\{a^3 a'^3\}, &&h=4, \\
&C_4'^{(0)}: &&\{b, ba'a^3, ba'^2 a^2, ba'^3 a\}, &&h=2, \\
&C_4'^{(1)}: &&\{ba', ba, ba'^2 a^3, ba'^3 a^2\}, &&h=8, \\
&C_4'^{(2)}: &&\{ba'^2, ba'a, ba^2, ba'^3 a^3\}, &&h=4, \\
&C_4'^{(3)}: &&\{ba'^3, ba'^2 a, ba'a^2, ba^3\}, &&h=8, \\
&C_2^{(1,0)}: &&\{a, a'\}, &&h=4, \\
&C_2^{(2,0)}: &&\{a^2, a'^2\}, &&h=2, \\
&C_2^{(2,1)}: &&\{a^2 a', aa'^2\}, &&h=4, \\
&C_2^{(3,0)}: &&\{a^3, a'^3\}, &&h=4, \\
&C_2^{(3,1)}: &&\{a^3 a', aa'^3\}, &&h=4, \\
&C_2^{(3,2)}: &&\{a^3 a'^2, a^2 a'^3\}, &&h=4,
\end{aligned} \quad (9.21)$$

where h is the order of each element in the given conjugacy class.

$\Sigma(32)$ has eight singlets $\mathbf{1}_{\pm,n}$ with $n = 0, 1, 2, 3$, and six doublets $\mathbf{2}_{p,q}$ with $(p, q) = (1, 0), (2, 0), (3, 0), (2, 1), (3, 1), (3, 2)$. The characters are shown in Table 9.3.

9.3 $\Sigma(32)$

Table 9.3 Characters of $\Sigma(32)$ representations

	h	$\chi_{1_{+0}}$	$\chi_{1_{+1}}$	$\chi_{1_{+2}}$	$\chi_{1_{+3}}$	$\chi_{1_{-0}}$	$\chi_{1_{-1}}$	$\chi_{1_{-2}}$	$\chi_{1_{-3}}$	$\chi_{2_{1,0}}$	$\chi_{2_{2,0}}$	$\chi_{2_{2,1}}$	$\chi_{2_{3,0}}$	$\chi_{2_{3,1}}$	$\chi_{2_{3,2}}$
C_1	1	1	1	1	1	1	1	1	1	2	2	2	2	2	2
$C_1^{(1)}$	4	1	-1	1	-1	1	-1	1	-1	$2i$	-2	$-2i$	$-2i$	2	$2i$
$C_1^{(2)}$	2	1	1	1	1	1	1	1	1	-2	2	-2	-2	2	-2
$C_1^{(3)}$	4	1	-1	1	-1	1	-1	1	-1	$-2i$	-2	$2i$	$2i$	2	$-2i$
$C_4'^{(0)}$	2	1	1	1	1	-1	-1	-1	-1	0	0	0	0	0	0
$C_4'^{(1)}$	8	1	i	-1	$-i$	-1	$-i$	1	i	0	0	0	0	0	0
$C_4'^{(2)}$	4	1	-1	1	-1	-1	1	-1	1	0	0	0	0	0	0
$C_4'^{(3)}$	8	1	$-i$	-1	i	-1	i	1	$-i$	0	0	0	0	0	0
$C_2^{(1,0)}$	4	1	i	-1	$-i$	1	i	-1	$-i$	$1+i$	0	$-1+i$	$1-i$	0	$-1-i$
$C_2^{(2,0)}$	2	1	-1	1	-1	1	-1	1	-1	0	2	0	0	-2	0
$C_2^{(2,1)}$	4	1	$-i$	-1	i	1	$-i$	-1	i	$-1+i$	0	$1+i$	$-1-i$	0	$1-i$
$C_2^{(3,0)}$	4	1	$-i$	-1	i	1	$-i$	-1	i	$1-i$	0	$-1-i$	$1+i$	0	$-1+i$
$C_2^{(3,1)}$	4	1	1	1	1	1	1	1	1	0	-2	0	0	-2	0
$C_2^{(3,2)}$	4	1	i	-1	$-i$	1	i	-1	$-i$	$-1-i$	0	$1-i$	$-1+i$	0	$1+i$

The tensor products between doublets are as follows:

$$\begin{pmatrix} x_3 \\ x_2 \end{pmatrix}_{2_{3,2}} \otimes \begin{pmatrix} y_3 \\ y_2 \end{pmatrix}_{2_{3,2}} = (x_2 y_3 + x_3 y_2)_{1_{+,1}} \oplus (x_2 y_3 - x_3 y_2)_{1_{-,1}} \oplus \begin{pmatrix} x_3 y_3 \\ x_2 y_2 \end{pmatrix}_{2_{2,0}}, \tag{9.22}$$

$$\begin{pmatrix} x_3 \\ x_1 \end{pmatrix}_{2_{3,1}} \otimes \begin{pmatrix} y_3 \\ y_1 \end{pmatrix}_{2_{3,1}} = (x_1 y_3 + x_3 y_1)_{1_{+,0}} \oplus (x_1 y_3 - x_3 y_1)_{1_{-,0}}$$
$$\oplus (x_3 y_3 + x_1 y_1)_{1_{+,2}} \oplus (x_3 y_3 - x_1 y_1)_{1_{-,2}}, \tag{9.23}$$

$$\begin{pmatrix} x_3 \\ x_0 \end{pmatrix}_{2_{3,0}} \otimes \begin{pmatrix} y_3 \\ y_0 \end{pmatrix}_{2_{3,0}} = (x_0 y_3 + x_3 y_0)_{1_{+,3}} \oplus (x_0 y_3 - x_3 y_0)_{1_{-,3}} \oplus \begin{pmatrix} x_3 y_3 \\ x_0 y_0 \end{pmatrix}_{2_{2,0}}, \tag{9.24}$$

$$\begin{pmatrix} x_2 \\ x_1 \end{pmatrix}_{2_{2,1}} \otimes \begin{pmatrix} y_2 \\ y_1 \end{pmatrix}_{2_{2,1}} = (x_1 y_2 + x_2 y_1)_{1_{+,3}} \oplus (x_1 y_2 - x_2 y_1)_{1_{-,3}} \oplus \begin{pmatrix} x_1 y_1 \\ x_2 y_2 \end{pmatrix}_{2_{2,0}}, \tag{9.25}$$

$$\begin{pmatrix} x_2 \\ x_0 \end{pmatrix}_{2_{2,0}} \otimes \begin{pmatrix} y_2 \\ y_0 \end{pmatrix}_{2_{2,0}} = (x_0 y_2 + x_2 y_0)_{1_{+,2}} \oplus (x_0 y_2 - x_2 y_0)_{1_{-,2}}$$
$$\oplus (x_2 y_2 + x_0 y_0)_{1_{+,0}} \oplus (x_2 y_2 - x_0 y_0)_{1_{-,0}}, \tag{9.26}$$

$$\begin{pmatrix} x_1 \\ x_0 \end{pmatrix}_{2_{1,0}} \otimes \begin{pmatrix} y_1 \\ y_0 \end{pmatrix}_{2_{1,0}} = (x_0 y_1 + x_1 y_0)_{1_{+,1}} \oplus (x_0 y_1 - x_1 y_0)_{1_{-,1}} \oplus \begin{pmatrix} x_1 y_1 \\ x_0 y_0 \end{pmatrix}_{2_{2,0}}, \tag{9.27}$$

$$\begin{pmatrix} x_3 \\ x_2 \end{pmatrix}_{2_{3,2}} \otimes \begin{pmatrix} y_3 \\ y_1 \end{pmatrix}_{2_{3,1}} = \begin{pmatrix} x_2 y_3 \\ x_3 y_1 \end{pmatrix}_{2_{1,0}} \oplus \begin{pmatrix} x_2 y_1 \\ x_3 y_3 \end{pmatrix}_{2_{3,2}}, \qquad (9.28)$$

$$\begin{pmatrix} x_3 \\ x_2 \end{pmatrix}_{2_{3,2}} \otimes \begin{pmatrix} y_3 \\ y_0 \end{pmatrix}_{2_{3,0}} = (x_3 y_3 + x_2 y_0)_{1_{+,2}} \oplus (x_3 y_3 - x_2 y_0)_{1_{-,2}} \oplus \begin{pmatrix} x_3 y_0 \\ x_2 y_3 \end{pmatrix}_{2_{3,1}}, \qquad (9.29)$$

$$\begin{pmatrix} x_3 \\ x_2 \end{pmatrix}_{2_{3,2}} \otimes \begin{pmatrix} y_2 \\ y_1 \end{pmatrix}_{2_{2,1}} = (x_2 y_2 + x_3 y_1)_{1_{+,0}} \oplus (x_2 y_2 - x_3 y_1)_{1_{-,0}} \oplus \begin{pmatrix} x_2 y_1 \\ x_3 y_2 \end{pmatrix}_{2_{3,1}}, \qquad (9.30)$$

$$\begin{pmatrix} x_3 \\ x_2 \end{pmatrix}_{2_{3,2}} \otimes \begin{pmatrix} y_2 \\ y_0 \end{pmatrix}_{2_{2,0}} = \begin{pmatrix} x_3 y_0 \\ x_2 y_2 \end{pmatrix}_{2_{3,1}} \oplus \begin{pmatrix} x_2 y_0 \\ x_3 y_2 \end{pmatrix}_{2_{2,1}}, \qquad (9.31)$$

$$\begin{pmatrix} x_3 \\ x_2 \end{pmatrix}_{2_{3,2}} \otimes \begin{pmatrix} y_1 \\ y_0 \end{pmatrix}_{2_{1,0}} = (x_2 y_1 + x_3 y_0)_{1_{+,3}} \oplus (x_2 y_1 - x_3 y_0)_{1_{-,3}} \oplus \begin{pmatrix} x_2 y_0 \\ x_3 y_1 \end{pmatrix}_{2_{2,0}}, \qquad (9.32)$$

$$\begin{pmatrix} x_3 \\ x_1 \end{pmatrix}_{2_{3,1}} \otimes \begin{pmatrix} y_3 \\ y_0 \end{pmatrix}_{2_{3,0}} = \begin{pmatrix} x_3 y_0 \\ x_1 y_3 \end{pmatrix}_{2_{3,0}} \oplus \begin{pmatrix} x_3 y_3 \\ x_1 y_0 \end{pmatrix}_{2_{2,1}}, \qquad (9.33)$$

$$\begin{pmatrix} x_3 \\ x_1 \end{pmatrix}_{2_{3,1}} \otimes \begin{pmatrix} y_2 \\ y_1 \end{pmatrix}_{2_{2,1}} = \begin{pmatrix} x_1 y_2 \\ x_3 y_1 \end{pmatrix}_{2_{3,0}} \oplus \begin{pmatrix} x_1 y_1 \\ x_3 y_2 \end{pmatrix}_{2_{2,1}}, \qquad (9.34)$$

$$\begin{pmatrix} x_3 \\ x_1 \end{pmatrix}_{2_{3,1}} \otimes \begin{pmatrix} y_2 \\ y_0 \end{pmatrix}_{2_{2,0}} = (x_1 y_2 + x_3 y_0)_{1_{+,3}} \oplus (x_1 y_2 - x_3 y_0)_{1_{-,3}}$$
$$\oplus (x_3 y_2 + x_1 y_0)_{1_{+,1}} \oplus (x_3 y_2 - x_1 y_0)_{1_{-,1}}, \qquad (9.35)$$

$$\begin{pmatrix} x_3 \\ x_1 \end{pmatrix}_{2_{3,1}} \otimes \begin{pmatrix} y_1 \\ y_0 \end{pmatrix}_{2_{1,0}} = \begin{pmatrix} x_3 y_0 \\ x_1 y_1 \end{pmatrix}_{2_{3,2}} \oplus \begin{pmatrix} x_1 y_0 \\ x_3 y_1 \end{pmatrix}_{2_{1,0}}, \qquad (9.36)$$

$$\begin{pmatrix} x_3 \\ x_0 \end{pmatrix}_{2_{3,0}} \otimes \begin{pmatrix} y_2 \\ y_1 \end{pmatrix}_{2_{2,1}} = (x_3 y_2 + x_0 y_1)_{1_{+,1}} \oplus (x_3 y_2 - x_0 y_1)_{1_{-,1}} \oplus \begin{pmatrix} x_0 y_2 \\ x_3 y_1 \end{pmatrix}_{2_{2,0}}, \qquad (9.37)$$

$$\begin{pmatrix} x_3 \\ x_0 \end{pmatrix}_{2_{3,0}} \otimes \begin{pmatrix} y_2 \\ y_0 \end{pmatrix}_{2_{2,0}} = \begin{pmatrix} x_3 y_0 \\ x_0 y_2 \end{pmatrix}_{2_{3,2}} \oplus \begin{pmatrix} x_3 y_2 \\ x_0 y_0 \end{pmatrix}_{2_{1,0}}, \qquad (9.38)$$

$$\begin{pmatrix} x_3 \\ x_0 \end{pmatrix}_{2_{3,0}} \otimes \begin{pmatrix} y_1 \\ y_0 \end{pmatrix}_{2_{1,0}} = (x_3 y_1 + x_0 y_0)_{1_{+,0}} \oplus (x_3 y_1 - x_0 y_0)_{1_{-,0}} \oplus \begin{pmatrix} x_3 y_0 \\ x_0 y_1 \end{pmatrix}_{2_{3,1}}, \qquad (9.39)$$

$$\begin{pmatrix} x_2 \\ x_1 \end{pmatrix}_{2_{2,1}} \otimes \begin{pmatrix} y_2 \\ y_0 \end{pmatrix}_{2_{2,0}} = (x_2 y_2 + x_1 y_0)_{1_{+,1}} \oplus (x_2 y_2 - x_1 y_0)_{1_{-,1}} \oplus \begin{pmatrix} x_1 y_2 \\ x_2 y_0 \end{pmatrix}_{2_{3,2}}, \qquad (9.40)$$

9.3 $\Sigma(32)$

$$\begin{pmatrix} x_2 \\ x_1 \end{pmatrix}_{2_{2,1}} \otimes \begin{pmatrix} y_1 \\ y_0 \end{pmatrix}_{2_{1,0}} = (x_1 y_1 + x_2 y_0)_{1_{+,2}} \oplus (x_1 y_1 - x_2 y_0)_{1_{-,2}} \oplus \begin{pmatrix} x_2 y_1 \\ x_1 y_0 \end{pmatrix}_{2_{3,1}},$$
(9.41)

$$\begin{pmatrix} x_2 \\ x_0 \end{pmatrix}_{2_{2,0}} \otimes \begin{pmatrix} y_1 \\ y_0 \end{pmatrix}_{2_{1,0}} = \begin{pmatrix} x_2 y_1 \\ x_0 y_0 \end{pmatrix}_{2_{2,0}} \oplus \begin{pmatrix} x_2 y_0 \\ x_0 y_1 \end{pmatrix}_{2_{2,1}}.$$
(9.42)

The tensor products between singlets are:

$$\begin{aligned}
&1_{\pm,0} \otimes 1_{\pm,0} = 1_{+,0}, & &1_{\pm,1} \otimes 1_{\pm,1} = 1_{+,2}, & &1_{\pm,2} \otimes 1_{\pm,2} = 1_{+,0}, \\
&1_{\pm,3} \otimes 1_{\pm,3} = 1_{+,2}, & &1_{\pm,3} \otimes 1_{\pm,2} = 1_{+,1}, & &1_{\pm,3} \otimes 1_{\pm,1} = 1_{+,0}, \\
&1_{\pm,3} \otimes 1_{\pm,0} = 1_{+,3}, & &1_{\pm,2} \otimes 1_{\pm,1} = 1_{+,3}, & &1_{\pm,2} \otimes 1_{\pm,0} = 1_{+,2}, \\
&1_{\pm,1} \otimes 1_{\pm,0} = 1_{+,1}, & &1_{\mp,0} \otimes 1_{\pm,0} = 1_{-,0}, & &1_{\mp,1} \otimes 1_{\pm,1} = 1_{-,2}, \quad (9.43) \\
&1_{\mp,2} \otimes 1_{\pm,2} = 1_{-,0}, & &1_{\mp,3} \otimes 1_{\pm,3} = 1_{-,2}, & &1_{\mp,3} \otimes 1_{\pm,2} = 1_{-,1}, \\
&1_{\mp,3} \otimes 1_{\pm,1} = 1_{-,0}, & &1_{\mp,3} \otimes 1_{\pm,0} = 1_{-,3}, & &1_{\mp,2} \otimes 1_{\pm,1} = 1_{-,3}, \\
&1_{\mp,2} \otimes 1_{\pm,0} = 1_{-,2}, & &1_{\mp,1} \otimes 1_{\pm,0} = 1_{-,1}.
\end{aligned}$$

Finally, the tensor products between singlets and doublets are:

$$(y)_{1_{\pm,0}} \otimes \begin{pmatrix} x_3 \\ x_2 \end{pmatrix}_{2_{3,2}} = \begin{pmatrix} y x_3 \\ y x_2 \end{pmatrix}_{2_{3,2}}, \quad (y)_{1_{\pm,1}} \otimes \begin{pmatrix} x_3 \\ x_2 \end{pmatrix}_{2_{3,2}} = \begin{pmatrix} y x_2 \\ y x_3 \end{pmatrix}_{2_{3,0}},$$

$$(y)_{1_{\pm,2}} \otimes \begin{pmatrix} x_3 \\ x_2 \end{pmatrix}_{2_{3,2}} = \begin{pmatrix} y x_3 \\ y x_2 \end{pmatrix}_{2_{1,0}}, \quad (y)_{1_{\pm,3}} \otimes \begin{pmatrix} x_3 \\ x_2 \end{pmatrix}_{2_{3,2}} = \begin{pmatrix} y x_3 \\ y x_2 \end{pmatrix}_{2_{2,1}},$$

$$(y)_{1_{\pm,0}} \otimes \begin{pmatrix} x_3 \\ x_1 \end{pmatrix}_{2_{3,1}} = \begin{pmatrix} y x_3 \\ y x_1 \end{pmatrix}_{2_{3,1}}, \quad (y)_{1_{\pm,1}} \otimes \begin{pmatrix} x_3 \\ x_1 \end{pmatrix}_{2_{3,1}} = \begin{pmatrix} y x_1 \\ y x_3 \end{pmatrix}_{2_{2,0}},$$

$$(y)_{1_{\pm,2}} \otimes \begin{pmatrix} x_3 \\ x_1 \end{pmatrix}_{2_{3,1}} = \begin{pmatrix} y x_1 \\ y x_3 \end{pmatrix}_{2_{3,1}}, \quad (y)_{1_{\pm,3}} \otimes \begin{pmatrix} x_3 \\ x_1 \end{pmatrix}_{2_{3,1}} = \begin{pmatrix} y x_3 \\ y x_1 \end{pmatrix}_{2_{2,0}},$$

$$(y)_{1_{\pm,0}} \otimes \begin{pmatrix} x_3 \\ x_0 \end{pmatrix}_{2_{3,0}} = \begin{pmatrix} y x_3 \\ y x_0 \end{pmatrix}_{2_{3,0}}, \quad (y)_{1_{\pm,1}} \otimes \begin{pmatrix} x_3 \\ x_0 \end{pmatrix}_{2_{3,0}} = \begin{pmatrix} y x_0 \\ y x_3 \end{pmatrix}_{2_{1,0}},$$

$$(y)_{1_{\pm,2}} \otimes \begin{pmatrix} x_3 \\ x_0 \end{pmatrix}_{2_{3,0}} = \begin{pmatrix} y x_0 \\ y x_3 \end{pmatrix}_{2_{2,1}}, \quad (y)_{1_{\pm,3}} \otimes \begin{pmatrix} x_3 \\ x_0 \end{pmatrix}_{2_{3,0}} = \begin{pmatrix} y x_0 \\ y x_3 \end{pmatrix}_{2_{3,2}},$$

$$(y)_{1_{\pm,0}} \otimes \begin{pmatrix} x_2 \\ x_1 \end{pmatrix}_{2_{2,1}} = \begin{pmatrix} y x_2 \\ y x_1 \end{pmatrix}_{2_{2,1}}, \quad (y)_{1_{\pm,1}} \otimes \begin{pmatrix} x_2 \\ x_1 \end{pmatrix}_{2_{2,1}} = \begin{pmatrix} y x_2 \\ y x_1 \end{pmatrix}_{2_{3,2}},$$

$$(y)_{1_{\pm,2}} \otimes \begin{pmatrix} x_2 \\ x_1 \end{pmatrix}_{2_{2,1}} = \begin{pmatrix} y x_1 \\ y x_2 \end{pmatrix}_{2_{3,0}}, \quad (y)_{1_{\pm,3}} \otimes \begin{pmatrix} x_2 \\ x_1 \end{pmatrix}_{2_{2,1}} = \begin{pmatrix} y x_2 \\ y x_1 \end{pmatrix}_{2_{1,0}},$$

$$(y)1_{\pm,0} \otimes \begin{pmatrix} x_2 \\ x_0 \end{pmatrix}_{2_{2,0}} = \begin{pmatrix} yx_2 \\ yx_0 \end{pmatrix}_{2_{2,0}}, \quad (y)1_{\pm,1} \otimes \begin{pmatrix} x_2 \\ x_0 \end{pmatrix}_{2_{2,0}} = \begin{pmatrix} yx_2 \\ yx_0 \end{pmatrix}_{2_{3,1}},$$

$$(y)1_{\pm,2} \otimes \begin{pmatrix} x_2 \\ x_0 \end{pmatrix}_{2_{2,0}} = \begin{pmatrix} yx_0 \\ yx_2 \end{pmatrix}_{2_{2,0}}, \quad (y)1_{\pm,3} \otimes \begin{pmatrix} x_2 \\ x_0 \end{pmatrix}_{2_{2,0}} = \begin{pmatrix} yx_0 \\ yx_2 \end{pmatrix}_{2_{3,1}},$$

$$(y)1_{\pm,0} \otimes \begin{pmatrix} x_1 \\ x_0 \end{pmatrix}_{2_{1,0}} = \begin{pmatrix} yx_1 \\ yx_0 \end{pmatrix}_{2_{1,0}}, \quad (y)1_{\pm,1} \otimes \begin{pmatrix} x_1 \\ x_0 \end{pmatrix}_{2_{1,0}} = \begin{pmatrix} yx_1 \\ yx_0 \end{pmatrix}_{2_{2,1}},$$

$$(y)1_{\pm,2} \otimes \begin{pmatrix} x_1 \\ x_0 \end{pmatrix}_{2_{1,0}} = \begin{pmatrix} yx_1 \\ yx_0 \end{pmatrix}_{2_{3,2}}, \quad (y)1_{\pm,3} \otimes \begin{pmatrix} x_1 \\ x_0 \end{pmatrix}_{2_{1,0}} = \begin{pmatrix} yx_0 \\ yx_1 \end{pmatrix}_{2_{3,0}}.$$

(9.44)

9.4 $\Sigma(50)$

This group has fifty elements, $b^k a^m a'^n$ for $k = 0, 1$, and $m, n = 0, 1, 2, 3, 4$, where a, a', and b satisfy the same conditions as (9.1) for the case $N = 5$. These elements are classified into twenty conjugacy classes:

$$\begin{aligned}
&C_1: &&\{e\}, &&h=1,\\
&C_1^{(1)}: &&\{aa'\}, &&h=5,\\
&C_1^{(2)}: &&\{a^2 a'^2\}, &&h=5,\\
&C_1^{(3)}: &&\{a^3 a'^3\}, &&h=5,\\
&C_1^{(4)}: &&\{a^3 a'^3\}, &&h=5,\\
&C_5'^{(0)}: &&\{b, ba'^2 a^3, ba'^3 a^2, ba'^4 a, ba'a^4\}, &&h=2,\\
&C_5'^{(1)}: &&\{ba', ba, ba'^3 a^3, ba'^4 a^2, ba'^2 a^4\}, &&h=10,\\
&C_5'^{(2)}: &&\{ba'^2, ba'a, ba^2, ba'^4 a^3, ba'^3 a^4\}, &&h=10,\\
&C_5'^{(3)}: &&\{ba'^3, ba'^2 a, ba'a^2, ba^3, ba'^4 a^4\}, &&h=10,\\
&C_5'^{(4)}: &&\{ba'^4, ba'^2 a^2, ba'a^3, ba'^3 a, ba^4\}, &&h=10,\\
&C_2^{(1,0)}: &&\{a, a'\}, &&h=5,\\
&C_2^{(2,0)}: &&\{a^2, a'^2\}, &&h=5,\\
&C_2^{(2,1)}: &&\{a^2 a', aa'^2\}, &&h=5,\\
&C_2^{(3,0)}: &&\{a^3, a'^3\}, &&h=5,\\
&C_2^{(3,1)}: &&\{a^3 a', aa'^3\}, &&h=5,\\
&C_2^{(3,2)}: &&\{a^3 a'^2, a^2 a'^3\}, &&h=5,\\
&C_2^{(4,0)}: &&\{a^4, a'^4\}, &&h=5,\\
&C_2^{(4,1)}: &&\{a^4 a', aa'^4\}, &&h=5,\\
&C_2^{(4,2)}: &&\{a^4 a'^2, a^2 a'^4\}, &&h=5,\\
&C_2^{(4,3)}: &&\{a^4 a'^3, a^3 a'^4\}, &&h=5,\\
\end{aligned}$$

(9.45)

where h is the order of each element in the given conjugacy class.

9.4 $\Sigma(50)$

Table 9.4 Characters of $\Sigma(50)$ representations, where $\rho = e^{2i\pi/5}$

	h	$\chi_{1\pm 0}$	$\chi_{1\pm 1}$	$\chi_{1\pm 2}$	$\chi_{1\pm 3}$	$\chi_{1\pm 4}$
C_1	1	1	1	1	1	1
$C_1^{(1)}$	5	1	ρ^2	ρ^4	ρ	ρ^3
$C_1^{(2)}$	5	1	ρ^4	ρ^3	ρ^2	ρ
$C_1^{(3)}$	5	1	ρ	ρ^2	ρ^3	ρ^4
$C_1^{(4)}$	5	1	ρ^3	ρ	ρ^4	ρ^2
$C_5'^{(0)}$	2	± 1	± 1	± 1	± 1	± 1
$C_5'^{(1)}$	10	± 1	$\pm\rho$	$\pm\rho^2$	$\pm\rho^3$	$\pm\rho^4$
$C_5'^{(2)}$	10	± 1	$\pm\rho^2$	$\pm\rho^4$	$\pm\rho$	$\pm\rho^3$
$C_5'^{(3)}$	10	± 1	$\pm\rho^3$	$\pm\rho$	$\pm\rho^4$	$\pm\rho^2$
$C_5'^{(4)}$	10	± 1	$\pm\rho^4$	$\pm\rho^3$	$\pm\rho^2$	$\pm\rho$
$C_2^{(1,0)}$	5	1	ρ	ρ^2	ρ^3	ρ^3
$C_2^{(2,0)}$	5	1	ρ^2	ρ^4	ρ	ρ^3
$C_2^{(2,1)}$	5	1	ρ^3	ρ	ρ^4	ρ^2
$C_2^{(3,0)}$	5	1	ρ^3	ρ	ρ^4	ρ^2
$C_2^{(3,1)}$	5	1	ρ^4	ρ^3	ρ^2	ρ
$C_2^{(3,2)}$	5	1	1	1	1	1
$C_2^{(4,0)}$	5	1	ρ^4	ρ^3	ρ^2	ρ
$C_2^{(4,1)}$	5	1	1	1	1	1
$C_2^{(4,2)}$	5	1	ρ	ρ^2	ρ^3	ρ^4
$C_2^{(4,3)}$	5	1	ρ^2	ρ^4	ρ	ρ^3

Table 9.5 Characters of $\Sigma(50)$ representations, where $\rho = e^{2i\pi/5}$

	h	$\chi_{2_{1,0}}$	$\chi_{2_{2,0}}$	$\chi_{2_{2,1}}$	$\chi_{2_{3,0}}$	$\chi_{2_{3,1}}$	$\chi_{2_{3,2}}$	$\chi_{2_{4,0}}$	$\chi_{2_{4,1}}$	$\chi_{2_{4,2}}$	$\chi_{2_{4,3}}$
C_1	1	2	2	2	2	2	2	2	2	2	2
$C_1^{(1)}$	5	2ρ	$2\rho^2$	$2\rho^3$	$2\rho^3$	$2\rho^4$	2	$2\rho^4$	2	2ρ	$2\rho^2$
$C_1^{(2)}$	5	$2\rho^2$	$2\rho^4$	2ρ	2ρ	$2\rho^3$	2	$2\rho^3$	2	$2\rho^2$	$2\rho^4$
$C_1^{(3)}$	5	$2\rho^3$	2ρ	$2\rho^4$	$2\rho^4$	$2\rho^2$	2	$2\rho^2$	2	$2\rho^3$	2ρ
$C_1^{(4)}$	5	$2\rho^4$	$2\rho^3$	$2\rho^2$	$2\rho^2$	2ρ	2	2ρ	2	$2\rho^4$	$2\rho^3$
$C_5'^{(0)}$	2	0	0	0	0	0	0	0	0	0	0
$C_5'^{(1-4)}$	10	0	0	0	0	0	0	0	0	0	0
$C_2^{(1,0)}$	5	$1+\rho$	$1+\rho^2$	$\rho+\rho^2$	$1+\rho^3$	$\rho+\rho^3$	$\rho^2+\rho^3$	$1+\rho^4$	$\rho+\rho^4$	$\rho^2+\rho^4$	$\rho^3+\rho^4$
$C_2^{(2,0)}$	5	$1+\rho^2$	$1+\rho^4$	$\rho^2+\rho^4$	$1+\rho$	$\rho+\rho^2$	$\rho+\rho^4$	$1+\rho^3$	$\rho^2+\rho^3$	$\rho^3+\rho^4$	$\rho+\rho^3$
$C_2^{(2,1)}$	5	$\rho+\rho^2$	$\rho^2+\rho^4$	$1+\rho^4$	$\rho+\rho^3$	$1+\rho^2$	$\rho^2+\rho^3$	$\rho^3+\rho^4$	$\rho+\rho^4$	$1+\rho^3$	$1+\rho$
$C_2^{(3,0)}$	5	$1+\rho^3$	$1+\rho$	$\rho+\rho^3$	$1+\rho^4$	$\rho^3+\rho^4$	$\rho+\rho^4$	$1+\rho^2$	$\rho^2+\rho^3$	$\rho+\rho^2$	$\rho^2+\rho^4$
$C_2^{(3,1)}$	5	$\rho+\rho^3$	$\rho+\rho^2$	$1+\rho^2$	$\rho^3+\rho^4$	$1+\rho$	$\rho+\rho^4$	$\rho^2+\rho^4$	$\rho^2+\rho^3$	$1+\rho^4$	$1+\rho^3$
$C_2^{(3,2)}$	5	$\rho^2+\rho^3$	$\rho+\rho^4$	$\rho^2+\rho^3$	$\rho+\rho^4$	$\rho+\rho^4$	$\rho^2+\rho^3$	$\rho^2+\rho^3$	$\rho+\rho^4$	$\rho+\rho^4$	$\rho^2+\rho^3$
$C_2^{(4,0)}$	5	$1+\rho^4$	$1+\rho^3$	$\rho^3+\rho^4$	$1+\rho^2$	$\rho^2+\rho^4$	$\rho^2+\rho^3$	$1+\rho$	$\rho+\rho^4$	$\rho+\rho^3$	$\rho+\rho^2$
$C_2^{(4,1)}$	5	$\rho+\rho^4$	$\rho^2+\rho^3$	$\rho+\rho^4$	$\rho^2+\rho^3$	$\rho^2+\rho^3$	$\rho+\rho^4$	$\rho+\rho^4$	$\rho^2+\rho^3$	$\rho^2+\rho^3$	$\rho+\rho^4$
$C_2^{(4,2)}$	5	$\rho^2+\rho^4$	$\rho^3+\rho^4$	$1+\rho^3$	$\rho+\rho^2$	$1+\rho^4$	$1+\rho$	$\rho+\rho^3$	$\rho^2+\rho^3$	$1+\rho$	$1+\rho^2$
$C_2^{(4,3)}$	5	$\rho^3+\rho^4$	$\rho+\rho^3$	$1+\rho$	$\rho^2+\rho^4$	$1+\rho^3$	$\rho^2+\rho^3$	$\rho+\rho^2$	$\rho+\rho^4$	$1+\rho^2$	$1+\rho^4$

The group $\Sigma(50)$ has ten singlets $\mathbf{1}_{\pm,n}$ with $n = 0, 1, 2, 3, 4$, and ten doublets $\mathbf{2}_{p,q}$ with $(p, q) = (1, 0), (2, 0), (3, 0), (4, 0), (2, 1), (3, 1), (4, 1), (3, 1), (3, 2), (4, 3)$. The characters are shown in Tables 9.4 and 9.5.

We omit the explicit expressions of the tensor products since they can be obtained in the same way as for $\Sigma(18)$ and $\Sigma(32)$.

Chapter 10
$\Delta(3N^2)$

In this chapter, we investigate the discrete group $\Delta(3N^2)$, which is isomorphic to $(Z_N \times Z'_N) \rtimes Z_3$ (see also [1]). The generators of Z_N and Z'_N are denoted by a and a', respectively, and the Z_3 generator is written b. These generators satisfy

$$a^N = a'^N = b^3 = e, \qquad aa' = a'a,$$
$$bab^{-1} = a^{-1}(a')^{-1}, \qquad ba'b^{-1} = a. \tag{10.1}$$

Therefore, all the elements of $\Delta(3N^2)$ can be written in the form

$$g = b^k a^m a'^n, \tag{10.2}$$

for $k = 0, 1, 2$, and $m, n = 0, 1, 2, \ldots, N-1$.

Since the generators, a, a', and b, are represented, e.g., by

$$b = \begin{pmatrix} 0 & 1 & 0 \\ 0 & 0 & 1 \\ 1 & 0 & 0 \end{pmatrix}, \quad a = \begin{pmatrix} \rho & 0 & 0 \\ 0 & 1 & 0 \\ 0 & 0 & \rho^{-1} \end{pmatrix}, \quad a' = \begin{pmatrix} \rho^{-1} & 0 & 0 \\ 0 & \rho & 0 \\ 0 & 0 & 1 \end{pmatrix}, \tag{10.3}$$

where $\rho = e^{2\pi i/N}$, all elements of $\Delta(3N^2)$ can be written in the form

$$\begin{pmatrix} \rho^m & 0 & 0 \\ 0 & \rho^n & 0 \\ 0 & 0 & \rho^{-m-n} \end{pmatrix}, \quad \begin{pmatrix} 0 & \rho^m & 0 \\ 0 & 0 & \rho^n \\ \rho^{-m-n} & 0 & 0 \end{pmatrix}, \quad \begin{pmatrix} 0 & 0 & \rho^m \\ \rho^n & 0 & 0 \\ 0 & \rho^{-m-n} & 0 \end{pmatrix}, \tag{10.4}$$

for $m, n = 0, 1, 2, \ldots, N-1$.

10.1 $\Delta(3N^2)$ with $N/3 \neq$ Integer

The groups $\Delta(3N^2)$ have different features depending on whether $N/3$ is an integer or not. First, we study $\Delta(3N^2)$ when $N/3 \neq$ integer.

10.1.1 Conjugacy Classes

The conjugacy classes of $\Delta(3N^2)$ with $N/3 \neq$ integer are found to be

$$ba^{\ell}a'^{m}b^{-1} = a^{-\ell+m}a'^{-\ell}, \qquad b^2 a^{\ell}a'^{m}b^{-2} = a^{-m}a'^{\ell-m}. \tag{10.5}$$

Thus, these elements $a^{\ell}a'^{m}$, $a^{-\ell+m}a'^{-\ell}$, and $a^{-m}a'^{\ell-m}$, must belong to the same conjugacy class. They are independent elements of $\Delta(3N^2)$ unless $N/3$ integer and $3\ell \equiv \ell + m \equiv 0 \pmod{N}$. As a result, the elements $a^{\ell}a'^{m}$ are classified into the following conjugacy classes:

$$C_3^{(\ell,m)} = \{a^{\ell}a'^{m}, a^{-\ell+m}a'^{-\ell}, a^{-m}a'^{\ell-m}\}, \tag{10.6}$$

for $N/3 \neq$ integer.

In the same way, we can obtain conjugacy classes containing $ba^{\ell}a'^{m}$. Let us consider the conjugates of the simplest element b among $ba^{\ell}a'^{m}$. We find

$$a^p a'^q (b) a^{-p} a'^{-q} = b a^{-p-q} a'^{p-2q} = b a^{-n+3q} a'^{n}, \tag{10.7}$$

where we have defined $n \equiv p - 2q$ for convenience. We also get

$$b(ba^{-n+3q}a'^{n})b^{-1} = ba^{2n-3q}a'^{n-3q}, \tag{10.8}$$

$$b^2(ba^{-n+3q}a'^{n})b^{-2} = ba^{-n}a'^{-2n+3q}. \tag{10.9}$$

The important thing to note here is that q appears only in the combination $3q$. Thus, if $N/3 \neq$ integer, the element b is conjugate to all elements $ba^{\ell}a'^{m}$. That is, all of them belong to the same conjugacy class $C_{N^2}^{(1)}$. Similarly, all elements $b^2 a^{\ell}a'^{m}$ belong to the same conjugacy class $C_{N^2}^{(2)}$ when $N/3 \neq$ integer.

We summarize the conjugacy classes of $\Delta(3N^2)$ for $N/3 \neq$ integer:

$$\begin{aligned}
&C_1: &&\{e\}, &&h=1,\\
&C_3^{(\ell,m)}: &&\{a^{\ell}a'^{m}, a^{-\ell+m}a'^{-\ell}, a^{-m}a'^{\ell-m}\}, &&h = N/\gcd(N,\ell,m),\\
&C_{N^2}^{(1)}: &&\{ba^{\ell}a'^{m} | \ell, m = 0, 1, \ldots, N-1\}, &&h=3,\\
&C_{N^2}^{(2)}: &&\{b^2 a^{\ell}a'^{m} | \ell, m = 0, 1, \ldots, N-1\}, &&h=3.
\end{aligned} \tag{10.10}$$

The number of conjugacy classes $C_3^{(\ell,m)}$ is $(N^2-1)/3$, whence the total number of conjugacy classes is $3 + (N^2-1)/3$. The relations (2.18) and (2.19) for $\Delta(3n^2)$ with $N/3 \neq$ integer give

$$m_1 + 2^2 m_2 + 3^2 m_3 + \cdots = 3N^2, \tag{10.11}$$

$$m_1 + m_2 + m_3 + \cdots = 3 + (N^2-1)/3, \tag{10.12}$$

which implies that $(m_1, m_3) = (3, (N^2-1)/3)$. Therefore, we find three singlets and $(N^2-1)/3$ triplets.

10.1.2 Characters and Representations

There are 3 singlets in the group $\Delta(3N^2)$ with $N/3 \neq$ integer. Since $b^3 = e$ is satisfied in this group, the characters of the three singlets have three possible values $\chi_{1k}(b) = \omega^k$ with $k = 0, 1, 2$, which correspond to three singlets $\mathbf{1}_k$. Because $\chi_{1k}(b) = \chi_{1k}(ba) = \chi_{1k}(ba')$, we find that $\chi_{1k}(a) = \chi_{1k}(a') = 1$. These characters are given in Table 10.1.

Now consider the triplet representations. As can be seen from (10.3), we have (3×3) matrices corresponding to one of the triplet representations. Therefore, (3×3) matrix representations for the generic triplets are found by replacing

$$a \to a^\ell a'^m, \qquad a' \to b^2 a b^{-2} = a^{-m} a'^{\ell-m}. \tag{10.13}$$

However, we may note that the two types of replacement

$$(a, a') \to (a^{-\ell+m} a'^{-\ell}, a^\ell a'^m), \qquad (a, a') \to (a^{-m} a'^{\ell-m}, a^{-\ell+m} a'^{-\ell}), \tag{10.14}$$

also lead to a representation equivalent to the above (10.13), because the three elements $a^\ell a'^m$, $a^{-\ell+m} a'^{-\ell}$, and $a^{-m} a'^{\ell-m}$ belong to the same conjugacy class, viz., $C_3^{(\ell,m)}$. Thus, the generators of $\Delta(3N^2)$ with $N/3 \neq$ integer are represented by

$$b = \begin{pmatrix} 0 & 1 & 0 \\ 0 & 0 & 1 \\ 1 & 0 & 0 \end{pmatrix}, \quad a = \begin{pmatrix} \rho^\ell & 0 & 0 \\ 0 & \rho^k & 0 \\ 0 & 0 & \rho^{-k-\ell} \end{pmatrix}, \quad a' = \begin{pmatrix} \rho^{-k-\ell} & 0 & 0 \\ 0 & \rho^\ell & 0 \\ 0 & 0 & \rho^k \end{pmatrix}, \tag{10.15}$$

on the triplet $\mathbf{3}_{[k][\ell]}$, where the notation $[k][\ell]$ is defined by[1]

$$[k][\ell] = (k, \ell), \ (-k-\ell, k), \ \text{or} \ (\ell, -k-\ell). \tag{10.16}$$

We denote the vector $\mathbf{3}_{[k][\ell]}$ by

$$\mathbf{3}_{[k][\ell]} = \begin{pmatrix} x_{\ell, -k-\ell} \\ x_{k, \ell} \\ x_{-k-\ell, k} \end{pmatrix}, \tag{10.17}$$

for $k, \ell = 0, 1, \ldots, N - 1$, where k and ℓ correspond to the Z_N and Z'_N charges, respectively. When $(k, \ell) = (0, 0)$, the matrices a and a' are the identity matrices. Thus, we exclude the case with $(k, \ell) = (0, 0)$. The characters are given in Table 10.1.

10.1.3 Tensor Products

First consider the tensor products of triplets. Taking into account the $Z_N \times Z'_N$ charges, these are given by

---

[1] $[k][\ell]$ corresponds to $\widetilde{(k, \ell)}$ in [1].

Table 10.1 Characters of $\Delta(3N^2)$ for $N/3 \neq$ integer

h		χ_{1_0}	χ_{1_1}	χ_{1_2}	$\chi_{3_{[n][m]}}$
C_1	1	1	1	1	3
$C_3^{(k,\ell)}$	$\dfrac{N}{\gcd(N,k,\ell)}$	1	1	1	$\rho^{mk-n\ell-m\ell} + \rho^{nk+m\ell} + \rho^{-nk-mk-n\ell}$
$C_{N^2}^{(1)}$	3	1	ω	ω^2	0
$C_{N^2}^{(2)}$	3	1	ω^2	ω	0

$$\begin{pmatrix} x_{\ell,-k-\ell} \\ x_{k,\ell} \\ x_{-k-\ell,k} \end{pmatrix}_{3_{[k][\ell]}} \otimes \begin{pmatrix} y_{\ell',-k'-\ell'} \\ y_{k',\ell'} \\ y_{-k'-\ell',k'} \end{pmatrix}_{3_{[k'][\ell']}}$$

$$= \begin{pmatrix} x_{\ell,-k-\ell}y_{\ell',-k'-\ell'} \\ x_{k,\ell}y_{k',\ell'} \\ x_{-k-\ell,k}y_{-k'-\ell',k'} \end{pmatrix}_{3_{[k+k'][\ell+\ell']}} \oplus \begin{pmatrix} x_{k,\ell}y_{-k'-\ell',k'} \\ x_{-k-\ell,k}y_{\ell',-k'-\ell'} \\ x_{\ell,-k-\ell}y_{k',\ell'} \end{pmatrix}_{3_{[-k-\ell+\ell'][k-k'-\ell']}}$$

$$\oplus \begin{pmatrix} x_{-k-\ell,k}y_{k',\ell'} \\ x_{\ell,-k-\ell}y_{-k'-\ell',k'} \\ x_{k,\ell}y_{\ell',-k'-\ell'} \end{pmatrix}_{3_{[\ell-k'-\ell'][-k-\ell+k']}} \quad (10.18)$$

for $-(k,\ell) \neq [k'][\ell']$, and

$$\begin{pmatrix} x_{\ell,-k-\ell} \\ x_{k,\ell} \\ x_{-k-\ell,k} \end{pmatrix}_{3_{[k][\ell]}} \otimes \begin{pmatrix} y_{-\ell,k+\ell} \\ y_{-k,-\ell} \\ y_{k+\ell,-k} \end{pmatrix}_{3_{-[k][\ell]}}$$

$$= (x_{\ell,-k-\ell}y_{-\ell,k+\ell} + x_{k,\ell}y_{-k,-\ell} + x_{-k-\ell,k}y_{k+\ell,-k})_{1_0}$$

$$\oplus \left(x_{\ell,-k-\ell}y_{-\ell,k+\ell} + \omega^2 x_{k,\ell}y_{-k,-\ell} + \omega x_{-k-\ell,k}y_{k+\ell,-k} \right)_{1_1}$$

$$\oplus \left(x_{\ell,-k-\ell}y_{-\ell,k+\ell} + \omega x_{k,\ell}y_{-k,-\ell} + \omega^2 x_{-k-\ell,k}y_{k+\ell,-k} \right)_{1_2}$$

$$\oplus \begin{pmatrix} x_{k,\ell}y_{k+\ell,-k} \\ x_{-k-\ell,k}y_{-\ell,k+\ell} \\ x_{\ell,-k-\ell}y_{-k,-\ell} \end{pmatrix}_{3_{[-k-2\ell][2k+\ell]}} \oplus \begin{pmatrix} x_{-k-\ell,k}y_{-k,-\ell} \\ x_{\ell,-k-\ell}y_{k+\ell,-k} \\ x_{k,\ell}y_{-\ell,k+\ell} \end{pmatrix}_{3_{[k+2\ell][-2k-\ell]}} \quad (10.19)$$

The product of $3_{[k][\ell]}$ and 1_r is

$$\begin{pmatrix} x_{\ell,-k-\ell} \\ x_{k,\ell} \\ x_{-k-\ell,k} \end{pmatrix}_{[k][\ell]} \otimes (z)_{1_r} = \begin{pmatrix} x_{\ell,-k-\ell}z \\ \omega^r x_{k,\ell}z \\ \omega^{2r} x_{-k-\ell,k}z \end{pmatrix}_{[k][\ell]} \quad (10.20)$$

The tensor products of singlets 1_k and $1_{k'}$ are simply given by

$$1_k \otimes 1_{k'} = 1_{k+k'} \bmod 3. \quad (10.21)$$

10.2 $\Delta(3N^2)$ with $N/3$ Integer

In this section, we study groups $\Delta(3N^2)$ with $N/3$ an integer.

10.2.1 Conjugacy Classes

The algebraic relation (10.5) holds true when $N/3$ is an integer, as well as for the case $N/3 \neq$ integer. Thus, the elements $a^\ell a'^m$, $a^{-\ell+m} a'^{-\ell}$, and $a^{-m} a'^{\ell-m}$, must belong to the same conjugacy class. When $N/3$ is an integer and $3\ell = \ell + m \equiv 0 \pmod{N}$, the above elements are the same, i.e., $a^\ell a'^{-\ell}$. As a result, the elements $a^\ell a'^m$ are classified into the following conjugacy classes:

$$C_1^{(k)} = \{a^k a'^{-k}\}, \quad k = \frac{N}{3}, \frac{2N}{3},$$

$$C_3^{(\ell,m)} = \{a^\ell a'^m, \ a^{-\ell+m} a'^{-\ell}, \ a^{-m} a'^{\ell-m}\}, \quad (\ell,m) \neq \left(\frac{N}{3}, \frac{2N}{3}\right), \left(\frac{2N}{3}, \frac{N}{3}\right),$$
(10.22)

for $N/3$ integer.

Similarly, we find the conjugacy classes containing $b a^\ell a'^m$. One can obtain the conjugates of b among $b a^\ell a'^m$ using (10.7), (10.8), and (10.9). When $N/3 \neq$ integer, the element b is conjugate to all elements of the form $b a^\ell a'^m$, as shown in the last section. However, the situation is different when $N/3$ integer. The elements conjugate to b do not include ba. The conjugates to ba are also obtained as

$$a^p a'^q (ba) a^{-p} a'^{-q} = b a^{1-p-q} a'^{p-2q} = b a^{1-n+3q} a'^n, \quad (10.23)$$

$$b(b a^{1-n+3q} a'^n) b^{-1} = b a^{-1+2n-3q} a'^{-1+n-3q}, \quad (10.24)$$

$$b^2 (b a^{1-n+3q} a'^n) b^{-2} = b a^{-n} a'^{1-2n+3q}, \quad (10.25)$$

where we note that these elements conjugate to ba, as well as conjugates of b, do not include ba^2 when $N/3 =$ integer. Therefore, for $N/3$ integer, the elements $b a^\ell a'^m$ are classified into three conjugacy classes $C_{N^2/3}^{(\ell)}$ for $\ell = 0, 1, 2$ i.e.,

$$C_{N^2/3}^{(\ell)} = \left\{ b a^{\ell-n-3m} a'^n \ \middle| \ m = 0, 1, \ldots, \frac{N-3}{3}; \ n = 0, \ldots, N-1 \right\}. \quad (10.26)$$

In the same way, the elements $b^2 a^\ell a'^m$ are classified into three conjugacy classes $C_{N^2/3}^{(\ell)}$ for $\ell = 0, 1, 2$, i.e.,

$$C_{N^2/3}^{(\ell)} = \left\{ b^2 a^{\ell-n-3m} a'^n \ \middle| \ m = 0, 1, \ldots, \frac{N-3}{3}; \ n = 0, \ldots, N-1 \right\}. \quad (10.27)$$

We summarize the conjugacy classes of $\Delta(3N^2)$ for $N/3$ integer as follows:

$C_1:$ $\{e\}$,

$C_1^{(k)}:$ $\{a^k a'^{-k}\}$, $k = \frac{N}{3}, \frac{2N}{3}$,

$C_3^{(\ell,m)}:$ $\{a^\ell a'^m, a^{-\ell+m} a'^{-\ell}, a^{-m} a'^{\ell-m}\}$, $(\ell, m) \neq \left(\frac{N}{3}, \frac{2N}{3}\right), \left(\frac{2N}{3}, \frac{N}{3}\right)$,

$C_{N^2/3}^{(1,p)}:$ $\left\{ ba^{p-n-3m} a'^n \,\middle|\, m = 0, 1, \ldots, \frac{N-3}{3}, n = 0, 1, \ldots, N-1 \right\}$, $p = 0, 1, 2$,

$C_{N^2/3}^{(2,p)}:$ $\left\{ b^2 a^{p-n-3m} a'^n \,\middle|\, m = 0, 1, \ldots, \frac{N-3}{3}, n = 0, 1, \ldots, N-1 \right\}$, $p = 0, 1, 2$.

(10.28)

The order h of each element in the given conjugacy classes, i.e., such that $g^h = e$, are given as follows:

$C_1:$ $h = 1$,

$C_1^{(k)}:$ $h = 3$,

$C_3^{(\ell,m)}:$ $h = N/\gcd(N, \ell, m)$, (10.29)

$C_{N^2/3}^{(1,p)}:$ $h = 3$,

$C_{N^2/3}^{(2,p)}:$ $h = 3$.

The numbers of conjugacy classes $C_1^{(k)}, C_3^{(\ell,m)}, C_{N^2/3}^{(1,p)}$, and $C_{N^2/3}^{(2,p)}$ are 2, $(N^2-3)/3$, 3, and 3, respectively. The total number of conjugacy classes is therefore equal to $9 + (N^2 - 3)/3$. The relations (2.18) and (2.19) for $\Delta(3N^2)$ with $N/3$ integer imply

$$m_1 + 2^2 m_2 + 3^2 m_3 + \cdots = 3N^2, \quad (10.30)$$
$$m_1 + m_2 + m_3 + \cdots = 9 + (N^2 - 3)/3, \quad (10.31)$$

which have the solution $(m_1, m_3) = (9, (N^2 - 3)/3)$. That is, there are nine singlets and $(N^2 - 3)/3$ triplets.

10.2.2 Characters and Representations

There are nine singlet representations of the group $\Delta(3N^2)$ with $N/3$ integer. Their characters must satisfy $\chi_\alpha(b) = \omega^k$ ($k = 0, 1, 2$) similarly to the case $N/3 \neq$ integer. In addition, it is found that $\chi_\alpha(a) = \chi_\alpha(a') = \omega^\ell$ ($\ell = 0, 1, 2$). Thus, the nine singlets can be specified by combinations of $\chi_\alpha(b)$ and $\chi_\alpha(a)$, i.e., $\mathbf{1}_{k,\ell}$ ($k, \ell = 0, 1, 2$) with $\chi_\alpha(b) = \omega^k$ and $\chi_\alpha(a) = \chi_\alpha(a') = \omega^\ell$. These characters are shown in Table 10.2.

10.2 $\Delta(3N^2)$ with $N/3$ Integer

Table 10.2 Characters of $\Delta(3N^2)$ for $N/3$ integer

	h	$\chi_{1_{r,s}}$	$\chi_{3_{[n][m]}}$
C_1	1	1	3
$C_1^{(k)}$	3	1	$\rho^{nk+2mk} + \rho^{nk+mk} + \rho^{-2nk-mk}$
$C_3^{(k,\ell)}$	$\dfrac{N}{\gcd(N,k,\ell)}$	$\omega^{s(\ell+m)}$	$\rho^{mk-n\ell-m\ell} + \rho^{nk-m\ell} + \rho^{-nk-mk+n\ell}$
$C_{N^2/3}^{(1,p)}$	3	ω^{r+sp}	0
$C_{N^2/3}^{(2,p)}$	3	ω^{2r+sp}	0

The triplet representations are also given similarly to the case with $N/3 \neq$ integer. That is, the generators of $\Delta(3N^2)$ with $N/3$ integer are represented by

$$b = \begin{pmatrix} 0 & 1 & 0 \\ 0 & 0 & 1 \\ 1 & 0 & 0 \end{pmatrix}, \quad a = \begin{pmatrix} \rho^\ell & 0 & 0 \\ 0 & \rho^k & 0 \\ 0 & 0 & \rho^{-k-\ell} \end{pmatrix}, \quad a' = \begin{pmatrix} \rho^{-k-\ell} & 0 & 0 \\ 0 & \rho^\ell & 0 \\ 0 & 0 & \rho^k \end{pmatrix}, \tag{10.32}$$

on the triplet $3_{[k][\ell]}$. It should be noted that the matrices a and a' are trivial for the case $(k, \ell) = (0, 0), (N/3, N/3), (2N/3, 2N/3)$. Thus, we exclude such values of (k, ℓ). These characters are shown in Table 10.2.

10.2.3 Tensor Products

For $N/3$ integer, we present the tensor products of two triplets:

$$\begin{pmatrix} x_{\ell, -k-\ell} \\ x_{k, \ell} \\ x_{-k-\ell, k} \end{pmatrix}_{3_{[k][\ell]}} \quad \text{and} \quad \begin{pmatrix} y_{\ell', -k'-\ell'} \\ y_{k', \ell'} \\ y_{-k'-\ell', k'} \end{pmatrix}_{3_{[k'][\ell']}}. \tag{10.33}$$

Unless $(k', \ell') = [k + mN/3][\ell + mN/3]$ for $m = 0, 1, 2$, their tensor products are the same as (10.18).

For $(k', \ell') = [-k + mN/3][-\ell + mN/3]$ ($m = 0, 1$ or 2), the tensor products of the above triplets are:

$$\begin{pmatrix} x_{\ell, -k-\ell} \\ x_{k, \ell} \\ x_{-k-\ell, k} \end{pmatrix}_{3_{[k][\ell]}} \otimes \begin{pmatrix} y_{-\ell+mN/3, k+\ell-2mN/3} \\ y_{-k+mN/3, -\ell+mN/3} \\ y_{k+\ell-2mN/3, -k+mN/3} \end{pmatrix}_{3_{[-k+mN/3][-\ell+mN/3]}}$$
$$= (x_{\ell,-k-\ell}y_{-\ell+mN/3,k+\ell-2mN/3} + x_{k,\ell}y_{-k+mN/3,-\ell+mN/3}$$
$$+ x_{-k-\ell,k}y_{k+\ell-2mN/3,-k+mN/3})_{1_{0,m}}$$
$$\oplus \left(x_{\ell,-k-\ell}y_{-\ell+mN/3,k+\ell-2mN/3} + \omega^2 x_{k,\ell}y_{-k+mN/3,-\ell+mN/3} \right.$$
$$\left. + \omega x_{-k-\ell,k}y_{k+\ell-2mN/3,-k+mN/3} \right)_{1_{1,m}}$$
$$\oplus \left(x_{\ell,-k-\ell}y_{-\ell+mN/3,k+\ell-2mN/3} + \omega x_{k,\ell}y_{-k+mN/3,-\ell+mN/3} \right.$$

$$+ \omega^2 x_{-k-\ell,k} y_{k+\ell-2mN/3,-k+mN/3})_{1_{2,m}}$$

$$\oplus \begin{pmatrix} x_{k,\ell} y_{k+\ell-2mN/3,-k+mN/3} \\ x_{-k-\ell,k} y_{-\ell+mN/3,k+\ell-2mN/3} \\ x_{\ell,-k-\ell} y_{-k+mN/3,-\ell+mN/3} \end{pmatrix}_{3_{[-k-2\ell+mN/3][2k+\ell-2mN/3]}}$$

$$\oplus \begin{pmatrix} x_{-k-\ell,k} y_{-k+mN/3,-\ell+mN/3} \\ x_{\ell,-k-\ell} y_{k+\ell-2mN/3,-k+mN/3} \\ x_{k,\ell} y_{-\ell+mN/3,k+\ell-2mN/3} \end{pmatrix}_{3_{[k+2\ell-2mN/3][-2k-\ell+mN/3]}} . \quad (10.34)$$

The product of $3_{[k][\ell]}$ and $1_{r,s}$ is

$$\begin{pmatrix} x_{\ell,-k-\ell} \\ x_{k,\ell} \\ x_{-k-\ell,k} \end{pmatrix}_{3_{[k][\ell]}} \otimes (z)_{1_{r,s}} = \begin{pmatrix} x_{\ell,-k-\ell} z \\ \omega^r x_{k,\ell} z \\ \omega^{2r} x_{-k-\ell,k} z \end{pmatrix}_{3_{[k+sN/3][\ell+sN/3]}} . \quad (10.35)$$

The tensor products of the singlets $1_{k,\ell}$ and $1_{k',\ell'}$ are

$$1_{k,\ell} \otimes 1_{k',\ell'} = 1_{k+k' \bmod 3, \ell+\ell' \bmod 3}. \quad (10.36)$$

10.3 $\Delta(27)$

We discuss here a simple example of the group $\Delta(3N^2)$. The simplest such group, viz., $\Delta(3)$, is nothing but Z_3. The next, viz., $\Delta(12)$, is isomorphic to A_4. Thus, the simplest non-trivial example is $\Delta(27)$.

The conjugacy classes of $\Delta(27)$ are:

$$\begin{array}{lll}
C_1: & \{e\}, & h=1, \\
C_1^{(1)}: & \{aa'^2\}, & h=3, \\
C_1^{(2)}: & \{a^2 a'\}, & h=3, \\
C_3^{(0,1)}: & \{a', a, a^2 a'^2\}, & h=3, \\
C_3^{(0,2)}: & \{a'^2, a^2, aa'\}, & h=3, \\
C_3^{(1,p)}: & \{ba^p, ba^{p-1}a', ba^{p-2}a'^2\}, & h=3, \\
C_3^{(2,p)}: & \{b^2 a^p, b^2 a^{p-1} a', b^2 a^{p-2} a'^2\}, & h=3.
\end{array} \quad (10.37)$$

$\Delta(27)$ has nine singlets $1_{k,\ell}$ ($k,\ell = 0, 1, 2$) and two triplets $3_{[0][1]}$ and $3_{[0][2]}$. The characters are shown in Table 10.3.

The tensor products between triplets are:

$$\begin{pmatrix} x_{1,-1} \\ x_{0,1} \\ x_{-1,0} \end{pmatrix}_{3_{[0][1]}} \otimes \begin{pmatrix} y_{1,-1} \\ y_{0,1} \\ y_{-1,0} \end{pmatrix}_{3_{[0][1]}}$$

$$= \begin{pmatrix} x_{1,-1} y_{1,-1} \\ x_{0,1} y_{0,1} \\ x_{-1,0} y_{-1,0} \end{pmatrix}_{3_{[0][2]}} \oplus \begin{pmatrix} x_{0,1} y_{-1,0} \\ x_{-1,0} y_{1,-1} \\ x_{1,-1} y_{0,1} \end{pmatrix}_{3_{[0][2]}} \oplus \begin{pmatrix} x_{-1,0} y_{0,1} \\ x_{1,-1} y_{-1,0} \\ x_{0,1} y_{1,-1} \end{pmatrix}_{3_{[0][2]}},$$

$$(10.38)$$

Table 10.3 Characters of $\Delta(27)$

	h	$\chi_{1_{k,\ell}}$	$\chi_{3_{[0,1]}}$	$\chi_{3_{[0,2]}}$
C_1	1	1	3	3
$C_1^{(1)}$	1	1	$3\omega^2$	3ω
$C_1^{(2)}$	1	1	3ω	$3\omega^2$
$C_1^{(0,1)}$	3	ω^ℓ	0	0
$C_1^{(0,2)}$	3	$\omega^{2\ell}$	0	0
$C_3^{(1,p)}$	3	$\omega^{k+\ell p}$	0	0
$C_3^{(2,p)}$	3	$\omega^{2k+\ell p}$	0	0

$$\begin{pmatrix} x_{2,-2} \\ x_{0,2} \\ x_{-2,0} \end{pmatrix}_{3_{[0][2]}} \otimes \begin{pmatrix} y_{2,-2} \\ y_{0,2} \\ y_{-2,0} \end{pmatrix}_{3_{[0][2]}}$$

$$= \begin{pmatrix} x_{2,-2}y_{2,-2} \\ x_{0,2}y_{0,2} \\ x_{-2,0}y_{-2,0} \end{pmatrix}_{3_{[0][1]}} \oplus \begin{pmatrix} x_{0,2}y_{-2,0} \\ x_{-2,0}y_{2,-2} \\ x_{2,-2}y_{0,2} \end{pmatrix}_{3_{[0][1]}} \oplus \begin{pmatrix} x_{-2,0}y_{0,2} \\ x_{2,-2}y_{-2,0} \\ x_{0,2}y_{2,-2} \end{pmatrix}_{3_{[0][1]}},$$

$$(10.39)$$

$$\begin{pmatrix} x_{1,-1} \\ x_{0,1} \\ x_{-1,0} \end{pmatrix}_{3_{[0][1]}} \otimes \begin{pmatrix} y_{-1,1} \\ y_{0,-1} \\ y_{1,0} \end{pmatrix}_{3_{[0][2]}}$$

$$= \sum_r \left(x_{1,-1}y_{-1,1} + \omega^{2r} x_{0,1}y_{0,-1} + \omega^r x_{-1,0}y_{1,0} \right)_{1_{(r,0)}}$$

$$\oplus \sum_r \left(x_{1,-1}y_{0,-1} + \omega^{2r} x_{0,1}y_{1,0} + \omega^r x_{-1,0}y_{-1,1} \right)_{1_{(r,1)}}$$

$$\oplus \sum_r \left(x_{1,-1}y_{1,0} + \omega^{2r} x_{0,1}y_{-1,1} + \omega^r x_{-1,0}y_{0,-1} \right)_{1_{(r,2)}}. \quad (10.40)$$

The tensor products between singlets and triplets are:

$$\begin{pmatrix} x_{(1,-1)} \\ x_{(0,1)} \\ x_{(-1,0)} \end{pmatrix}_{3_{[0][1]}} \otimes (z)_{1_{k,\ell}} = \begin{pmatrix} x_{(1,-1)}z \\ \omega^r x_{(0,1)}z \\ \omega^{2r} x_{(-1,0)}z \end{pmatrix}_{3_{[\ell][1+\ell]}},$$

$$\begin{pmatrix} x_{(2,-2)} \\ x_{(0,2)} \\ x_{(-2,0)} \end{pmatrix}_{3_{[0][2]}} \otimes (z)_{1_{k,\ell}} = \begin{pmatrix} x_{(2,-2)}z \\ \omega^r x_{(0,2)}z \\ \omega^{2r} x_{(-2,0)}z \end{pmatrix}_{3_{[\ell][2+\ell]}}. \quad (10.41)$$

The tensor products of singlets are easily obtained from (10.36).

References

1. Luhn, C., Nasri, S., Ramond, P.: J. Math. Phys. **48**, 073501 (2007). arXiv:hep-th/0701188

Chapter 11
T_N

11.1 Generic Aspects

We now study the group T_N, which is isomorphic to $Z_N \rtimes Z_3$ (see, e.g., [1–3]). Here we focus on the case where N is any prime number except 3 or any power of such a prime number, i.e., $N = p^q$ with $p \neq 3$ a prime number and q a positive number. We denote the generators of Z_N and Z_3 by a and b, respectively. These satisfy

$$a^N = e, \qquad b^3 = e. \tag{11.1}$$

Because of the semi-direct product structure, we impose

$$ba = a^m b, \tag{11.2}$$

with $m \neq 0$. When $m = 1 \mod N$, a and b commute and the group is just the direct product $Z_N \times Z_3$. Thus, we impose $m \neq 1 \mod N$. For example, the case with $N = 2$ is excluded, because we have only $m = 1 \mod N$ for $N = 2$, except if $m = 0$.

We find

$$a = b^2 a^m b, \quad a^m = (b^2 a^m b) \cdots (b^2 a^m b) = b^2 a^{m^2} b = bab^2, \tag{11.3}$$

$$a = ba^{m^2} b^2, \quad a^m = (ba^{m^2} b^2) \cdots (ba^{m^2} b^2) = ba^{m^3} b^2 = bab^2. \tag{11.4}$$

The consistency of these equations implies that

$$a^{m^3} = a, \tag{11.5}$$

and hence,

$$m^3 - 1 = (m-1)(m^2 + m + 1) = 0 \mod N. \tag{11.6}$$

Here, we focus on the case where the condition

$$m^2 + m + 1 = 0 \mod N \tag{11.7}$$

is satisfied with $1 < m < N$.

Suppose now that $m = 3\ell$ with integer ℓ. Thus, it follows that

$$m^2 + m + 1 = 3\ell(3\ell + 1) + 1, \tag{11.8}$$

Table 11.1 m for $N \leq 50$

N	7	13	19	31	43	49 $(= 7 \times 7)$
m	2	3	7	5	6	18

and $3\ell(3\ell + 1)$ is always a multiple of 6, i.e., $3\ell(3\ell + 1) = 6k$ and

$$m^2 + m + 1 = 6k + 1.$$

Similarly, when $m = 3\ell + 2$, we obtain

$$m^2 + m + 1 = 3\ell(3\ell + 5) + 7. \tag{11.9}$$

Then, it is found that $m^2 + m + 1 = 6k + 1$. These results show that the possible values of N have the form $N = 6k + 1$.

On the other hand, when $m = 3\ell + 1$, we find

$$m^2 + m + 1 = 3[3\ell(\ell + 1) + 1]. \tag{11.10}$$

Here, $\ell(\ell + 1)$ is always even, i.e., $\ell(\ell + 1) = 2\ell'$, and we can write

$$m^2 + m + 1 = 3(6\ell' + 1). \tag{11.11}$$

This implies that, for $N = p^q$ ($p \neq 3$), the possible values of N would be $N = 6\ell' + 1$. If we took $N = 3$, we could not find non-trivial m, because $m^3 - 1 = 7$ for $m = 2$. We thus find that the possible values of N are also $N = 6k + 1$ for $m = 3\ell + 1$.

Indeed, explicit computation with (11.7) leads to the following possible values of N:

$$N = (7, 13, 19, 31, 43, 49(= 7 \times 7)) \quad \text{for } N \leq 50. \tag{11.12}$$

These values and the corresponding values of m are shown in Table 11.1.

All elements of T_N can be expressed in terms of the two generators a and b as follows:

$$g = b^k a^\ell, \tag{11.13}$$

for $k = 0, 1, 2$, and $\ell = 0, \ldots, N - 1$. The generators b and a are represented, e.g., by

$$b = \begin{pmatrix} 0 & 1 & 0 \\ 0 & 0 & 1 \\ 1 & 0 & 0 \end{pmatrix}, \quad a = \begin{pmatrix} \rho & 0 & 0 \\ 0 & \rho^m & 0 \\ 0 & 0 & \rho^{m^2} \end{pmatrix}, \tag{11.14}$$

where we define $\rho = e^{2\pi i/N}$.

11.1.1 Conjugacy Classes

All elements $b^k a^\ell$ of T_N are classified into $3 + (N - 1)/3$ conjugacy classes:

11.1 Generic Aspects

$$
\begin{array}{lll}
C_1: & \{e\}, & h = 1, \\
C_N^{(1)}: & \{b, ba, \ldots, ba^{N-2}, ba^{N-1}\}, & h = 3, \\
C_N^{(2)}: & \{b^2, b^2 a, \ldots, b^2 a^{N-2}, b^2 a^{N-1}\}, & h = 3, \\
C_{3[k]}: & \{a^k, a^{km}, a^{km^2}\}, & h = N/\gcd(N, k),
\end{array}
\qquad (11.15)
$$

where $\gcd(N, k) = N$ when N is a prime number.
The relations (2.18) and (2.19) lead to

$$m_1 + m_2 + m_3 = 3 + (N - 1)/3, \qquad (11.16)$$

$$m_1 + 4m_2 + 9m_3 = 3N. \qquad (11.17)$$

For specific values of N in (11.12), their solutions are found to be $m_1 = 3$, $m_2 = 0$, and $m_3 = (N - 1)/3$.

11.1.2 Characters and Representations

The group T_N has 3 singlets. Since $b^3 = e$, the characters of the three singlets have three possible values $\chi_{1k}(b) = \omega^k$ with $k = 0, 1, 2$, and they correspond to three singlets $\mathbf{1}_k$. Note that $\chi_{1k}(a) = 1$, because $\chi_{1k}(b) = \chi_{1k}(ba)$. These characters are shown in Table 11.2.

Now consider the triplets. We use the notation $[k]$ defined by

$$[k] = k, \; km, \; \text{or} \; km^2 \; (\text{mod } N). \qquad (11.18)$$

We also define $\xi_{[k]} = \rho^k + \rho^{km} + \rho^{km^2}$. The notation $\bar{\xi}_{[k]}$ is defined as the complex conjugate of $\xi_{[k]}$, which is $\xi_{[N-k]}$. The two generators b and a are represented, e.g., by

$$
b = \begin{pmatrix} 0 & 1 & 0 \\ 0 & 0 & 1 \\ 1 & 0 & 0 \end{pmatrix}, \qquad
a = \begin{pmatrix} \rho^k & 0 & 0 \\ 0 & \rho^{km} & 0 \\ 0 & 0 & \rho^{km^2} \end{pmatrix}, \qquad (11.19)
$$

on the triplets $\mathbf{3}_{[k]}$. We also define the vector $\mathbf{3}_{[k]}$ by

$$
\mathbf{3}_{[k]} \equiv \begin{pmatrix} x_k \\ x_{km} \\ x_{km^2} \end{pmatrix}, \qquad (11.20)
$$

for $k \in N - 1$, where k corresponds to the Z_N charge. Thus we have $(N - 1)/3$ different triplets. The characters are shown in Table 11.2. Note also that we use the notation $\bar{\mathbf{3}}_{[k]} = \mathbf{3}_{[N-k]}$.

11.1.3 Tensor Products

The tensor products between triplets are obtained in general as follows:

Table 11.2 Characters of T_N

	h	χ_{1_0}	χ_{1_1}	χ_{1_2}	$\chi_{3_{[\ell]}}$
C_1	1	1	1	1	3
$C_N^{(1)}$	3	1	ω	ω^2	0
$C_N^{(2)}$	3	1	ω^2	ω	0
$C_{3_{[k]}}$	$N/\gcd(N,k)$	1	1	1	$\xi_{[k\ell]}$

$$\begin{pmatrix} x_k \\ x_{km} \\ x_{km^2} \end{pmatrix}_{3_{[k]}} \otimes \begin{pmatrix} y_\ell \\ y_{\ell m} \\ y_{\ell m^2} \end{pmatrix}_{3_{[\ell]}} = \begin{pmatrix} x_k y_\ell \\ x_{km} y_{\ell m} \\ x_{km^2} y_{\ell m^2} \end{pmatrix}_{3_{[k+\ell]}} \oplus \begin{pmatrix} x_k y_{\ell m} \\ x_{km} y_{\ell m^2} \\ x_{km^2} y_\ell \end{pmatrix}_{3_{[k+\ell m]}}$$

$$\oplus \begin{pmatrix} x_k y_{\ell m^2} \\ x_{km} y_\ell \\ x_{km^2} y_{\ell m} \end{pmatrix}_{3_{[k+\ell m^2]}}, \quad (11.21)$$

for $[k] \neq [N-\ell]$. When $[k] = [N-\ell]$, one of $3_{[k+\ell]}$, $3_{[k+\ell m]}$, or $3_{[k+\ell m^2]}$ can be reduced to three singlets. For example, if $k+\ell = 0 \bmod N$, the triplet $3_{[k+\ell]}$ can be reduced as follows:

$$\begin{pmatrix} x_k y_\ell \\ x_{km} y_{\ell m} \\ x_{km^2} y_{\ell m^2} \end{pmatrix}_{3_{[k+\ell]}} = \oplus \sum_{k'=0,1,2} \left(x_k y_\ell + \omega^{2k'} x_{km} y_{\ell m} + \omega^{k'} x_{km^2} y_{\ell m^2} \right)_{1_{k'}}. \quad (11.22)$$

Similarly, when $k + \ell m^n = 0 \bmod N$ for $n = 1, 2$, the triplet $3_{[k+\ell m^n]}$ can be reduced to three singlets.

In addition, the tensor products including singlets are simply

$$1_k \otimes 3_{[\ell]} = 3_{[\ell]}, \qquad 1_k \otimes 1_{k'} = 1_{k+k'}. \quad (11.23)$$

11.2 T_7

Here we study the smallest group T_N, that is, T_7. We denote the generator of Z_7 by a and that of Z_3 by b. They satisfy

$$a^7 = 1, \qquad ba = a^2 b. \quad (11.24)$$

Using these, all elements of T_7 can be written in the form

$$g = b^k a^\ell, \quad (11.25)$$

with $k = 0, 1, 2$, and $\ell = 0, \ldots, 6$.

The generators a and b are represented, e.g., by

$$b = \begin{pmatrix} 0 & 1 & 0 \\ 0 & 0 & 1 \\ 1 & 0 & 0 \end{pmatrix}, \qquad a = \begin{pmatrix} \rho & 0 & 0 \\ 0 & \rho^2 & 0 \\ 0 & 0 & \rho^4 \end{pmatrix}, \quad (11.26)$$

where $\rho = e^{2i\pi/7}$. These elements are classified into five conjugacy classes:

11.2 T_7

Table 11.3 Characters of T_7

	h	χ_{1_0}	χ_{1_1}	χ_{1_2}	χ_3	$\chi_{\bar{3}}$
C_1	1	1	1	1	3	3
$C_7^{(1)}$	3	1	ω	ω^2	0	0
$C_7^{(2)}$	3	1	ω^2	ω	0	0
C_3	7	1	1	1	ξ	$\bar{\xi}$
$C_{\bar{3}}$	7	1	1	1	$\bar{\xi}$	ξ

$$
\begin{aligned}
C_1 &: \{e\}, & h &= 1, \\
C_7^{(1)} &: \{b, ba, ba^2, ba^3, ba^4, ba^5, ba^6\}, & h &= 3, \\
C_7^{(2)} &: \{b^2, b^2a, b^2a^2, b^2a^3, b^2a^4, b^2a^5, b^2a^6\}, & h &= 3, \\
C_3 &: \{a, a^2, a^4\}, & h &= 7, \\
C_{\bar{3}} &: \{a^3, a^5, a^6\}, & h &= 7,
\end{aligned}
\quad (11.27)
$$

T_7 has three singlet representations $\mathbf{1}_k$ with $k = 0, 1, 2$, and two triplets $\mathbf{3}$ and $\bar{\mathbf{3}}$. The characters are shown in Table 11.3, where $\xi = (-1 + i\sqrt{7})/2$.

Using the order of ρ in a, we define the triplets $\mathbf{3}$ and $\bar{\mathbf{3}}$ by

$$
\mathbf{3} \equiv \begin{pmatrix} x_1 \\ x_2 \\ x_4 \end{pmatrix}, \quad \bar{\mathbf{3}} \equiv \begin{pmatrix} x_{-1} \\ x_{-2} \\ x_{-4} \end{pmatrix} = \begin{pmatrix} x_6 \\ x_5 \\ x_3 \end{pmatrix}. \quad (11.28)
$$

The tensor products between triplets are

$$
\begin{pmatrix} x_1 \\ x_2 \\ x_4 \end{pmatrix}_3 \otimes \begin{pmatrix} y_1 \\ y_2 \\ y_4 \end{pmatrix}_3 = \begin{pmatrix} x_2 y_4 \\ x_4 y_1 \\ x_1 y_2 \end{pmatrix}_{\bar{3}} \oplus \begin{pmatrix} x_4 y_2 \\ x_1 y_4 \\ x_2 y_1 \end{pmatrix}_{\bar{3}} \oplus \begin{pmatrix} x_4 y_4 \\ x_1 y_1 \\ x_2 y_2 \end{pmatrix}_3, \quad (11.29)
$$

$$
\begin{pmatrix} x_6 \\ x_5 \\ x_3 \end{pmatrix}_{\bar{3}} \otimes \begin{pmatrix} y_6 \\ y_5 \\ y_3 \end{pmatrix}_{\bar{3}} = \begin{pmatrix} x_5 y_3 \\ x_3 y_6 \\ x_6 y_5 \end{pmatrix}_3 \oplus \begin{pmatrix} x_3 y_5 \\ x_6 y_3 \\ x_5 y_6 \end{pmatrix}_3 \oplus \begin{pmatrix} x_3 y_3 \\ x_6 y_6 \\ x_5 y_5 \end{pmatrix}_{\bar{3}}, \quad (11.30)
$$

$$
\begin{pmatrix} x_1 \\ x_2 \\ x_4 \end{pmatrix}_3 \otimes \begin{pmatrix} y_6 \\ y_5 \\ y_3 \end{pmatrix}_{\bar{3}} = \begin{pmatrix} x_2 y_6 \\ x_4 y_5 \\ x_1 y_3 \end{pmatrix}_3 \oplus \begin{pmatrix} x_1 y_5 \\ x_2 y_3 \\ x_4 y_6 \end{pmatrix}_{\bar{3}}
$$

$$
\oplus \sum_{k=0,1,2} \left(x_1 y_6 + \omega^{2k} x_2 y_5 + \omega^k x_4 y_3 \right)_{\mathbf{1}_k}. \quad (11.31)
$$

The tensor products between singlets are

$$
\begin{aligned}
(x)_{1_0}(y)_{1_0} &= (x)_{1_1}(y)_{1_2} = (x)_{1_2}(y)_{1_1} = (xy)_{1_0}, \\
(x)_{1_1}(y)_{1_1} &= (xy)_{1_2}, \quad (x)_{1_2}(y)_{1_2} = (xy)_{1_1}.
\end{aligned}
\quad (11.32)
$$

The tensor products between triplets and singlets are

$$
\mathbf{1}_k \otimes \mathbf{3} = \mathbf{3}, \quad \mathbf{1}_k \otimes \bar{\mathbf{3}} = \bar{\mathbf{3}}, \quad (11.33)
$$

where $k = 0, 1, 2$. Notice that singlets have no affect on the form of triplets.

Table 11.4 Characters of T_{13}. $\bar{\xi}_i$ is defined as the complex conjugate of ξ_i

	h	χ_{1_0}	χ_{1_1}	χ_{1_2}	χ_{3_1}	$\chi_{\bar{3}_1}$	χ_{3_2}	$\chi_{\bar{3}_2}$
C_1	1	1	1	1	3	3	3	3
$C_{13}^{(1)}$	3	1	ω	ω^2	0	0	0	0
$C_{13}^{(2)}$	3	1	ω^2	ω	0	0	0	0
C_{3_1}	13	1	1	1	ξ_1	$\bar{\xi}_1$	ξ_2	$\bar{\xi}_2$
$C_{\bar{3}_1}$	13	1	1	1	$\bar{\xi}_1$	ξ_1	$\bar{\xi}_2$	ξ_2
C_{3_2}	13	1	1	1	ξ_2	$\bar{\xi}_2$	ξ_1	$\bar{\xi}_1$
$C_{\bar{3}_2}$	13	1	1	1	$\bar{\xi}_2$	ξ_2	$\bar{\xi}_1$	ξ_1

11.3 T_{13}

The non-Abelian discrete group T_{13} is isomorphic to $Z_{13} \rtimes Z_3$ [3, 4]. This group is a subgroup of $SU(3)$, and known to be the minimal non-Abelian discrete group with two complex triplets as irreducible representations. We denote the generators of Z_{13} and Z_3 by a and b, respectively. They satisfy

$$a^{13} = 1, \qquad ba = a^3 b. \tag{11.34}$$

Using these, all elements of T_{13} can be expressed in the form

$$g = b^k a^\ell, \tag{11.35}$$

with $k = 0, 1, 2$, and $\ell = 0, \ldots, 12$.

The generators a and b are represented, e.g., by

$$b = \begin{pmatrix} 0 & 1 & 0 \\ 0 & 0 & 1 \\ 1 & 0 & 0 \end{pmatrix}, \qquad a = \begin{pmatrix} \rho & 0 & 0 \\ 0 & \rho^3 & 0 \\ 0 & 0 & \rho^9 \end{pmatrix}, \tag{11.36}$$

where $\rho = e^{2i\pi/13}$. These elements are classified into seven conjugacy classes:

$$\begin{aligned}
C_1 &: \{e\}, & h &= 1, \\
C_{13}^{(1)} &: \{b, ba, ba^2, \ldots, ba^{10}, ba^{11}, ba^{12}\}, & h &= 3, \\
C_{13}^{(2)} &: \{b^2, b^2 a, b^2 a^2, \ldots, b^2 a^{10}, b^2 a^{11}, b^2 a^{12}\}, & h &= 3, \\
C_{3_1} &: \{a, a^3, a^9\}, & h &= 13, \\
C_{\bar{3}_1} &: \{a^4, a^{10}, a^{12}\}, & h &= 13, \\
C_{3_2} &: \{a^2, a^5, a^6\}, & h &= 13, \\
C_{\bar{3}_2} &: \{a^7, a^8, a^{11}\}, & h &= 13.
\end{aligned} \tag{11.37}$$

T_{13} has three singlet representations $\mathbf{1}_k$ with $k = 0, 1, 2$, together with two complex triplets $\mathbf{3}_1$ and $\mathbf{3}_2$ and their conjugates as irreducible representations. The characters are shown in Table 11.4, where $\xi_1 \equiv \rho + \rho^3 + \rho^9$, $\xi_2 \equiv \rho^2 + \rho^5 + \rho^6$, and $\omega \equiv e^{2i\pi/3}$.

11.3 T_{13}

Next we consider the multiplication rules of the group T_{13}. We define the triplets by

$$\mathbf{3}_1 \equiv \begin{pmatrix} x_1 \\ x_3 \\ x_9 \end{pmatrix}, \quad \bar{\mathbf{3}}_1 \equiv \begin{pmatrix} \bar{x}_{12} \\ \bar{x}_{10} \\ \bar{x}_4 \end{pmatrix}, \quad \mathbf{3}_2 = \begin{pmatrix} y_2 \\ y_6 \\ y_5 \end{pmatrix}, \quad \bar{\mathbf{3}}_2 \equiv \begin{pmatrix} \bar{y}_{11} \\ \bar{y}_7 \\ \bar{y}_8 \end{pmatrix}, \quad (11.38)$$

where the subscripts denote the Z_{13} charge of each element.

The tensor products between triplets are:

$$\begin{pmatrix} x_1 \\ x_3 \\ x_9 \end{pmatrix}_{\mathbf{3}_1} \otimes \begin{pmatrix} y_1 \\ y_3 \\ y_9 \end{pmatrix}_{\mathbf{3}_1} = \begin{pmatrix} x_3 y_9 \\ x_9 y_1 \\ x_1 y_3 \end{pmatrix}_{\bar{\mathbf{3}}_1} \oplus \begin{pmatrix} x_9 y_3 \\ x_1 y_9 \\ x_3 y_1 \end{pmatrix}_{\bar{\mathbf{3}}_1} \oplus \begin{pmatrix} x_1 y_1 \\ x_3 y_3 \\ x_9 y_9 \end{pmatrix}_{\mathbf{3}_2}, \quad (11.39)$$

$$\begin{pmatrix} \bar{x}_{12} \\ \bar{x}_{10} \\ \bar{x}_4 \end{pmatrix}_{\bar{\mathbf{3}}_1} \otimes \begin{pmatrix} \bar{y}_{12} \\ \bar{y}_{10} \\ \bar{y}_4 \end{pmatrix}_{\bar{\mathbf{3}}_1} = \begin{pmatrix} \bar{x}_{10}\bar{y}_4 \\ \bar{x}_4\bar{y}_{12} \\ \bar{x}_{12}\bar{y}_{10} \end{pmatrix}_{\mathbf{3}_1} \oplus \begin{pmatrix} \bar{x}_4\bar{y}_{10} \\ \bar{x}_{12}\bar{y}_4 \\ \bar{x}_{10}\bar{y}_{12} \end{pmatrix}_{\mathbf{3}_1} \oplus \begin{pmatrix} \bar{x}_{12}\bar{y}_{12} \\ \bar{x}_{10}\bar{y}_{10} \\ \bar{x}_4\bar{y}_4 \end{pmatrix}_{\bar{\mathbf{3}}_2},$$
$$(11.40)$$

$$\begin{pmatrix} x_1 \\ x_3 \\ x_9 \end{pmatrix}_{\mathbf{3}_1} \otimes \begin{pmatrix} \bar{y}_{12} \\ \bar{y}_{10} \\ \bar{y}_4 \end{pmatrix}_{\bar{\mathbf{3}}_1} = \sum_{k=0,1,2} \left(x_1\bar{y}_{12} + \omega^{2k} x_3\bar{y}_{10} + \omega^k x_9\bar{y}_4 \right)_{\mathbf{1}_k}$$
$$\oplus \begin{pmatrix} x_3\bar{y}_{12} \\ x_9\bar{y}_{10} \\ x_1\bar{y}_4 \end{pmatrix}_{\mathbf{3}_2} \oplus \begin{pmatrix} x_1\bar{y}_{10} \\ x_3\bar{y}_4 \\ x_9\bar{y}_{12} \end{pmatrix}_{\bar{\mathbf{3}}_2}, \quad (11.41)$$

$$\begin{pmatrix} x_2 \\ x_6 \\ x_5 \end{pmatrix}_{\mathbf{3}_2} \otimes \begin{pmatrix} y_2 \\ y_6 \\ y_5 \end{pmatrix}_{\mathbf{3}_2} = \begin{pmatrix} x_5 y_6 \\ x_2 y_5 \\ x_6 y_2 \end{pmatrix}_{\bar{\mathbf{3}}_2} \oplus \begin{pmatrix} x_6 y_5 \\ x_5 y_2 \\ x_2 y_6 \end{pmatrix}_{\bar{\mathbf{3}}_2} \oplus \begin{pmatrix} x_6 y_6 \\ x_5 y_5 \\ x_2 y_2 \end{pmatrix}_{\bar{\mathbf{3}}_1}, \quad (11.42)$$

$$\begin{pmatrix} \bar{x}_{11} \\ \bar{x}_7 \\ \bar{x}_8 \end{pmatrix}_{\bar{\mathbf{3}}_2} \otimes \begin{pmatrix} \bar{y}_{11} \\ \bar{y}_7 \\ \bar{y}_8 \end{pmatrix}_{\bar{\mathbf{3}}_2} = \begin{pmatrix} \bar{x}_8\bar{y}_7 \\ \bar{x}_{11}\bar{y}_8 \\ \bar{x}_7\bar{y}_{11} \end{pmatrix}_{\mathbf{3}_2} \oplus \begin{pmatrix} \bar{x}_7\bar{y}_8 \\ \bar{x}_8\bar{y}_{11} \\ \bar{x}_{11}\bar{y}_7 \end{pmatrix}_{\mathbf{3}_2} \oplus \begin{pmatrix} \bar{x}_7\bar{y}_7 \\ \bar{x}_8\bar{y}_8 \\ \bar{x}_{11}\bar{y}_{11} \end{pmatrix}_{\mathbf{3}_1}, \quad (11.43)$$

$$\begin{pmatrix} x_2 \\ x_6 \\ x_5 \end{pmatrix}_{\mathbf{3}_2} \otimes \begin{pmatrix} \bar{y}_{11} \\ \bar{y}_7 \\ \bar{y}_8 \end{pmatrix}_{\bar{\mathbf{3}}_2} = \sum_{k=0,1,2} \left(x_2\bar{y}_{11} + \omega^{2k} x_6\bar{y}_7 + \omega^k x_5\bar{y}_8 \right)_{\mathbf{1}_k}$$
$$\oplus \begin{pmatrix} x_6\bar{y}_8 \\ x_5\bar{y}_{11} \\ x_2\bar{y}_7 \end{pmatrix}_{\bar{\mathbf{3}}_1} \oplus \begin{pmatrix} x_5\bar{y}_7 \\ x_2\bar{y}_8 \\ x_6\bar{y}_{11} \end{pmatrix}_{\bar{\mathbf{3}}_1}, \quad (11.44)$$

$$\begin{pmatrix} x_1 \\ x_3 \\ x_9 \end{pmatrix}_{\mathbf{3}_1} \otimes \begin{pmatrix} y_2 \\ y_6 \\ y_5 \end{pmatrix}_{\mathbf{3}_2} = \begin{pmatrix} x_9 y_6 \\ x_1 y_5 \\ x_3 y_2 \end{pmatrix}_{\mathbf{3}_2} \oplus \begin{pmatrix} x_9 y_2 \\ x_1 y_6 \\ x_3 y_5 \end{pmatrix}_{\bar{\mathbf{3}}_2} \oplus \begin{pmatrix} x_9 y_5 \\ x_1 y_2 \\ x_3 y_6 \end{pmatrix}_{\bar{\mathbf{3}}_1}, \quad (11.45)$$

$$\begin{pmatrix} x_1 \\ x_3 \\ x_9 \end{pmatrix}_{\mathbf{3}_1} \otimes \begin{pmatrix} \bar{y}_{11} \\ \bar{y}_7 \\ \bar{y}_8 \end{pmatrix}_{\bar{\mathbf{3}}_2} = \begin{pmatrix} x_1\bar{y}_{11} \\ x_3\bar{y}_7 \\ x_9\bar{y}_8 \end{pmatrix}_{\bar{\mathbf{3}}_1} \oplus \begin{pmatrix} x_3\bar{y}_8 \\ x_9\bar{y}_{11} \\ x_1\bar{y}_7 \end{pmatrix}_{\bar{\mathbf{3}}_2} \oplus \begin{pmatrix} x_3\bar{y}_{11} \\ x_9\bar{y}_7 \\ x_1\bar{y}_8 \end{pmatrix}_{\bar{\mathbf{3}}_1}, \quad (11.46)$$

$$\begin{pmatrix} x_2 \\ x_6 \\ x_5 \end{pmatrix}_{3_2} \otimes \begin{pmatrix} \bar{y}_{12} \\ \bar{y}_{10} \\ \bar{y}_4 \end{pmatrix}_{\bar{3}_1} = \begin{pmatrix} x_2 \bar{y}_{12} \\ x_6 \bar{y}_{10} \\ x_5 \bar{y}_4 \end{pmatrix}_{\bar{3}_1} \oplus \begin{pmatrix} x_2 \bar{y}_{10} \\ x_6 \bar{y}_4 \\ x_5 \bar{y}_{12} \end{pmatrix}_{\bar{3}_1} \oplus \begin{pmatrix} x_5 \bar{y}_{10} \\ x_2 \bar{y}_4 \\ x_6 \bar{y}_{12} \end{pmatrix}_{3_2}, \quad (11.47)$$

$$\begin{pmatrix} \bar{x}_{12} \\ \bar{x}_{10} \\ \bar{x}_4 \end{pmatrix}_{\bar{3}_1} \otimes \begin{pmatrix} \bar{y}_{11} \\ \bar{y}_7 \\ \bar{y}_8 \end{pmatrix}_{\bar{3}_2} = \begin{pmatrix} \bar{x}_4 \bar{y}_8 \\ \bar{x}_{12} \bar{y}_{11} \\ \bar{x}_{10} \bar{y}_7 \end{pmatrix}_{\bar{3}_1} \oplus \begin{pmatrix} \bar{x}_4 \bar{y}_7 \\ \bar{x}_{12} \bar{y}_8 \\ \bar{x}_{10} \bar{y}_{11} \end{pmatrix}_{\bar{3}_2} \oplus \begin{pmatrix} \bar{x}_4 \bar{y}_{11} \\ \bar{x}_{12} \bar{y}_7 \\ \bar{x}_{10} \bar{y}_8 \end{pmatrix}_{\bar{3}_2}.$$

$$(11.48)$$

The tensor products between singlets are

$$\begin{aligned} (x)_{1_0}(y)_{1_0} &= (x)_{1_1}(y)_{1_2} = (x)_{1_2}(y)_{1_1} = (xy)_{1_0}, \\ (x)_{1_1}(y)_{1_1} &= (xy)_{1_2}, \quad (x)_{1_2}(y)_{1_2} = (xy)_{1_1}. \end{aligned} \quad (11.49)$$

The tensor products between triplets and singlets are

$$\mathbf{1}_j \otimes \mathbf{3}_k = \mathbf{3}_k, \quad (11.50)$$

where j and k run over all singlets and triplets, respectively. Note that singlets do not affect the form of triplets.

11.4 T_{19}

The non-Abelian discrete group T_{19} is isomorphic to $Z_{19} \rtimes Z_3$. This group is a subgroup of $SU(3)$, and known to be the minimal non-Abelian discrete group with three complex triplets as irreducible representations. We denote the generators of Z_{19} and Z_3 by a and b, respectively. They satisfy

$$a^{19} = 1, \quad ba = a^7 b. \quad (11.51)$$

Using these, all elements of T_{19} can be expressed in the form

$$g = b^k a^\ell, \quad (11.52)$$

with $k = 0, 1, 2$, and $\ell = 0, \ldots, 18$.

The generators a and b are represented, e.g., by

$$b = \begin{pmatrix} 0 & 1 & 0 \\ 0 & 0 & 1 \\ 1 & 0 & 0 \end{pmatrix}, \quad a = \begin{pmatrix} \rho & 0 & 0 \\ 0 & \rho^7 & 0 \\ 0 & 0 & \rho^{11} \end{pmatrix}, \quad (11.53)$$

11.4 T_{19}

Table 11.5 Characters of T_{19}. $\bar{\xi}_i$ is defined as the complex conjugate of ξ_i

	h	χ_{1_0}	χ_{1_1}	χ_{1_2}	χ_{3_1}	$\chi_{\bar{3}_1}$	χ_{3_2}	$\chi_{\bar{3}_2}$	χ_{3_3}	$\chi_{\bar{3}_3}$
C_1	1	1	1	1	3	3	3	3	3	3
$C_{19}^{(1)}$	3	1	ω	ω^2	0	0	0	0	0	0
$C_{19}^{(2)}$	3	1	ω^2	ω	0	0	0	0	0	0
C_{3_1}	19	1	1	1	ξ_1	$\bar{\xi}_1$	ξ_2	$\bar{\xi}_2$	ξ_3	$\bar{\xi}_3$
$C_{\bar{3}_1}$	19	1	1	1	$\bar{\xi}_1$	ξ_1	$\bar{\xi}_2$	ξ_2	$\bar{\xi}_3$	ξ_3
C_{3_2}	19	1	1	1	ξ_2	$\bar{\xi}_2$	ξ_3	$\bar{\xi}_3$	ξ_1	$\bar{\xi}_1$
$C_{\bar{3}_2}$	19	1	1	1	$\bar{\xi}_2$	ξ_2	$\bar{\xi}_3$	ξ_3	$\bar{\xi}_1$	ξ_1
C_{3_3}	19	1	1	1	ξ_3	$\bar{\xi}_3$	ξ_1	$\bar{\xi}_1$	ξ_2	$\bar{\xi}_2$
$C_{\bar{3}_3}$	19	1	1	1	$\bar{\xi}_3$	ξ_3	$\bar{\xi}_1$	ξ_1	$\bar{\xi}_2$	ξ_2

where $\rho = e^{2i\pi/19}$. These elements are classified into nine conjugacy classes:

$$\begin{aligned}
C_1 &: \{e\}, & h &= 1, \\
C_{19}^{(1)} &: \{b, ba, ba^2, \ldots, ba^{16}, ba^{17}, ba^{18}\}, & h &= 3, \\
C_{19}^{(2)} &: \{b^2, b^2 a, b^2 a^2, \ldots, b^2 a^{16}, b^2 a^{17}, b^2 a^{18}\}, & h &= 3, \\
C_{3_1} &: \{a, a^7, a^{11}\}, & h &= 19, \\
C_{\bar{3}_1} &: \{a^8, a^{12}, a^{18}\}, & h &= 19, \quad (11.54) \\
C_{3_2} &: \{a^2, a^3, a^{14}\}, & h &= 19, \\
C_{\bar{3}_2} &: \{a^5, a^{16}, a^{17}\}, & h &= 19, \\
C_{3_3} &: \{a^4, a^6, a^9\}, & h &= 19, \\
C_{\bar{3}_3} &: \{a^{10}, a^{13}, a^{15}\}, & h &= 19.
\end{aligned}$$

T_{19} has three singlet representations $\mathbf{1}_k$ with $k = 0, 1, 2$, and three complex triplets $\mathbf{3}_1$, $\mathbf{3}_2$, and $\mathbf{3}_3$, together with their conjugates, as irreducible representations. The characters are shown in Table 11.5, where $\xi_1 \equiv \rho + \rho^7 + \rho^{11}$, $\xi_2 \equiv \rho^2 + \rho^3 + \rho^{14}$, $\xi_3 \equiv \rho^4 + \rho^6 + \rho^9$, and $\omega \equiv e^{2i\pi/3}$.

Next we consider the multiplication rules of the T_{19} group. We define the triplets by

$$\mathbf{3}_1 \equiv \begin{pmatrix} x_1 \\ x_7 \\ x_{11} \end{pmatrix}, \quad \bar{\mathbf{3}}_1 \equiv \begin{pmatrix} \bar{x}_{18} \\ \bar{x}_{12} \\ \bar{x}_8 \end{pmatrix}, \quad \mathbf{3}_2 = \begin{pmatrix} y_2 \\ y_{14} \\ y_3 \end{pmatrix},$$

$$\bar{\mathbf{3}}_2 \equiv \begin{pmatrix} \bar{y}_{17} \\ \bar{y}_5 \\ \bar{y}_{16} \end{pmatrix}, \quad \mathbf{3}_3 = \begin{pmatrix} z_4 \\ z_9 \\ z_6 \end{pmatrix}, \quad \bar{\mathbf{3}}_3 \equiv \begin{pmatrix} \bar{z}_{15} \\ \bar{z}_{10} \\ \bar{z}_{13} \end{pmatrix}, \qquad (11.55)$$

where the subscripts denote the Z_{19} charge of each element.

The tensor products between triplets are:

$$\begin{pmatrix} x_1 \\ x_7 \\ x_{11} \end{pmatrix}_{3_1} \otimes \begin{pmatrix} y_1 \\ y_7 \\ y_{11} \end{pmatrix}_{3_1} = \begin{pmatrix} x_7 y_{11} \\ x_{11} y_1 \\ x_1 y_7 \end{pmatrix}_{\bar{3}_1} \oplus \begin{pmatrix} x_{11} y_7 \\ x_1 y_{11} \\ x_7 y_1 \end{pmatrix}_{\bar{3}_1} \oplus \begin{pmatrix} x_1 y_1 \\ x_7 y_7 \\ x_{11} y_{11} \end{pmatrix}_{\bar{3}_2}, \quad (11.56)$$

$$\begin{pmatrix} \bar{x}_{18} \\ \bar{x}_{12} \\ \bar{x}_8 \end{pmatrix}_{\bar{3}_1} \otimes \begin{pmatrix} \bar{y}_{18} \\ \bar{y}_{12} \\ \bar{y}_8 \end{pmatrix}_{\bar{3}_1} = \begin{pmatrix} \bar{x}_8 \bar{y}_{12} \\ \bar{x}_{18} \bar{y}_8 \\ \bar{x}_{12} \bar{y}_{18} \end{pmatrix}_{3_1} \oplus \begin{pmatrix} \bar{x}_{12} \bar{y}_8 \\ \bar{x}_8 \bar{y}_{18} \\ \bar{x}_{18} \bar{y}_{12} \end{pmatrix}_{3_1} \oplus \begin{pmatrix} \bar{x}_{18} \bar{y}_{18} \\ \bar{x}_{12} \bar{y}_{12} \\ \bar{x}_8 \bar{y}_8 \end{pmatrix}_{\bar{3}_2},$$
$$(11.57)$$

$$\begin{pmatrix} x_1 \\ x_7 \\ x_{11} \end{pmatrix}_{3_1} \otimes \begin{pmatrix} \bar{y}_{18} \\ \bar{y}_{12} \\ \bar{y}_8 \end{pmatrix}_{\bar{3}_1} = \sum_{k=0,1,2} (x_1 \bar{y}_{18} + \omega^{2k} x_7 \bar{y}_{12} + \omega^k x_{11} \bar{y}_8)_{1_k}$$
$$\oplus \begin{pmatrix} x_{11} \bar{y}_{12} \\ x_1 \bar{y}_8 \\ x_7 \bar{y}_{18} \end{pmatrix}_{3_3} \oplus \begin{pmatrix} x_7 \bar{y}_8 \\ x_{11} \bar{y}_{18} \\ x_1 \bar{y}_{12} \end{pmatrix}_{\bar{3}_3}, \quad (11.58)$$

$$\begin{pmatrix} x_2 \\ x_{14} \\ x_3 \end{pmatrix}_{3_2} \otimes \begin{pmatrix} y_2 \\ y_{14} \\ y_3 \end{pmatrix}_{3_2} = \begin{pmatrix} x_2 y_2 \\ x_{14} y_{14} \\ x_3 y_3 \end{pmatrix}_{3_3} \oplus \begin{pmatrix} x_3 y_{14} \\ x_2 y_3 \\ x_{14} y_2 \end{pmatrix}_{\bar{3}_2} \oplus \begin{pmatrix} x_{14} y_3 \\ x_3 y_2 \\ x_2 y_{14} \end{pmatrix}_{\bar{3}_2}, \quad (11.59)$$

$$\begin{pmatrix} \bar{x}_{17} \\ \bar{x}_5 \\ \bar{x}_{16} \end{pmatrix}_{\bar{3}_2} \otimes \begin{pmatrix} \bar{y}_{17} \\ \bar{y}_5 \\ \bar{y}_{16} \end{pmatrix}_{\bar{3}_2} = \begin{pmatrix} \bar{x}_{17} \bar{y}_{17} \\ \bar{x}_5 \bar{y}_5 \\ \bar{x}_{16} \bar{y}_{16} \end{pmatrix}_{\bar{3}_3} \oplus \begin{pmatrix} \bar{x}_5 \bar{y}_{16} \\ \bar{x}_{16} \bar{y}_{17} \\ \bar{x}_{17} \bar{y}_5 \end{pmatrix}_{3_2} \oplus \begin{pmatrix} \bar{x}_{16} \bar{y}_5 \\ \bar{x}_{17} \bar{y}_{16} \\ \bar{x}_5 \bar{y}_{17} \end{pmatrix}_{3_2},$$
$$(11.60)$$

$$\begin{pmatrix} x_2 \\ x_{14} \\ x_3 \end{pmatrix}_{3_2} \otimes \begin{pmatrix} \bar{y}_{17} \\ \bar{y}_5 \\ \bar{y}_{16} \end{pmatrix}_{\bar{3}_2} = \sum_{k=0,1,2} (x_2 \bar{y}_{17} + \omega^{2k} x_{14} \bar{y}_5 + \omega^k x_3 \bar{y}_{16})_{1_k}$$
$$\oplus \begin{pmatrix} x_3 \bar{y}_{17} \\ x_2 \bar{y}_5 \\ x_{14} \bar{y}_{16} \end{pmatrix}_{3_1} \oplus \begin{pmatrix} x_2 \bar{y}_{16} \\ x_{14} \bar{y}_{17} \\ x_3 \bar{y}_5 \end{pmatrix}_{\bar{3}_1}, \quad (11.61)$$

$$\begin{pmatrix} x_4 \\ x_9 \\ x_6 \end{pmatrix}_{3_3} \otimes \begin{pmatrix} y_4 \\ y_9 \\ y_6 \end{pmatrix}_{3_3} = \begin{pmatrix} x_9 y_9 \\ x_6 y_6 \\ x_4 y_4 \end{pmatrix}_{\bar{3}_1} \oplus \begin{pmatrix} x_6 y_9 \\ x_4 y_6 \\ x_9 y_4 \end{pmatrix}_{\bar{3}_3} \oplus \begin{pmatrix} x_9 y_6 \\ x_6 y_4 \\ x_4 y_9 \end{pmatrix}_{\bar{3}_3}, \quad (11.62)$$

$$\begin{pmatrix} \bar{x}_{15} \\ \bar{x}_{10} \\ \bar{x}_{13} \end{pmatrix}_{\bar{3}_3} \otimes \begin{pmatrix} \bar{y}_{15} \\ \bar{y}_{10} \\ \bar{y}_{13} \end{pmatrix}_{\bar{3}_3} = \begin{pmatrix} \bar{x}_{10} \bar{y}_{10} \\ \bar{x}_{13} \bar{y}_{13} \\ \bar{x}_{15} \bar{y}_{15} \end{pmatrix}_{3_1} \oplus \begin{pmatrix} \bar{x}_{10} \bar{y}_{13} \\ \bar{x}_{13} \bar{y}_{15} \\ \bar{x}_{15} \bar{y}_{10} \end{pmatrix}_{3_3} \oplus \begin{pmatrix} \bar{x}_{13} \bar{y}_{10} \\ \bar{x}_{15} \bar{y}_{13} \\ \bar{x}_{10} \bar{y}_{15} \end{pmatrix}_{3_3},$$
$$(11.63)$$

$$\begin{pmatrix} x_4 \\ x_9 \\ x_6 \end{pmatrix}_{3_3} \otimes \begin{pmatrix} \bar{y}_{15} \\ \bar{y}_{10} \\ \bar{y}_{13} \end{pmatrix}_{\bar{3}_3} = \sum_{k=0,1,2} (x_4 \bar{y}_{15} + \omega^{2k} x_9 \bar{y}_{10} + \omega^k x_6 \bar{y}_{13})_{1_k}$$
$$\oplus \begin{pmatrix} x_6 \bar{y}_{15} \\ x_4 \bar{y}_{10} \\ x_9 \bar{y}_{13} \end{pmatrix}_{3_2} \oplus \begin{pmatrix} x_4 \bar{y}_{13} \\ x_9 \bar{y}_{15} \\ x_6 \bar{y}_{10} \end{pmatrix}_{\bar{3}_2}, \quad (11.64)$$

11.4 T_{19}

$$\begin{pmatrix} x_1 \\ x_7 \\ x_{11} \end{pmatrix}_{3_1} \otimes \begin{pmatrix} y_2 \\ y_{14} \\ y_3 \end{pmatrix}_{3_2} = \begin{pmatrix} x_7 y_{14} \\ x_{11} y_3 \\ x_1 y_2 \end{pmatrix}_{3_2} \oplus \begin{pmatrix} x_1 y_3 \\ x_7 y_2 \\ x_{11} y_{14} \end{pmatrix}_{3_3} \oplus \begin{pmatrix} x_1 y_{14} \\ x_7 y_3 \\ x_{11} y_2 \end{pmatrix}_{\bar{3}_3},$$
(11.65)

$$\begin{pmatrix} x_1 \\ x_7 \\ x_{11} \end{pmatrix}_{3_1} \otimes \begin{pmatrix} \bar{y}_{17} \\ \bar{y}_5 \\ \bar{y}_{16} \end{pmatrix}_{\bar{3}_2} = \begin{pmatrix} x_1 \bar{y}_{17} \\ x_7 \bar{y}_5 \\ x_{11} \bar{y}_{16} \end{pmatrix}_{\bar{3}_1} \oplus \begin{pmatrix} x_7 \bar{y}_{16} \\ x_{11} \bar{y}_{17} \\ x_1 \bar{y}_5 \end{pmatrix}_{3_3} \oplus \begin{pmatrix} x_1 \bar{y}_{16} \\ x_7 \bar{y}_{17} \\ x_{11} \bar{y}_5 \end{pmatrix}_{\bar{3}_2},$$
(11.66)

$$\begin{pmatrix} x_1 \\ x_7 \\ x_{11} \end{pmatrix}_{3_1} \otimes \begin{pmatrix} y_4 \\ y_9 \\ y_6 \end{pmatrix}_{3_3} = \begin{pmatrix} x_{11} y_6 \\ x_1 y_4 \\ x_7 y_9 \end{pmatrix}_{\bar{3}_2} \oplus \begin{pmatrix} x_{11} y_4 \\ x_1 y_9 \\ x_7 y_6 \end{pmatrix}_{\bar{3}_3} \oplus \begin{pmatrix} x_{11} y_9 \\ x_1 y_6 \\ x_7 y_4 \end{pmatrix}_{3_1},$$
(11.67)

$$\begin{pmatrix} x_1 \\ x_7 \\ x_{11} \end{pmatrix}_{3_1} \otimes \begin{pmatrix} \bar{y}_{15} \\ \bar{y}_{10} \\ \bar{y}_{13} \end{pmatrix}_{\bar{3}_3} = \begin{pmatrix} x_7 \bar{y}_{10} \\ x_{11} \bar{y}_{13} \\ x_1 \bar{y}_{15} \end{pmatrix}_{\bar{3}_2} \oplus \begin{pmatrix} x_7 \bar{y}_{13} \\ x_{11} \bar{y}_{15} \\ x_1 \bar{y}_{10} \end{pmatrix}_{3_1} \oplus \begin{pmatrix} x_{11} \bar{y}_{10} \\ x_1 \bar{y}_{13} \\ x_7 \bar{y}_{15} \end{pmatrix}_{3_2},$$
(11.68)

$$\begin{pmatrix} x_2 \\ x_{14} \\ x_3 \end{pmatrix}_{3_2} \otimes \begin{pmatrix} \bar{y}_{18} \\ \bar{y}_{12} \\ \bar{y}_8 \end{pmatrix}_{\bar{3}_1} = \begin{pmatrix} x_2 \bar{y}_{18} \\ x_{14} \bar{y}_{12} \\ x_3 \bar{y}_8 \end{pmatrix}_{3_1} \oplus \begin{pmatrix} x_3 \bar{y}_{18} \\ x_2 \bar{y}_{12} \\ x_{14} \bar{y}_8 \end{pmatrix}_{3_2} \oplus \begin{pmatrix} x_3 \bar{y}_{12} \\ x_2 \bar{y}_8 \\ x_{14} \bar{y}_{18} \end{pmatrix}_{\bar{3}_3},$$
(11.69)

$$\begin{pmatrix} \bar{x}_{18} \\ \bar{x}_{12} \\ \bar{x}_8 \end{pmatrix}_{\bar{3}_1} \otimes \begin{pmatrix} \bar{y}_{17} \\ \bar{y}_5 \\ \bar{y}_{16} \end{pmatrix}_{\bar{3}_2} = \begin{pmatrix} \bar{x}_{12} \bar{y}_5 \\ \bar{x}_8 \bar{y}_{16} \\ \bar{x}_{18} \bar{y}_{17} \end{pmatrix}_{\bar{3}_2} \oplus \begin{pmatrix} \bar{x}_{18} \bar{y}_5 \\ \bar{x}_{12} \bar{y}_{16} \\ \bar{x}_8 \bar{y}_{17} \end{pmatrix}_{3_3} \oplus \begin{pmatrix} \bar{x}_{18} \bar{y}_{16} \\ \bar{x}_{12} \bar{y}_{17} \\ \bar{x}_8 \bar{y}_5 \end{pmatrix}_{\bar{3}_3},$$
(11.70)

$$\begin{pmatrix} x_4 \\ x_9 \\ x_6 \end{pmatrix}_{3_3} \otimes \begin{pmatrix} \bar{y}_{18} \\ \bar{y}_{12} \\ \bar{y}_8 \end{pmatrix}_{\bar{3}_1} = \begin{pmatrix} x_9 \bar{y}_{12} \\ x_6 \bar{y}_8 \\ x_4 \bar{y}_{18} \end{pmatrix}_{3_2} \oplus \begin{pmatrix} x_9 \bar{y}_8 \\ x_6 \bar{y}_{18} \\ x_4 \bar{y}_{12} \end{pmatrix}_{\bar{3}_2} \oplus \begin{pmatrix} x_6 \bar{y}_{12} \\ x_4 \bar{y}_8 \\ x_9 \bar{y}_{18} \end{pmatrix}_{\bar{3}_1},$$
(11.71)

$$\begin{pmatrix} \bar{x}_{18} \\ \bar{x}_{12} \\ \bar{x}_8 \end{pmatrix}_{\bar{3}_1} \otimes \begin{pmatrix} \bar{y}_{15} \\ \bar{y}_{10} \\ \bar{y}_{13} \end{pmatrix}_{\bar{3}_3} = \begin{pmatrix} \bar{x}_8 \bar{y}_{13} \\ \bar{x}_{18} \bar{y}_{15} \\ \bar{x}_{12} \bar{y}_{10} \end{pmatrix}_{3_2} \oplus \begin{pmatrix} \bar{x}_8 \bar{y}_{15} \\ \bar{x}_{18} \bar{y}_{10} \\ \bar{x}_{12} \bar{y}_{13} \end{pmatrix}_{3_3} \oplus \begin{pmatrix} \bar{x}_8 \bar{y}_{10} \\ \bar{x}_{18} \bar{y}_{13} \\ \bar{x}_{12} \bar{y}_{15} \end{pmatrix}_{\bar{3}_1},$$
(11.72)

$$\begin{pmatrix} x_2 \\ x_{14} \\ x_3 \end{pmatrix}_{3_2} \otimes \begin{pmatrix} y_4 \\ y_9 \\ y_6 \end{pmatrix}_{3_3} = \begin{pmatrix} x_{14} y_9 \\ x_3 y_6 \\ x_2 y_4 \end{pmatrix}_{3_3} \oplus \begin{pmatrix} x_{14} y_6 \\ x_3 y_4 \\ x_2 y_9 \end{pmatrix}_{3_1} \oplus \begin{pmatrix} x_{14} y_4 \\ x_3 y_9 \\ x_2 y_6 \end{pmatrix}_{\bar{3}_1},$$
(11.73)

$$\begin{pmatrix} x_2 \\ x_{14} \\ x_3 \end{pmatrix}_{3_2} \otimes \begin{pmatrix} \bar{y}_{15} \\ \bar{y}_{10} \\ \bar{y}_{13} \end{pmatrix}_{\bar{3}_3} = \begin{pmatrix} x_2 \bar{y}_{15} \\ x_{14} \bar{y}_{10} \\ x_3 \bar{y}_{13} \end{pmatrix}_{\bar{3}_2} \oplus \begin{pmatrix} x_3 \bar{y}_{15} \\ x_2 \bar{y}_{10} \\ x_{14} \bar{y}_{13} \end{pmatrix}_{3_1} \oplus \begin{pmatrix} x_2 \bar{y}_{13} \\ x_{14} \bar{y}_{15} \\ x_3 \bar{y}_{10} \end{pmatrix}_{\bar{3}_3},$$
(11.74)

$$\begin{pmatrix} x_4 \\ x_9 \\ x_6 \end{pmatrix}_{3_3} \otimes \begin{pmatrix} \bar{y}_{17} \\ \bar{y}_5 \\ \bar{y}_{16} \end{pmatrix}_{\bar{3}_2} = \begin{pmatrix} x_4\bar{y}_{17} \\ x_9\bar{y}_5 \\ x_6\bar{y}_{16} \end{pmatrix}_{3_2} \oplus \begin{pmatrix} x_6\bar{y}_{17} \\ x_4\bar{y}_5 \\ x_9\bar{y}_{16} \end{pmatrix}_{3_3} \oplus \begin{pmatrix} x_4\bar{y}_{16} \\ x_9\bar{y}_{17} \\ x_6\bar{y}_5 \end{pmatrix}_{3_1}, \quad (11.75)$$

$$\begin{pmatrix} \bar{x}_{17} \\ \bar{x}_5 \\ \bar{x}_{16} \end{pmatrix}_{\bar{3}_2} \otimes \begin{pmatrix} \bar{y}_{15} \\ \bar{y}_{10} \\ \bar{y}_{13} \end{pmatrix}_{\bar{3}_3} = \begin{pmatrix} \bar{x}_5\bar{y}_{10} \\ \bar{x}_{16}\bar{y}_{13} \\ \bar{x}_{17}\bar{y}_{15} \end{pmatrix}_{\bar{3}_3} \oplus \begin{pmatrix} \bar{x}_5\bar{y}_{13} \\ \bar{x}_{16}\bar{y}_{15} \\ \bar{x}_{17}\bar{y}_{10} \end{pmatrix}_{\bar{3}_1} \oplus \begin{pmatrix} \bar{x}_5\bar{y}_{15} \\ \bar{x}_{16}\bar{y}_{10} \\ \bar{x}_{17}\bar{y}_{13} \end{pmatrix}_{\bar{3}_1}.$$
(11.76)

The tensor products between singlets are

$$(x)_{1_0}(y)_{1_0} = (x)_{1_1}(y)_{1_2} = (x)_{1_2}(y)_{1_1} = (xy)_{1_0},$$
$$(x)_{1_1}(y)_{1_1} = (xy)_{1_2}, \quad (x)_{1_2}(y)_{1_2} = (xy)_{1_1}.$$
(11.77)

The tensor products between triplets and singlets are

$$\mathbf{1}_j \otimes \mathbf{3}_k = \mathbf{3}_k, \tag{11.78}$$

where j and k run over all the singlets and triplets, respectively. Notice that singlets do not affect the form of triplets.

References

1. Bovier, A., Luling, M., Wyler, D.: J. Math. Phys. **22**, 1536 (1981)
2. Bovier, A., Luling, M., Wyler, D.: J. Math. Phys. **22**, 1543 (1981)
3. Fairbairn, W.M., Fulton, T.: J. Math. Phys. **23**, 1747 (1982)
4. King, S.F., Luhn, C.: J. High Energy Phys. **0910**, 093 (2009). arXiv:0908.1897 [hep-ph]

Chapter 12
$\Sigma(3N^3)$

12.1 Generic Aspects

In this chapter, we study $\Sigma(3N^3)$ [1]. This discrete group is defined as a closed algebra of three Abelian symmetries, namely, Z_N, Z'_N, and Z''_N, which commute with each other, and their Z_3 permutations. Let us denote the generators of Z_N, Z'_N, and Z''_N by a, a', and a'', respectively, and the Z_3 generator by b. All elements of $\Sigma(3N^3)$ can be expressed in the form

$$g = b^k a^m a'^n a''^\ell, \tag{12.1}$$

with $k = 0, 1, 2$, and $m, n, \ell = 0, \ldots, N-1$, where a, a', a'', and b satisfy

$$a^N = a'^N = a''^N = 1, \quad aa' = a'a, \quad aa'' = a''a, \quad a''a' = a'a'', \tag{12.2}$$
$$b^3 = 1, \quad b^2 ab = a'', \quad b^2 a'b = a, \quad b^2 a''b = a'.$$

These generators a, a', a'', and b are represented, e.g., by

$$b = \begin{pmatrix} 0 & 1 & 0 \\ 0 & 0 & 1 \\ 1 & 0 & 0 \end{pmatrix}, \quad a = \begin{pmatrix} 1 & 0 & 0 \\ 0 & 1 & 0 \\ 0 & 0 & \rho \end{pmatrix},$$
$$a' = \begin{pmatrix} 1 & 0 & 0 \\ 0 & \rho & 0 \\ 0 & 0 & 1 \end{pmatrix}, \quad a'' = \begin{pmatrix} \rho & 0 & 0 \\ 0 & 1 & 0 \\ 0 & 0 & 1 \end{pmatrix}, \tag{12.3}$$

where $\rho = e^{2i\pi/N}$. Then, all elements of $\Sigma(3N^3)$ can be expressed as follows:

$$\begin{pmatrix} 0 & \rho^n & 0 \\ 0 & 0 & \rho^m \\ \rho^\ell & 0 & 0 \end{pmatrix}, \quad \begin{pmatrix} \rho^\ell & 0 & 0 \\ 0 & \rho^m & 0 \\ 0 & 0 & \rho^n \end{pmatrix}, \quad \begin{pmatrix} 0 & 0 & \rho^m \\ \rho^\ell & 0 & 0 \\ 0 & \rho^n & 0 \end{pmatrix}. \tag{12.4}$$

For the case $N = 2$, the element $aa'a''$ commutes with all the elements. In addition, when we define $\tilde{a} = aa''$ and $\tilde{a}' = a'a''$, we find the closed algebra among $\tilde{a}$, $\tilde{a}'$, and b, which corresponds to $\Delta(12)$. Since the element $aa'a''$ is not contained in this closed algebra, this group is isomorphic to $Z_2 \times \Delta(12)$.

The situation for $N = 3$ is different. Once again, the element $aa'a''$ commutes with all the elements. When we define $\tilde{a} = a^2 a''$ and $\tilde{a}' = a'a''^2$, the closed algebra among $\tilde{a}$, $\tilde{a}'$, and b corresponds to $\Delta(27)$. On the other hand, the element $aa'a''$ can be written $aa'a'' = \tilde{a}^2 \tilde{a}'$ in this case. Thus, $aa'a''$ is one of elements of $\Delta(27)$. That is, the group $\Sigma(81)$ is not $Z_3 \times \Delta(27)$, but isomorphic to $(Z_3 \times Z_3' \times Z_3'') \rtimes Z_3$.

Similarly, for a generic value of N, the element $aa'a''$ commutes with all the elements. When we define $\tilde{a} = a^{N-1} a''$ and $\tilde{a}' = a'a''^{N-1}$, the closed algebra generated by $\tilde{a}$, $\tilde{a}'$, and b corresponds to $\Delta(3N^2)$. For the case $N/3 \neq$ integer, the element $aa'a''$ is not included in $\Delta(3N^2)$. Thus, we find that this group is isomorphic to $Z_N \times \Delta(3N^2)$. On the other hand, when $N/3$ is an integer, the element $(aa'a'')^{Nk/3}$ with $k = 0, 1, 2$, is included in $\Delta(3N^2)$. That is, the group $\Sigma(3N^3)$ cannot be $Z_N \times \Delta(3N^2)$.

12.1.1 Conjugacy Classes

Here we summarize the conjugacy classes of $\Sigma(3N^3)$:

$$C_1: \quad \{e\},$$
$$C_1^{(k)}: \quad \{a^k a'^k a''^k\}, \quad k = 1, \ldots, N-1,$$
$$C_3^{(\ell,m,n)}: \quad \{a^\ell a'^m a''^n, a^m a'^n a''^\ell, a^n a'^\ell a''^m\}, \quad \ell \neq m \neq n,$$
$$C_{N^2}^{(p)}: \quad \{ba^\ell a'^m a''^{p-\ell-m} | \ell, m = 0, \ldots, N-1\}, \quad p = 0, \ldots, N-1,$$
$$C_{N^2}^{\prime(p)}: \quad \{b^2 a^\ell a'^m a''^{p'-\ell-m} | \ell, m = 0, \ldots, N-1\}, \quad p = 0, \ldots, N-1,$$
(12.5)

and the order h of the elements in each conjugacy class, viz.,

$$\begin{aligned} C_1: &\quad h = 1, \\ C_1^{(k)}: &\quad h = N/\gcd(N, k), \\ C_3^{(\ell,m,n)}: &\quad h = N/\gcd(N, \ell, m, n), \\ C_{N^2}^{(p)}: &\quad h = 3N/\gcd(N, p), \\ C_{N^2}^{\prime(p)}: &\quad h = 3N/\gcd(N, p). \end{aligned}$$
(12.6)

For example, the conjugacy classes $C_{N^2}^{(p)}$ and $C_{N^2}^{\prime(p)}$ are obtained with the help of the following relations:

12.1 Generic Aspects

$$a^p a'^q a'''^r (b a^\ell a'^m a''^n) a^{-p} a'^{-q} a''^{-r} = (b a^\ell a'^m a''^n) a^{q-p} a'^{r-q} a''^{p-r},$$

$$b a^p a'^q a'''^r (b a^\ell a'^m a''^n) a^{-p} a'^{-q} a''^{-r} b^{-1} = b a^{n+p-r} a'^{\ell+q-p} a''^{m+r-q}, \quad (12.7)$$

$$b^2 a^p a'^q a'''^r (b a^\ell a'^m a''^n) a^{-p} a'^{-q} a''^{-r} b^{-2} = b a^{m+r-q} a'^{n+p-r} a''^{\ell+q-p},$$

where we have used $a^p a'^q a'''^r b = b a^q a'^r a''^p$, $a^p a'^q a'''^r b^2 = b^2 a^r a'^p a''^q$. Note that the sum of the factors of a, a', and a'' is the same. Since the numbers of each of the classes $C_1^{(\ell)}$, $C_3^{(\ell,m,n)}$, $C_{N^2}^{(p)}$, and $C_{N^2}^{\prime(p)}$ are $(N-1)$, $(N^3-N)/3$, N, and N, respectively, the total number of conjugacy classes is

$$1 + (N-1) + (N^3-N)/3 + N + N = \frac{1}{3}N(N^2+8). \quad (12.8)$$

The number of irreducible representations can be determined by using the relations

$$m_1 + 4m_2 + 9m_3 + \cdots = 3N^3, \qquad m_1 + m_2 + m_3 + \cdots = \frac{1}{3}N(N^2+8). \quad (12.9)$$

Their solutions are $(m_1, m_2, m_3) = (3N, 0, N(N^2-1)/3)$. Hence the group $\Sigma(3N^3)$ has $N(N^2+8)/3$ conjugacy classes, $3N$ singlets, and $N(N^2-1)/3$ triplets.

12.1.2 Characters and Representations

The number of singlets is $3N$. We denote them by $\mathbf{1}_{k,\ell}$ with $k = 0, 1, 2$, and $\ell = 0, \ldots, N-1$. For the representations of singlets, all operators a, a', a'', and b are mutually commuting. Then, from the algebraic relations it is found that the characters for a, a', and a'' must be the same. Thus we can represent the $3N$ singlets by $\chi_{1_{k,\ell}}(a) = \chi_{1_{k,\ell}}(a') = \chi_{1_{k,\ell}}(a'') = \rho^\ell$ and $\chi_{1_{k,\ell}}(b) = \omega^k$, as shown in Table 12.1.

The number of triplets is $N(N^2-1)/3$. We represent a, a', a'', and b by

$$a = \begin{pmatrix} \rho^\ell & 0 & 0 \\ 0 & \rho^m & 0 \\ 0 & 0 & \rho^n \end{pmatrix}, \quad a' = \begin{pmatrix} \rho^m & 0 & 0 \\ 0 & \rho^n & 0 \\ 0 & 0 & \rho^\ell \end{pmatrix}, \quad a'' = \begin{pmatrix} \rho^n & 0 & 0 \\ 0 & \rho^\ell & 0 \\ 0 & 0 & \rho^m \end{pmatrix},$$

$$(12.10)$$

and

$$b = \begin{pmatrix} 0 & 1 & 0 \\ 0 & 0 & 1 \\ 1 & 0 & 0 \end{pmatrix}, \quad (12.11)$$

on the triplet $\mathbf{3}_{[\ell][m][n]}$,

$$\mathbf{3}_{[\ell][m][n]} = \begin{pmatrix} x_\ell \\ x_m \\ x_n \end{pmatrix}, \quad (12.12)$$

Table 12.1 Characters of $\Sigma(3N^3)$

	h	$\chi 1_{\ell,m}$	$\chi 3_{[\ell][m][n]}$
C_1	1	1	3
$C_1^{(p)}$	$N/\gcd(N,p)$	ρ^{3pm}	$3\rho^{p(\ell+m+n)}$
$C_3^{(p,q,r)}$	$N/\gcd(N,p,q,r)$	$\rho^{(p+q+r)m}$	$\rho^{p\ell+qm+rn} + \rho^{pm+qn+r\ell} + \rho^{pn+q\ell+rm}$
$C_{N^2}^{(p)}$	$3N/\gcd(N,p)$	$\omega^\ell \rho^{pm}$	0
$C'^{(p)}_{N^2}$	$3N/\gcd(N,p)$	$\omega^{2\ell} \rho^{pm}$	0

with $[\ell][m][n] = (\ell,m,n)$, (m,n,ℓ), or (n,ℓ,m). These characters are shown in Table 12.1. The subscripts of the components give the Z_N charge.

12.1.3 Tensor Products

The tensor products between two triplets are given by

$$\begin{pmatrix} x_\ell \\ x_m \\ x_n \end{pmatrix}_{3_{[\ell][m][n]}} \otimes \begin{pmatrix} y_{\ell'} \\ y_{m'} \\ y_{n'} \end{pmatrix}_{3_{[\ell'][m'][n']}}$$

$$= \begin{pmatrix} x_\ell y_{\ell'} \\ x_m y_{m'} \\ x_n y_{n'} \end{pmatrix}_{3_{[\ell+\ell'][m+m'][n+n']}} \oplus \begin{pmatrix} x_m y_{n'} \\ x_n y_{\ell'} \\ x_\ell y_{m'} \end{pmatrix}_{3_{[m+n'][n+\ell'][\ell+m']}}$$

$$\oplus \begin{pmatrix} x_n y_{m'} \\ x_\ell y_{n'} \\ x_m y_{\ell'} \end{pmatrix}_{3_{[n+m'][\ell+n'][m+\ell']}} . \tag{12.13}$$

If all the subscripts become equal, the triplet can be decomposed into singlets as

$$(x_a, x_b, x_c)_{3_{[k][k][k]}} = (x_a + x_b + x_c)_{1_{0,k}} + (x_a + \omega^2 x_b + \omega x_c)_{1_{1,k}}$$
$$+ (x_a + \omega x_b + \omega^2 x_c)_{1_{2,k}}.$$

The tensor products between singlets and triplets are

$$\begin{pmatrix} x_\ell \\ x_m \\ x_n \end{pmatrix}_{3_{[\ell][m][n]}} \otimes (y)_{1_{k',k}} = \begin{pmatrix} x_\ell y \\ \omega^{k'} x_m y \\ \omega^{2k'} x_n y \end{pmatrix}_{3_{[\ell+k][m+k][n+k]}} . \tag{12.14}$$

The product of singlets is

$$(x)_{1_{k,\ell}} \otimes (y)_{1_{k',\ell'}} = (xy)_{1_{k+k',\ell+\ell'}} . \tag{12.15}$$

12.2 $\Sigma(81)$

We now detail the case $N = 3$, that is, $\Sigma(81)$. It has eighty-one elements and these can be written in the form $b^k a^m a'^n a''^\ell$ for $k = 0, 1, 2$, and $m, n, \ell = 0, 1, 2$, where a, a', a'', and b satisfy $a^3 = a'^3 = a''^3 = 1$, $aa' = a'a$, $aa'' = a''a$, $a''a' = a'a''$, $b^3 = 1$, $b^2 ab = a''$, $b^2 a' b = a$, and $b^2 a'' b = a'$. These elements are classified into seventeen conjugacy classes as follows:

$$
\begin{aligned}
&C_1: &&\{e\}, &&h = 1,\\
&C_1^{(1)}: &&\{aa'a''\}, &&h = 3,\\
&C_1^{(2)}: &&\{(aa'a'')^2\}, &&h = 3,\\
&C_3^{(0)}: &&\{aa'^1 a''^2, a'' a^1 a'^2, a' a''^1 a^2\}, &&h = 3,\\
&C_3'^{(0)}: &&\{a^1 a' a''^2, a''^1 aa'^2, a'^1 a'' a^2\}, &&h = 3,\\
&C_3^{(1)}: &&\{aa' a''^1, a'' aa'^1, a' a'' a^1\}, &&h = 3,\\
&C_3'^{(1)}: &&\{aa'^2 a''^2, a'' a^2 a'^2, a' a''^2 a^2\}, &&h = 3,\\
&C_3''^{(1)}: &&\{a^1 a'^1 a''^2, a''^1 a^1 a'^2, a'^1 a''^1 a^2\}, &&h = 3,\\
&C_3^{(2)}: &&\{aa' a''^2, a'' aa'^2, a' a'' a^2\}, &&h = 3, &&(12.16)\\
&C_3'^{(2)}: &&\{aa'^1 a''^1, a'' a^1 a'^1, a' a''^1 a^1\}, &&h = 3,\\
&C_3''^{(2)}: &&\{a^1 a'^2 a''^2, a''^1 a^2 a'^2, a'^1 a''^2 a^2\}, &&h = 3,\\
&C_9^{(0)}: &&\{b a^p a'^q a''^{-p-q} | p, q = 0, \ldots, N-1\}, &&h = 3,\\
&C_9^{(1)}: &&\{b a^p a'^q a''^{1-p-q} | p, q = 0, \ldots, N-1\}, &&h = 9,\\
&C_9^{(2)}: &&\{b a^p a'^q a''^{2-p-q} | p, q = 0, \ldots, N-1\}, &&h = 9,\\
&C_9'^{(0)}: &&\{b^2 a^p a'^q a''^{-p-q} | p, q = 0, \ldots, N-1\}, &&h = 3,\\
&C_9'^{(1)}: &&\{b^2 a^p a'^q a''^{1-p-q} | p, q = 0, \ldots, N-1\}, &&h = 9,\\
&C_9'^{(2)}: &&\{b^2 a^p a'^q a''^{2-p-q} | p, q = 0, \ldots, N-1\}, &&h = 9,
\end{aligned}
$$

where h denotes the order of each element in the given conjugacy class.

The relations (2.18) and (2.19) for $\Sigma(81)$ give

$$m_1 + 2^2 m_2 + 3^2 m_3 + \cdots = 81, \qquad (12.17)$$

$$m_1 + m_2 + m_3 + \cdots = 17. \qquad (12.18)$$

Table 12.2 Characters of $\Sigma(81)$ for the 9 one-dimensional representations

	h	$\chi 1_{0,0}$	$\chi 1_{1,0}$	$\chi 1_{2,0}$	$\chi 1_{0,1}$	$\chi 1_{1,1}$	$\chi 1_{2,1}$	$\chi 1_{0,2}$	$\chi 1_{1,2}$	$\chi 1_{2,2}$
C_1	1	1	1	1	1	1	1	1	1	1
$C_1^{(1)}$	1	1	1	1	1	1	1	1	1	1
$C_1^{(2)}$	1	1	1	1	1	1	1	1	1	1
$C_3^{(0)}$	3	1	1	1	1	1	1	1	1	1
$C_3'^{(0)}$	3	1	1	1	1	1	1	1	1	1
$C_3^{(1)}$	3	1	1	1	ω	ω	ω	ω^2	ω^2	ω^2
$C_3'^{(1)}$	3	1	1	1	ω	ω	ω	ω^2	ω^2	ω^2
$C_3''^{(1)}$	3	1	1	1	ω	ω	ω	ω^2	ω^2	ω^2
$C_3^{(2)}$	3	1	1	1	ω^2	ω^2	ω^2	ω	ω	ω
$C_3'^{(2)}$	3	1	1	1	ω^2	ω^2	ω^2	ω	ω	ω
$C_3''^{(2)}$	3	1	1	1	ω^2	ω^2	ω^2	ω	ω	ω
$C_9^{(0)}$	3	1	ω	ω^2	1	ω	ω^2	1	ω	ω^2
$C_9^{(1)}$	9	1	ω	ω^2	ω	ω^2	1	ω^2	1	ω
$C_9^{(2)}$	9	1	ω	ω^2	ω^2	1	ω	ω	ω^2	1
$C_9'^{(0)}$	3	1	ω^2	ω	1	ω^2	ω	1	ω^2	ω
$C_9'^{(1)}$	9	1	ω^2	ω	ω	1	ω^2	ω^2	ω	1
$C_9'^{(2)}$	9	1	ω^2	ω	ω^2	ω	1	ω	1	ω^2

We thus find $(m_1, m_3) = (9, 8)$, whence there are nine singlets $\mathbf{1}_{k,\ell}$ with $k, \ell = 0, 1, 2$, and eight triplets, for which we use the notation

$$\mathbf{3}_A = \mathbf{3}_{[1][0][0]}, \quad \mathbf{3}_B = \mathbf{3}_{[0][2][2]}, \quad \mathbf{3}_C = \mathbf{3}_{[2][1][1]}, \quad \mathbf{3}_D = \mathbf{3}_{[2][0][1]},$$

$$\bar{\mathbf{3}}_A = \mathbf{3}_{[2][0][0]}, \quad \bar{\mathbf{3}}_B = \mathbf{3}_{[0][1][1]}, \quad \bar{\mathbf{3}}_C = \mathbf{3}_{[1][2][2]}, \quad \bar{\mathbf{3}}_D = \mathbf{3}_{[1][0][2]}.$$

The characters are given in Tables 12.2 and 12.3.

On all the triplets, the generator b is represented by

$$b = \begin{pmatrix} 0 & 1 & 0 \\ 0 & 0 & 1 \\ 1 & 0 & 0 \end{pmatrix}. \tag{12.19}$$

The generators a, a', and a'' are represented on each triplet as follows:

$$a = \begin{pmatrix} \omega & 0 & 0 \\ 0 & 1 & 0 \\ 0 & 0 & 1 \end{pmatrix}, \quad a' = \begin{pmatrix} 1 & 0 & 0 \\ 0 & 1 & 0 \\ 0 & 0 & \omega \end{pmatrix}, \quad a'' = \begin{pmatrix} 1 & 0 & 0 \\ 0 & \omega & 0 \\ 0 & 0 & 1 \end{pmatrix}, \tag{12.20}$$

12.2 $\Sigma(81)$

Table 12.3 Characters of $\Sigma(81)$ for the 8 three-dimensional representations

Class	n	h	χ_{3_A}	$\chi_{\bar{3}_A}$	χ_{3_B}	$\chi_{\bar{3}_B}$	χ_{3_C}	$\chi_{\bar{3}_C}$	χ_{3_D}	$\chi_{\bar{3}_D}$
C_1	1	1	3	3	3	3	3	3	3	3
$C_1^{(1)}$	1	3	3ω	$3\omega^2$	3ω	$3\omega^2$	3ω	$3\omega^2$	3	3
$C_1^{(2)}$	1	3	$3\omega^2$	3ω	$3\omega^2$	3ω	$3\omega^2$	3ω	3	3
$C_3^{(0)}$	3	3	0	0	0	0	0	0	$3\omega^2$	3ω
$C_3^{'(0)}$	3	3	0	0	0	0	0	0	3ω	$3\omega^2$
$C_3^{(1)}$	3	3	$-i\sqrt{3}\omega$	$i\sqrt{3}\omega^2$	$-i\sqrt{3}$	$i\sqrt{3}$	$-i\sqrt{3}\omega^2$	$i\sqrt{3}\omega$	0	0
$C_3^{'(1)}$	3	3	$-i\sqrt{3}$	$i\sqrt{3}$	$-i\sqrt{3}\omega^2$	$i\sqrt{3}\omega$	$-i\sqrt{3}\omega$	$i\sqrt{3}\omega^2$	0	0
$C_3^{''(1)}$	3	3	$-i\sqrt{3}\omega^2$	$i\sqrt{3}\omega$	$-i\sqrt{3}\omega$	$i\sqrt{3}\omega^2$	$-i\sqrt{3}$	$i\sqrt{3}$	0	0
$C_3^{(2)}$	3	3	$i\sqrt{3}\omega^2$	$-i\sqrt{3}\omega$	$i\sqrt{3}$	$-i\sqrt{3}$	$i\sqrt{3}\omega$	$-i\sqrt{3}\omega^2$	0	0
$C_3^{'(2)}$	3	3	$i\sqrt{3}$	$-i\sqrt{3}$	$i\sqrt{3}\omega$	$-i\sqrt{3}\omega^2$	$i\sqrt{3}\omega^2$	$-i\sqrt{3}\omega$	0	0
$C_3^{''(2)}$	3	3	$i\sqrt{3}\omega$	$-i\sqrt{3}\omega^2$	$i\sqrt{3}\omega^2$	$-i\sqrt{3}\omega$	$i\sqrt{3}$	$-i\sqrt{3}$	0	0
$C_9^{(0)}$	9	3	0	0	0	0	0	0	0	0
$C_9^{(1)}$	9	9	0	0	0	0	0	0	0	0
$C_9^{(2)}$	9	9	0	0	0	0	0	0	0	0
$C_9^{'(0)}$	9	3	0	0	0	0	0	0	0	0
$C_9^{'(1)}$	9	9	0	0	0	0	0	0	0	0
$C_9^{'(2)}$	9	9	0	0	0	0	0	0	0	0

on 3_A,

$$a = \begin{pmatrix} 1 & 0 & 0 \\ 0 & \omega^2 & 0 \\ 0 & 0 & \omega^2 \end{pmatrix}, \quad a' = \begin{pmatrix} \omega^2 & 0 & 0 \\ 0 & \omega^2 & 0 \\ 0 & 0 & 1 \end{pmatrix}, \quad a'' = \begin{pmatrix} \omega^2 & 0 & 0 \\ 0 & 1 & 0 \\ 0 & 0 & \omega^2 \end{pmatrix},$$
(12.21)

on 3_B,

$$a = \begin{pmatrix} \omega^2 & 0 & 0 \\ 0 & \omega & 0 \\ 0 & 0 & \omega \end{pmatrix}, \quad a' = \begin{pmatrix} \omega & 0 & 0 \\ 0 & \omega & 0 \\ 0 & 0 & \omega^2 \end{pmatrix}, \quad a'' = \begin{pmatrix} \omega & 0 & 0 \\ 0 & \omega^2 & 0 \\ 0 & 0 & \omega \end{pmatrix},$$
(12.22)

on 3_C, and

$$a = \begin{pmatrix} \omega^2 & 0 & 0 \\ 0 & 1 & 0 \\ 0 & 0 & \omega \end{pmatrix}, \quad a' = \begin{pmatrix} 1 & 0 & 0 \\ 0 & \omega & 0 \\ 0 & 0 & \omega^2 \end{pmatrix}, \quad a'' = \begin{pmatrix} \omega & 0 & 0 \\ 0 & \omega^2 & 0 \\ 0 & 0 & 1 \end{pmatrix},$$
(12.23)

on $\mathbf{3}_D$. The representations of a, a', and a'' on $\bar{\mathbf{3}}_A, \bar{\mathbf{3}}_B, \bar{\mathbf{3}}_C$, and $\bar{\mathbf{3}}_D$ are obtained as complex conjugates of the representations on $\mathbf{3}_A, \mathbf{3}_B, \mathbf{3}_C$, and $\mathbf{3}_D$, respectively.

On the other hand, for the singlet $\mathbf{1}_{k,\ell}$, these generators are represented by

$$b = \omega^k, \qquad a = a' = a'' = \omega^\ell.$$

The tensor products between triplets are:

$$\begin{pmatrix} x_1 \\ x_2 \\ x_3 \end{pmatrix}_{\mathbf{3}_A} \otimes \begin{pmatrix} y_1 \\ y_2 \\ y_3 \end{pmatrix}_{\mathbf{3}_A} = \begin{pmatrix} x_1 y_1 \\ x_2 y_2 \\ x_3 y_3 \end{pmatrix}_{\bar{\mathbf{3}}_A} \oplus \begin{pmatrix} x_2 y_3 \\ x_3 y_1 \\ x_1 y_2 \end{pmatrix}_{\bar{\mathbf{3}}_B} \oplus \begin{pmatrix} x_3 y_2 \\ x_1 y_3 \\ x_2 y_1 \end{pmatrix}_{\bar{\mathbf{3}}_B}, \quad (12.24)$$

$$\begin{pmatrix} x_1 \\ x_2 \\ x_3 \end{pmatrix}_{\mathbf{3}_A} \otimes \begin{pmatrix} y_1 \\ y_2 \\ y_3 \end{pmatrix}_{\bar{\mathbf{3}}_A} = \sum_{k=0,1,2} \left(x_1 y_1 + \omega^{2k} x_2 y_2 + \omega^k x_3 y_3 \right)_{\mathbf{1}_{k,0}}$$

$$\oplus \begin{pmatrix} x_2 y_3 \\ x_3 y_1 \\ x_1 y_2 \end{pmatrix}_{\mathbf{3}_D} \oplus \begin{pmatrix} x_3 y_2 \\ x_1 y_3 \\ x_2 y_1 \end{pmatrix}_{\mathbf{3}_D}, \quad (12.25)$$

$$\begin{pmatrix} x_1 \\ x_2 \\ x_3 \end{pmatrix}_{\mathbf{3}_A} \otimes \begin{pmatrix} y_1 \\ y_2 \\ y_3 \end{pmatrix}_{\mathbf{3}_B} = \begin{pmatrix} x_1 y_1 \\ x_2 y_2 \\ x_3 y_3 \end{pmatrix}_{\bar{\mathbf{3}}_C} \oplus \begin{pmatrix} x_3 y_2 \\ x_1 y_3 \\ x_2 y_1 \end{pmatrix}_{\bar{\mathbf{3}}_A} \oplus \begin{pmatrix} x_2 y_3 \\ x_3 y_1 \\ x_1 y_2 \end{pmatrix}_{\bar{\mathbf{3}}_A}, \quad (12.26)$$

$$\begin{pmatrix} x_1 \\ x_2 \\ x_3 \end{pmatrix}_{\mathbf{3}_A} \otimes \begin{pmatrix} y_1 \\ y_2 \\ y_3 \end{pmatrix}_{\bar{\mathbf{3}}_B} = \sum_{k=0,1,2} \left(x_1 y_1 + \omega^{2k} x_2 y_2 + \omega^k x_3 y_3 \right)_{\mathbf{1}_{k,1}}$$

$$\oplus \begin{pmatrix} x_2 y_3 \\ x_3 y_1 \\ x_1 y_2 \end{pmatrix}_{\mathbf{3}_D} \oplus \begin{pmatrix} x_3 y_2 \\ x_1 y_3 \\ x_2 y_1 \end{pmatrix}_{\mathbf{3}_D}, \quad (12.27)$$

$$\begin{pmatrix} x_1 \\ x_2 \\ x_3 \end{pmatrix}_{\mathbf{3}_A} \otimes \begin{pmatrix} y_1 \\ y_2 \\ y_3 \end{pmatrix}_{\mathbf{3}_C} = \begin{pmatrix} x_1 y_1 \\ x_2 y_2 \\ x_3 y_3 \end{pmatrix}_{\bar{\mathbf{3}}_B} \oplus \begin{pmatrix} x_2 y_3 \\ x_3 y_1 \\ x_1 y_2 \end{pmatrix}_{\bar{\mathbf{3}}_C} \oplus \begin{pmatrix} x_3 y_2 \\ x_1 y_3 \\ x_2 y_1 \end{pmatrix}_{\bar{\mathbf{3}}_C}, \quad (12.28)$$

$$\begin{pmatrix} x_1 \\ x_2 \\ x_3 \end{pmatrix}_{\mathbf{3}_A} \otimes \begin{pmatrix} y_1 \\ y_2 \\ y_3 \end{pmatrix}_{\bar{\mathbf{3}}_C} = \sum_{k=0,1,2} \left(x_1 y_1 + \omega^{2k} x_2 y_2 + \omega^k x_3 y_3 \right)_{\mathbf{1}_{k,2}}$$

$$\oplus \begin{pmatrix} x_2 y_3 \\ x_3 y_1 \\ x_1 y_2 \end{pmatrix}_{\mathbf{3}_D} \oplus \begin{pmatrix} x_3 y_2 \\ x_1 y_3 \\ x_2 y_1 \end{pmatrix}_{\mathbf{3}_D}, \quad (12.29)$$

$$\begin{pmatrix} x_1 \\ x_2 \\ x_3 \end{pmatrix}_{\mathbf{3}_A} \otimes \begin{pmatrix} y_1 \\ y_2 \\ y_3 \end{pmatrix}_{\mathbf{3}_D} = \begin{pmatrix} x_3 y_3 \\ x_1 y_1 \\ x_2 y_2 \end{pmatrix}_{\bar{\mathbf{3}}_A} \oplus \begin{pmatrix} x_3 y_2 \\ x_1 y_3 \\ x_2 y_1 \end{pmatrix}_{\bar{\mathbf{3}}_B} \oplus \begin{pmatrix} x_3 y_1 \\ x_1 y_2 \\ x_2 y_3 \end{pmatrix}_{\bar{\mathbf{3}}_C}, \quad (12.30)$$

12.2 $\Sigma(81)$

$$\begin{pmatrix} x_1 \\ x_2 \\ x_3 \end{pmatrix}_{\mathbf{3}_A} \otimes \begin{pmatrix} y_1 \\ y_2 \\ y_3 \end{pmatrix}_{\bar{\mathbf{3}}_D} = \begin{pmatrix} x_2 y_1 \\ x_3 y_2 \\ x_1 y_3 \end{pmatrix}_{\mathbf{3}_A} \oplus \begin{pmatrix} x_2 y_2 \\ x_3 y_3 \\ x_1 y_1 \end{pmatrix}_{\mathbf{3}_B} \oplus \begin{pmatrix} x_2 y_3 \\ x_3 y_1 \\ x_1 y_2 \end{pmatrix}_{\mathbf{3}_C}, \quad (12.31)$$

$$\begin{pmatrix} x_1 \\ x_2 \\ x_3 \end{pmatrix}_{\bar{\mathbf{3}}_A} \otimes \begin{pmatrix} y_1 \\ y_2 \\ y_3 \end{pmatrix}_{\bar{\mathbf{3}}_B} = \begin{pmatrix} x_1 y_1 \\ x_2 y_2 \\ x_3 y_3 \end{pmatrix}_{\mathbf{3}_C} \oplus \begin{pmatrix} x_3 y_2 \\ x_1 y_3 \\ x_2 y_1 \end{pmatrix}_{\mathbf{3}_A} \oplus \begin{pmatrix} x_2 y_3 \\ x_3 y_1 \\ x_1 y_2 \end{pmatrix}_{\mathbf{3}_A}, \quad (12.32)$$

$$\begin{pmatrix} x_1 \\ x_2 \\ x_3 \end{pmatrix}_{\bar{\mathbf{3}}_A} \otimes \begin{pmatrix} y_1 \\ y_2 \\ y_3 \end{pmatrix}_{\bar{\mathbf{3}}_C} = \begin{pmatrix} x_1 y_1 \\ x_2 y_2 \\ x_3 y_3 \end{pmatrix}_{\mathbf{3}_B} \oplus \begin{pmatrix} x_3 y_2 \\ x_1 y_3 \\ x_2 y_1 \end{pmatrix}_{\mathbf{3}_C} \oplus \begin{pmatrix} x_2 y_3 \\ x_3 y_1 \\ x_1 y_2 \end{pmatrix}_{\mathbf{3}_C}, \quad (12.33)$$

$$\begin{pmatrix} x_1 \\ x_2 \\ x_3 \end{pmatrix}_{\mathbf{3}_B} \otimes \begin{pmatrix} y_1 \\ y_2 \\ y_3 \end{pmatrix}_{\mathbf{3}_B} = \begin{pmatrix} x_1 y_1 \\ x_2 y_2 \\ x_3 y_3 \end{pmatrix}_{\bar{\mathbf{3}}_B} \oplus \begin{pmatrix} x_3 y_2 \\ x_1 y_3 \\ x_2 y_1 \end{pmatrix}_{\bar{\mathbf{3}}_C} \oplus \begin{pmatrix} x_2 y_3 \\ x_3 y_1 \\ x_1 y_2 \end{pmatrix}_{\bar{\mathbf{3}}_C}, \quad (12.34)$$

$$\begin{pmatrix} x_1 \\ x_2 \\ x_3 \end{pmatrix}_{\mathbf{3}_B} \otimes \begin{pmatrix} y_1 \\ y_2 \\ y_3 \end{pmatrix}_{\mathbf{3}_C} = \begin{pmatrix} x_1 y_1 \\ x_2 y_2 \\ x_3 y_3 \end{pmatrix}_{\bar{\mathbf{3}}_A} \oplus \begin{pmatrix} x_3 y_2 \\ x_1 y_3 \\ x_2 y_1 \end{pmatrix}_{\bar{\mathbf{3}}_B} \oplus \begin{pmatrix} x_2 y_3 \\ x_3 y_1 \\ x_1 y_2 \end{pmatrix}_{\bar{\mathbf{3}}_B}, \quad (12.35)$$

$$\begin{pmatrix} x_1 \\ x_2 \\ x_3 \end{pmatrix}_{\mathbf{3}_B} \otimes \begin{pmatrix} y_1 \\ y_2 \\ y_3 \end{pmatrix}_{\mathbf{3}_D} = \begin{pmatrix} x_3 y_3 \\ x_1 y_1 \\ x_2 y_2 \end{pmatrix}_{\mathbf{3}_B} \oplus \begin{pmatrix} x_3 y_2 \\ x_1 y_3 \\ x_2 y_1 \end{pmatrix}_{\mathbf{3}_C} \oplus \begin{pmatrix} x_3 y_1 \\ x_1 y_2 \\ x_2 y_3 \end{pmatrix}_{\mathbf{3}_A}, \quad (12.36)$$

$$\begin{pmatrix} x_1 \\ x_2 \\ x_3 \end{pmatrix}_{\bar{\mathbf{3}}_B} \otimes \begin{pmatrix} y_1 \\ y_2 \\ y_3 \end{pmatrix}_{\mathbf{3}_B} = \sum_{k=0,1,2} \left(x_1 y_1 + \omega^{2k} x_2 y_2 + \omega^k x_3 y_3 \right)_{\mathbf{1}_{k,0}}$$

$$\oplus \begin{pmatrix} x_1 y_2 \\ x_2 y_3 \\ x_3 y_1 \end{pmatrix}_{\mathbf{3}_D} \oplus \begin{pmatrix} x_2 y_1 \\ x_3 y_2 \\ x_1 y_3 \end{pmatrix}_{\bar{\mathbf{3}}_D}, \quad (12.37)$$

$$\begin{pmatrix} x_1 \\ x_2 \\ x_3 \end{pmatrix}_{\bar{\mathbf{3}}_B} \otimes \begin{pmatrix} y_1 \\ y_2 \\ y_3 \end{pmatrix}_{\mathbf{3}_C} = \sum_{k=0,1,2} \left(x_1 y_1 + \omega^{2k} x_2 y_2 + \omega^k x_3 y_3 \right)_{\mathbf{1}_{k,2}}$$

$$\oplus \begin{pmatrix} x_2 y_3 \\ x_3 y_1 \\ x_1 y_2 \end{pmatrix}_{\mathbf{3}_D} \oplus \begin{pmatrix} x_1 y_3 \\ x_2 y_1 \\ x_3 y_2 \end{pmatrix}_{\bar{\mathbf{3}}_D}, \quad (12.38)$$

$$\begin{pmatrix} x_1 \\ x_2 \\ x_3 \end{pmatrix}_{\bar{\mathbf{3}}_B} \otimes \begin{pmatrix} y_1 \\ y_2 \\ y_3 \end{pmatrix}_{\mathbf{3}_D} = \begin{pmatrix} x_2 y_2 \\ x_3 y_3 \\ x_1 y_1 \end{pmatrix}_{\bar{\mathbf{3}}_C} \oplus \begin{pmatrix} x_2 y_1 \\ x_3 y_2 \\ x_1 y_3 \end{pmatrix}_{\bar{\mathbf{3}}_B} \oplus \begin{pmatrix} x_2 y_3 \\ x_3 y_1 \\ x_1 y_2 \end{pmatrix}_{\bar{\mathbf{3}}_A}, \quad (12.39)$$

$$\begin{pmatrix} x_1 \\ x_2 \\ x_3 \end{pmatrix}_{\mathbf{3}_C} \otimes \begin{pmatrix} y_1 \\ y_2 \\ y_3 \end{pmatrix}_{\mathbf{3}_C} = \begin{pmatrix} x_1 y_1 \\ x_2 y_2 \\ x_3 y_3 \end{pmatrix}_{\bar{\mathbf{3}}_C} \oplus \begin{pmatrix} x_3 y_2 \\ x_1 y_3 \\ x_2 y_1 \end{pmatrix}_{\bar{\mathbf{3}}_A} \oplus \begin{pmatrix} x_2 y_3 \\ x_3 y_1 \\ x_1 y_2 \end{pmatrix}_{\bar{\mathbf{3}}_A}, \quad (12.40)$$

$$\begin{pmatrix} x_1 \\ x_2 \\ x_3 \end{pmatrix}_{3_C} \otimes \begin{pmatrix} y_1 \\ y_2 \\ y_3 \end{pmatrix}_{3_C} = \sum_{k=0,1,2} \left(x_1 y_1 + \omega^{2k} x_2 y_2 + \omega^k x_3 y_3 \right)_{1_{k,0}}$$

$$\oplus \begin{pmatrix} x_1 y_2 \\ x_2 y_3 \\ x_3 y_1 \end{pmatrix}_{3_D} \oplus \begin{pmatrix} x_2 y_1 \\ x_3 y_2 \\ x_1 y_3 \end{pmatrix}_{\bar{3}_D} , \quad (12.41)$$

$$\begin{pmatrix} x_1 \\ x_2 \\ x_3 \end{pmatrix}_{3_C} \otimes \begin{pmatrix} y_1 \\ y_2 \\ y_3 \end{pmatrix}_{3_D} = \begin{pmatrix} x_2 y_2 \\ x_3 y_3 \\ x_1 y_1 \end{pmatrix}_{\bar{3}_A} \oplus \begin{pmatrix} x_2 y_1 \\ x_3 y_2 \\ x_1 y_3 \end{pmatrix}_{\bar{3}_C} \oplus \begin{pmatrix} x_2 y_3 \\ x_3 y_1 \\ x_1 y_2 \end{pmatrix}_{\bar{3}_B} , \quad (12.42)$$

$$\begin{pmatrix} x_1 \\ x_2 \\ x_3 \end{pmatrix}_{3_D} \otimes \begin{pmatrix} y_1 \\ y_2 \\ y_3 \end{pmatrix}_{3_D} = \begin{pmatrix} x_1 y_1 \\ x_2 y_2 \\ x_3 y_3 \end{pmatrix}_{\bar{3}_D} \oplus \begin{pmatrix} x_2 y_3 \\ x_3 y_1 \\ x_1 y_2 \end{pmatrix}_{\bar{3}_D} \oplus \begin{pmatrix} x_3 y_2 \\ x_1 y_3 \\ x_2 y_1 \end{pmatrix}_{\bar{3}_D} , \quad (12.43)$$

$$\begin{pmatrix} x_1 \\ x_2 \\ x_3 \end{pmatrix}_{3_D} \otimes \begin{pmatrix} y_1 \\ y_2 \\ y_3 \end{pmatrix}_{\bar{3}_D} = \sum_{k=0,1,2} \left[\left(x_1 y_1 + \omega^{2k} x_2 y_2 + \omega^k x_3 y_3 \right)_{1_{k,0}} \right.$$

$$\oplus \left(x_2 y_3 + \omega^{2k} x_3 y_1 + \omega^k x_1 y_2 \right)_{1_{k,2}}$$

$$\left. \oplus \left(x_3 y_2 + \omega^{2k} x_1 y_3 + \omega^k x_2 y_1 \right)_{1_{k,1}} \right].$$

$$(12.44)$$

The tensor products between singlets are

$$\mathbf{1}_{k,\ell} \otimes \mathbf{1}_{k',\ell'} = \mathbf{1}_{k+k' \ (\mathrm{mod}\, 3),\, \ell+\ell' \ (\mathrm{mod}\, 3)}. \quad (12.45)$$

The tensor products between singlets and triplets are:

$$(x)_{1_{0,0}} \otimes \begin{pmatrix} y_1 \\ y_2 \\ y_3 \end{pmatrix}_{3(\bar{3})_A} = \begin{pmatrix} x y_1 \\ x y_2 \\ x y_3 \end{pmatrix}_{3(\bar{3})_A},$$

$$(x)_{1_{1,0}} \otimes \begin{pmatrix} y_1 \\ y_2 \\ y_3 \end{pmatrix}_{3(\bar{3})_A} = \begin{pmatrix} x y_1 \\ \omega x y_2 \\ \omega^2 x y_3 \end{pmatrix}_{3(\bar{3})_A}, \quad (12.46)$$

$$(x)_{1_{2,0}} \otimes \begin{pmatrix} y_1 \\ y_2 \\ y_3 \end{pmatrix}_{3(\bar{3})_A} = \begin{pmatrix} x y_1 \\ \omega^2 x y_2 \\ \omega x y_3 \end{pmatrix}_{3(\bar{3})_A},$$

12.2 $\Sigma(81)$

$$(x)\mathbf{1}_{0,1} \otimes \begin{pmatrix} y_1 \\ y_2 \\ y_3 \end{pmatrix}_{\mathbf{3}(\bar{\mathbf{3}})_A} = \begin{pmatrix} xy_1 \\ xy_2 \\ xy_3 \end{pmatrix}_{\mathbf{3}_C,(\bar{\mathbf{3}}_B)},$$

$$(x)\mathbf{1}_{1,1} \otimes \begin{pmatrix} y_1 \\ y_2 \\ y_3 \end{pmatrix}_{\mathbf{3}(\bar{\mathbf{3}})_A} = \begin{pmatrix} xy_1 \\ \omega xy_2 \\ \omega^2 xy_3 \end{pmatrix}_{\mathbf{3}_C,(\bar{\mathbf{3}}_B)}, \qquad (12.47)$$

$$(x)\mathbf{1}_{2,1} \otimes \begin{pmatrix} y_1 \\ y_2 \\ y_3 \end{pmatrix}_{\mathbf{3}(\bar{\mathbf{3}})_A} = \begin{pmatrix} xy_1 \\ \omega^2 xy_2 \\ \omega xy_3 \end{pmatrix}_{\mathbf{3}_C,(\bar{\mathbf{3}}_B)},$$

$$(x)\mathbf{1}_{0,2} \otimes \begin{pmatrix} y_1 \\ y_2 \\ y_3 \end{pmatrix}_{\mathbf{3}(\bar{\mathbf{3}})_A} = \begin{pmatrix} xy_1 \\ xy_2 \\ xy_3 \end{pmatrix}_{\mathbf{3}_B,(\bar{\mathbf{3}}_C)},$$

$$(x)\mathbf{1}_{1,2} \otimes \begin{pmatrix} y_1 \\ y_2 \\ y_3 \end{pmatrix}_{\mathbf{3}(\bar{\mathbf{3}})_A} = \begin{pmatrix} xy_1 \\ \omega xy_2 \\ \omega^2 xy_3 \end{pmatrix}_{\mathbf{3}_B,(\bar{\mathbf{3}}_C)}, \qquad (12.48)$$

$$(x)\mathbf{1}_{2,2} \otimes \begin{pmatrix} y_1 \\ y_2 \\ y_3 \end{pmatrix}_{\mathbf{3}(\bar{\mathbf{3}})_A} = \begin{pmatrix} xy_1 \\ \omega^2 xy_2 \\ \omega xy_3 \end{pmatrix}_{\mathbf{3}_B,(\bar{\mathbf{3}}_C)},$$

$$(x)\mathbf{1}_{0,0} \otimes \begin{pmatrix} y_1 \\ y_2 \\ y_3 \end{pmatrix}_{\mathbf{3}(\bar{\mathbf{3}})_B} = \begin{pmatrix} xy_1 \\ xy_2 \\ xy_3 \end{pmatrix}_{\mathbf{3}(\bar{\mathbf{3}})_B},$$

$$(x)\mathbf{1}_{1,0} \otimes \begin{pmatrix} y_1 \\ y_2 \\ y_3 \end{pmatrix}_{\mathbf{3}(\bar{\mathbf{3}})_B} = \begin{pmatrix} xy_1 \\ \omega xy_2 \\ \omega^2 xy_3 \end{pmatrix}_{\mathbf{3}(\bar{\mathbf{3}})_B}, \qquad (12.49)$$

$$(x)\mathbf{1}_{2,0} \otimes \begin{pmatrix} y_1 \\ y_2 \\ y_3 \end{pmatrix}_{\mathbf{3}(\bar{\mathbf{3}})_B} = \begin{pmatrix} xy_1 \\ \omega^2 xy_2 \\ \omega xy_3 \end{pmatrix}_{\mathbf{3}(\bar{\mathbf{3}})_B},$$

$$(x)\mathbf{1}_{0,1} \otimes \begin{pmatrix} y_1 \\ y_2 \\ y_3 \end{pmatrix}_{\mathbf{3}(\bar{\mathbf{3}})_B} = \begin{pmatrix} xy_1 \\ xy_2 \\ xy_3 \end{pmatrix}_{\mathbf{3}_A,(\bar{\mathbf{3}}_C)},$$

$$(x)\mathbf{1}_{1,1} \otimes \begin{pmatrix} y_1 \\ y_2 \\ y_3 \end{pmatrix}_{\mathbf{3}(\bar{\mathbf{3}})_B} = \begin{pmatrix} xy_1 \\ \omega xy_2 \\ \omega^2 xy_3 \end{pmatrix}_{\mathbf{3}_A,(\bar{\mathbf{3}}_C)}, \qquad (12.50)$$

$$(x)\mathbf{1}_{2,1} \otimes \begin{pmatrix} y_1 \\ y_2 \\ y_3 \end{pmatrix}_{\mathbf{3}(\bar{\mathbf{3}})_B} = \begin{pmatrix} xy_1 \\ \omega^2 xy_2 \\ \omega xy_3 \end{pmatrix}_{\mathbf{3}_A,(\bar{\mathbf{3}}_C)},$$

$$(x)\mathbf{1}_{0,2} \otimes \begin{pmatrix} y_1 \\ y_2 \\ y_3 \end{pmatrix}_{\mathbf{3}(\bar{\mathbf{3}})_B} = \begin{pmatrix} xy_1 \\ xy_2 \\ xy_3 \end{pmatrix}_{\mathbf{3}_C,(\bar{\mathbf{3}}_A)},$$

$$(x)\mathbf{1}_{1,2} \otimes \begin{pmatrix} y_1 \\ y_2 \\ y_3 \end{pmatrix}_{\mathbf{3}(\bar{\mathbf{3}})_B} = \begin{pmatrix} xy_1 \\ \omega xy_2 \\ \omega^2 xy_3 \end{pmatrix}_{\mathbf{3}_C,(\bar{\mathbf{3}}_A)}, \qquad (12.51)$$

$$(x)\mathbf{1}_{2,2} \otimes \begin{pmatrix} y_1 \\ y_2 \\ y_3 \end{pmatrix}_{\mathbf{3}(\bar{\mathbf{3}})_B} = \begin{pmatrix} xy_1 \\ \omega^2 xy_2 \\ \omega xy_3 \end{pmatrix}_{\mathbf{3}_C,(\bar{\mathbf{3}}_A)},$$

$$(x)\mathbf{1}_{0,0} \otimes \begin{pmatrix} y_1 \\ y_2 \\ y_3 \end{pmatrix}_{\mathbf{3}(\bar{\mathbf{3}})_C} = \begin{pmatrix} xy_1 \\ xy_2 \\ xy_3 \end{pmatrix}_{\mathbf{3}(\bar{\mathbf{3}})_C},$$

$$(x)\mathbf{1}_{1,0} \otimes \begin{pmatrix} y_1 \\ y_2 \\ y_3 \end{pmatrix}_{\mathbf{3}(\bar{\mathbf{3}})_C} = \begin{pmatrix} xy_1 \\ \omega xy_2 \\ \omega^2 xy_3 \end{pmatrix}_{\mathbf{3}(\bar{\mathbf{3}})_C}, \qquad (12.52)$$

$$(x)\mathbf{1}_{2,0} \otimes \begin{pmatrix} y_1 \\ y_2 \\ y_3 \end{pmatrix}_{\mathbf{3}(\bar{\mathbf{3}})_C} = \begin{pmatrix} xy_1 \\ \omega^2 xy_2 \\ \omega xy_3 \end{pmatrix}_{\mathbf{3}(\bar{\mathbf{3}})_C},$$

$$(x)\mathbf{1}_{0,1} \otimes \begin{pmatrix} y_1 \\ y_2 \\ y_3 \end{pmatrix}_{\mathbf{3}(\bar{\mathbf{3}})_C} = \begin{pmatrix} xy_1 \\ xy_2 \\ xy_3 \end{pmatrix}_{\mathbf{3}_B,(\bar{\mathbf{3}}_A)},$$

$$(x)\mathbf{1}_{1,1} \otimes \begin{pmatrix} y_1 \\ y_2 \\ y_3 \end{pmatrix}_{\mathbf{3}(\bar{\mathbf{3}})_C} = \begin{pmatrix} xy_1 \\ \omega xy_2 \\ \omega^2 xy_3 \end{pmatrix}_{\mathbf{3}_B,(\bar{\mathbf{3}}_A)}, \qquad (12.53)$$

$$(x)\mathbf{1}_{2,1} \otimes \begin{pmatrix} y_1 \\ y_2 \\ y_3 \end{pmatrix}_{\mathbf{3}(\bar{\mathbf{3}})_C} = \begin{pmatrix} xy_1 \\ \omega^2 xy_2 \\ \omega xy_3 \end{pmatrix}_{\mathbf{3}_B,(\bar{\mathbf{3}}_A)},$$

$$(x)\mathbf{1}_{0,2} \otimes \begin{pmatrix} y_1 \\ y_2 \\ y_3 \end{pmatrix}_{\mathbf{3}(\bar{\mathbf{3}})_C} = \begin{pmatrix} xy_1 \\ xy_2 \\ xy_3 \end{pmatrix}_{\mathbf{3}_A,(\bar{\mathbf{3}}_B)},$$

$$(x)\mathbf{1}_{1,2} \otimes \begin{pmatrix} y_1 \\ y_2 \\ y_3 \end{pmatrix}_{\mathbf{3}(\bar{\mathbf{3}})_C} = \begin{pmatrix} xy_1 \\ \omega xy_2 \\ \omega^2 xy_3 \end{pmatrix}_{\mathbf{3}_A,(\bar{\mathbf{3}}_B)}, \qquad (12.54)$$

$$(x)\mathbf{1}_{2,2} \otimes \begin{pmatrix} y_1 \\ y_2 \\ y_3 \end{pmatrix}_{\mathbf{3}(\bar{\mathbf{3}})_C} = \begin{pmatrix} xy_1 \\ \omega^2 xy_2 \\ \omega xy_3 \end{pmatrix}_{\mathbf{3}_A,(\bar{\mathbf{3}}_B)},$$

$$(x)_{\mathbf{1}_{k,\ell}} \otimes \begin{pmatrix} y_1 \\ y_2 \\ y_3 \end{pmatrix}_{\mathbf{3}(\bar{\mathbf{3}})_D} = \begin{pmatrix} xy_1 \\ xy_2 \\ xy_3 \end{pmatrix}_{\mathbf{3}(\bar{\mathbf{3}})_D}, \tag{12.55}$$

where $k, \ell = 0, 1, 2$.

References

1. Ishimori, H., Kobayashi, T.: arXiv:1201.3429 [hep-ph]

Chapter 13
$\Delta(6N^2)$

In this chapter, we investigate the discrete group $\Delta(6N^2)$, which is isomorphic to $(Z_N \times Z'_N) \rtimes S_3$ (see also [1]). Let us denote the generators of Z_N and Z'_N by a and a', respectively. We denote the S_3 generators by b and c, where b and c are the Z_3 and Z_2 generators of S_3, respectively. These satisfy

$$a^N = a'^N = b^3 = c^2 = (bc)^2 = e, \qquad aa' = a'a,$$
$$bab^{-1} = a^{-1}(a')^{-1}, \qquad ba'b^{-1} = a, \qquad (13.1)$$
$$cac^{-1} = (a')^{-1}, \qquad ca'c^{-1} = a^{-1}.$$

Using these, all elements of $\Delta(6N^2)$ can be expressed in the form

$$g = b^k c^\ell a^m a'^n, \qquad (13.2)$$

for $k = 0, 1, 2$, $\ell = 0, 1$, and $m, n = 0, 1, 2, \ldots, N - 1$.

Note that the group $\Delta(6N^2)$ includes as a subgroup $\Delta(3N^2)$, whose elements can be written $b^k a^m a'^n$. Thus, some group-theoretical aspects of $\Delta(6N^2)$ can be derived from those of $\Delta(3N^2)$.

13.1 $\Delta(6N^2)$ with $N/3 \neq$ Integer

13.1.1 Conjugacy Classes

With a view to identifying the conjugacy classes, we note that

$$aba^{-1} = ba^{-1}a', \qquad a'ba'^{-1} = ba^{-1}a'^{-2},$$
$$aca^{-1} = ca^{-1}a'^{-1}, \qquad a'ca'^{-1} = ca^{-1}a'^{-1}, \qquad (13.3)$$
$$cbc^{-1} = b^2, \qquad bcb^{-1} = b^2 c.$$

Using these relations, one can obtain the conjugacy classes of $\Delta(6N^2)$. Indeed, these relations are nothing but those in $\Delta(3N^2)$, except for the relations involving c. Hence, the conjugacy classes of $\Delta(3N^2)$ are useful to obtain those of $\Delta(6N^2)$.

First, we consider the elements $a^\ell a'^m$. As shown in Chap. 10, the element $a^\ell a'^m$ is conjugate to $a^{-\ell+m}a'^{-\ell}$ and $a^{-m}a'^{\ell-m}$ for any group $\Delta(3N^2)$ with $N/3 \neq$ integer. These elements must therefore be conjugate to each other in $\Delta(6N^2)$, too. In addition, it is found that

$$ca^\ell a'^m c^{-1} = a^{-m}a'^{-\ell}, \qquad ca^{-\ell+m}a'^{-\ell}c^{-1} = a^\ell a'^{\ell-m},$$
$$ca^{-m}a'^{\ell-m}c^{-1} = a^{-\ell+m}a'^m. \tag{13.4}$$

Thus, the following elements,

$$a^\ell a'^m, \qquad a^{-\ell+m}a'^{-\ell}, \qquad a^{-m}a'^{\ell-m},$$
$$a^{-m}a'^{-\ell}, \qquad a^\ell a'^{\ell-m}, \qquad a^{-\ell+m}a'^m, \tag{13.5}$$

are conjugate to each other in $\Delta(6N^2)$. However, the elements, $a^{-m}a'^{-\ell}$, $a^\ell a'^{\ell-m}$, and $a^{-\ell+m}a'^m$ are the same as $a^\ell a'^m$ when ℓ and m satisfy the conditions

$$\ell + m = 0 \mod(N), \qquad 2\ell - m = 0 \mod(N), \qquad \ell - 2m = 0 \mod(N), \tag{13.6}$$

respectively. Under these conditions, the above conjugate elements in (13.5) reduce to the three elements $a^\ell a'^{-\ell}$, $a^{-2\ell}a'^{-\ell}$, and $a^\ell a'^{2\ell}$.

As a result, the elements $a^\ell a'^m$ are classified into the following conjugacy classes:

$$C_3^{(k)} = \{a^k a'^{-k}, a^{-2k}a'^{-k}, a^k a'^{2k}\}, \quad k = 1, 2, \ldots, N-1, \tag{13.7}$$

$$C_6^{(\ell,m)} = \{a^\ell a'^m, a^{m-\ell}a'^{-\ell}, a^{-m}a'^{\ell-m}, a^{-m}a'^{-\ell}, a^{m-\ell}a'^m, a^\ell a'^{\ell-m}\}, \tag{13.8}$$

for $N/3 \neq$ integer. The parameters ℓ and m of $C_6^{(\ell,m)}$ run from 0 to $N-1$, but they do not satisfy the conditions (13.6). The numbers of each of the conjugacy classes $C_3^{(k)}$ and $C_6^{(\ell,m)}$ are $(N-1)$ and $(N^2 - 3N + 2)/6$, respectively.

Similarly, we can obtain conjugacy classes containing $ba^\ell a'^m$ and $b^2 a^\ell a'^m$. As shown in Chap. 10, all the elements $ba^\ell a'^m$ with $\ell, m = 0, \ldots, N-1$, belong to the same conjugacy classes in the group $\Delta(3N^2)$. In addition, all the elements $b^2 a^\ell a'^m$ with $\ell, m = 0, \ldots, N-1$, also belong to the same conjugacy classes in $\Delta(3N^2)$. Furthermore, we obtain

$$cba^\ell a'^m c^{-1} = b^2 a^{-m}a'^{-\ell}. \tag{13.9}$$

Thus, all the elements $b^k a^\ell a'^m$, for $k = 1, 2$, and $\ell, m = 0, \ldots, N-1$, belong to the same conjugacy class in $\Delta(6N^2)$.

Finally, we consider the conjugate elements involving c, which are not included in $\Delta(3N^2)$. Here we find that

$$(a^p a'^q)ca^\ell a'^m (a^p a'^q)^{-1} = ca^{\ell-p-q}a'^{m-p-q} = ca^{k+n}a'^n, \tag{13.10}$$

13.1 $\Delta(6N^2)$ with $N/3 \neq$ Integer

$$(ba^p a'^q) ca^\ell a'^m (ba^p a'^q)^{-1} = b^2 ca^{-\ell+m} a'^{-\ell+p+q} = b^2 ca^{-k} a'^{-k-n}, \quad (13.11)$$

$$(b^2 a^p a'^q) ca^\ell a'^m (b^2 a^p a'^q)^{-1} = bca^{-m+p+q} a'^{\ell-m} = bca^{-n} a'^k, \quad (13.12)$$

where $k = \ell - m$ and $n = m - p - q$. As a result, all the elements

$$ca^{k+n} a'^m, \quad b^2 ca^{-k} a'^{-k-n}, \quad bca^{-n} a'^k, \quad (13.13)$$

with $n = 0, 1, \ldots, N-1$, belong to the same conjugacy class.

Here, we summarize the conjugacy classes of $\Delta(6N^2)$. For $N/3 \neq$ integer, $\Delta(6N^2)$ has the following conjugacy classes:

C_1: $\{e\}$,

$C_3^{(k)}$: $\{a^k a'^{-k}, a^{-2k} a'^{-k}, a^k a'^{2k}\}, \quad k = 1, 2, \ldots, N-1$,

$C_6^{(\ell,m)}$: $\{a^\ell a'^m, a^{m-\ell} a'^{-\ell}, a^{-m} a'^{\ell-m}, a^{-m} a'^{-\ell}, a^{m-\ell} a'^m, a^\ell a'^{\ell-m}\}$,
$\ell, m = 0, 1, \ldots, N-1$, $\quad (13.14)$

C_{2N^2}: $\{b^k a^\ell a'^m \mid k = 1, 2, \ \ell, m = 0, 1, \ldots, N-1\}$,

$C_{3N}^{(\ell)}$: $\{ca^{\ell+n} a'^m, b^2 ca^{-\ell} a'^{-\ell-n}, bca^{-n} a'^\ell \mid n = 0, 1, \ldots, N-1\}$,
$\ell = 0, 1, \ldots, N-1$,

where ℓ and m in $C_6^{(\ell,m)}$ do not satisfy the conditions (13.6). The order h of each element in the given conjugacy class is:

$$\begin{aligned} C_1: & \quad h = 1, \\ C_3^{(k)}: & \quad h = N/\gcd(N, k), \\ C_6^{(\ell,m)}: & \quad h = N/\gcd(N, \ell, m), \\ C_{2N^2}: & \quad h = 3, \\ C_{3N}^{(\ell)}: & \quad h = 2N/\gcd(N, \ell). \end{aligned} \quad (13.15)$$

Since the numbers of each of the conjugacy classes $C_3^{(k)}$, $C_6^{(\ell,m)}$, and $C_{3N}^{(\ell)}$ are $N-1$, $(N^2 - 3N + 2)/6$, and N, respectively, the total number of conjugacy classes is equal to $(N^2 + 9N + 8)/6$. The relations (2.18) and (2.19) for $\Delta(6n^2)$ with $N/3 \neq$ integer give

$$m_1 + 2^2 m_2 + 3^2 m_3 + \cdots = 6N^2, \quad (13.16)$$

$$m_1 + m_2 + m_3 + \cdots = \frac{N^2 + 9N + 8}{6}. \quad (13.17)$$

These have the solution $(m_1, m_2, m_3, m_6) = (2, 1, 2(N-1), (N^2 - 3N + 2)/6)$.

13.1.2 Characters and Representations

The group $\Delta(6N^2)$ with $N/3 \neq$ integer has two singlets and one doublet. These are nothing but the irreducible representations of S_3. Thus, on these representations, the Z_N and Z'_N generators a and a' are identity matrices. Since $c^2 = e$, the characters of the two singlets have two possible values $\chi_{1_k}(c) = (-1)^k$ with $k = 0, 1$, and they correspond to two singlets $\mathbf{1}_k$. Note that $\chi_{1_k}(a) = \chi_{1_k}(a') = \chi_{1_k}(b) = 1$. Similarly to S_3, the characters of the doublet $\mathbf{2}$ are obtained as

$$\chi_2(C_{2N^2}) = -1, \qquad \chi_2\big(C_{3N}^{(k)}\big) = 0, \tag{13.18}$$

together with $\chi_2(a) = \chi_2(a') = 2$. The two-dimensional representations are, e.g.,

$$b(\mathbf{2}) = \begin{pmatrix} \omega & 0 \\ 0 & \omega^2 \end{pmatrix}, \quad c(\mathbf{2}) = \begin{pmatrix} 0 & 1 \\ 1 & 0 \end{pmatrix}, \quad a(\mathbf{2}) = a'(\mathbf{2}) = \begin{pmatrix} 1 & 0 \\ 0 & 1 \end{pmatrix}. \tag{13.19}$$

Next, we consider the sextet representations. We can obtain (6×6) matrix representations for the generic sextet. The group $\Delta(6N^2)$ is represented as follows:

$$b = \begin{pmatrix} b_1 & 0 \\ 0 & b_2 \end{pmatrix}, \quad c = \begin{pmatrix} 0 & 1 \\ 1 & 0 \end{pmatrix}, \quad a = \begin{pmatrix} a_1 & 0 \\ 0 & a_2 \end{pmatrix}, \quad a' = \begin{pmatrix} a'_1 & 0 \\ 0 & a'_2 \end{pmatrix}, \tag{13.20}$$

where

$$b_1 = \begin{pmatrix} 0 & 1 & 0 \\ 0 & 0 & 1 \\ 1 & 0 & 0 \end{pmatrix}, \quad b_2 = \begin{pmatrix} 0 & 0 & 1 \\ 1 & 0 & 0 \\ 0 & 1 & 0 \end{pmatrix}, \tag{13.21}$$

$$a_1 = a'_2{}^{-1} = \begin{pmatrix} \rho^l & 0 & 0 \\ 0 & \rho^k & 0 \\ 0 & 0 & \rho^{-l-k} \end{pmatrix},$$

$$a_2 = a'_1{}^{-1} = \begin{pmatrix} \rho^{l+k} & 0 & 0 \\ 0 & \rho^{-l} & 0 \\ 0 & 0 & \rho^{-k} \end{pmatrix}, \tag{13.22}$$

on the sextet $\mathbf{6}_{[[k],[\ell]]}$ with $(k, \ell) \neq (0, 0)$, where $[[k], [\ell]]$ is defined by[1]

$$[[k], [\ell]] = (k, \ell), (-k - \ell, k), (\ell, -k - \ell), (-\ell, -k), (k + \ell, -\ell), \text{ or}(-k, k + \ell). \tag{13.23}$$

---

[1] The notation $[[k], [\ell]]$ corresponds to $\widetilde{(k, \ell)}$ in [1].

13.1 $\Delta(6N^2)$ with $N/3 \neq$ Integer

We also denote the vector $\mathbf{6}_{[[k],[\ell]]}$ by

$$\mathbf{6}_{[[k],[\ell]]} = \begin{pmatrix} x_{\ell,-k-\ell} \\ x_{k,\ell} \\ x_{-k-\ell,k} \\ x_{k+\ell,-\ell} \\ x_{-\ell,-k} \\ x_{-k,k+\ell} \end{pmatrix}, \qquad (13.24)$$

for $k, \ell = 0, 1, \ldots, N-1$, where k and ℓ correspond to the Z_N and Z'_N charges, respectively.

In certain cases, the above representation becomes reducible. For $N/3 \neq$ integer, $\mathbf{6}_{[[k],[\ell]]}$ is reducible if $k + \ell = 0 \pmod{N}$, $k = 0, \ell \neq 0$, and $\ell = 0, k \neq 0$, and then the number of irreducible sextet representations is $[N^2 - N - 2(N-1)]/6$.

On the other hand, we can diagonalize the above reducible 6D representations so as to obtain $2(N-1)$ irreducible triplets $\mathbf{3}_{1k}$ and $\mathbf{3}_{2k}$, with $k = 1, \ldots, N-1$, where the generators are represented by

$$\begin{aligned} b(\mathbf{3}_{1k}) &= \begin{pmatrix} 0 & 1 & 0 \\ 0 & 0 & 1 \\ 1 & 0 & 0 \end{pmatrix}, & c(\mathbf{3}_{1k}) &= \begin{pmatrix} 0 & 0 & 1 \\ 0 & 1 & 0 \\ 1 & 0 & 0 \end{pmatrix}, \\ a(\mathbf{3}_{1k}) &= \begin{pmatrix} \rho^k & 0 & 0 \\ 0 & \rho^{-k} & 0 \\ 0 & 0 & 1 \end{pmatrix}, & a'(\mathbf{3}_{1k}) &= \begin{pmatrix} 1 & 0 & 0 \\ 0 & \rho^k & 0 \\ 0 & 0 & \rho^{-k} \end{pmatrix}, \end{aligned} \qquad (13.25)$$

$$\begin{aligned} b(\mathbf{3}_{2k}) &= b(\mathbf{3}_{1k}), & c(\mathbf{3}_{2k}) &= -c(\mathbf{3}_{1k}), \\ a(\mathbf{3}_{2k}) &= a(\mathbf{3}_{1k}), & a'(\mathbf{3}_{2k}) &= a'(\mathbf{3}_{1k}). \end{aligned} \qquad (13.26)$$

We also denote the vectors $\mathbf{3}_{1k}$ and $\mathbf{3}_{2k}$ by

$$\mathbf{3}_{1k} = \begin{pmatrix} x_{k,0} \\ x_{-k,k} \\ x_{0,-k} \end{pmatrix}, \qquad \mathbf{3}_{2k} = \begin{pmatrix} x_{k,0} \\ x_{-k,k} \\ x_{0,-k} \end{pmatrix}, \qquad (13.27)$$

for $k = 1, \ldots, N-1$.

The characters are shown in Table 13.1.

Table 13.1 Characters of $\Delta(6N^2)$ for $N/3 \neq$ integer

h		χ_{1_r}	χ_2	$\chi_{3_{1k}}$	$\chi_{3_{2k}}$	$\chi_{6_{[[k],[\ell]]}}$
C_1	1	1	2	3	3	6
$C_3^{(m)}$	$N/\gcd(N,m)$	1	2	$\rho^{-2mk}+2\rho^{mk}$	$\rho^{-2mk}+2\rho^{mk}$	$2\rho^{m(k-\ell)}+2\rho^{-m(2k+\ell)}$ $+2\rho^{m(k+2\ell)}$
$C_6^{(m,n)}$	$N/\gcd(N,m,n)$	1	2	$\rho^{mk}+\rho^{-nk}$ $+\rho^{(-m+n)k}$	$\rho^{mk}+\rho^{-nk}$ $+\rho^{(-m+n)k}$	$\rho^{mk+n\ell}+\rho^{(-m+n)k-m\ell}$ $+\rho^{-nk+(m-n)\ell}$ $+\rho^{-nk-m\ell}$ $+\rho^{mk+(-m+n)\ell}$ $+\rho^{(-m+n)k+n\ell}$
C_{2N^2}	3	1	-1	0	0	0
$C_{3N}^{(m)}$	$2N/\gcd(N,m)$	$(-1)^r$	0	ρ^{-mk}	$-\rho^{-mk}$	0

13.1.3 Tensor Products

Because of their $Z_N \times Z'_N$ charges, tensor products of sextets **6** can be obtained as follows:

$$\begin{pmatrix} x_{\ell,-k-\ell} \\ x_{k,\ell} \\ x_{-k-\ell,k} \\ x_{k+\ell,-\ell} \\ x_{-\ell,-k} \\ x_{-k,k+\ell} \end{pmatrix}_{6_{[[k],[\ell]]}} \otimes \begin{pmatrix} y_{\ell',-k'-\ell'} \\ y_{k',\ell'} \\ y_{-k'-\ell',k'} \\ y_{k'+\ell',-\ell'} \\ y_{-\ell',-k'} \\ y_{-k',k'+\ell'} \end{pmatrix}_{6_{[[k'],[\ell']]}}$$

$$= \begin{pmatrix} x_{\ell,-k-\ell}y_{\ell',-k'-\ell'} \\ x_{k,\ell}y_{k',\ell'} \\ x_{-k-\ell,k}y_{-k'-\ell',k'} \\ x_{k+\ell,-\ell}y_{k'+\ell',-\ell'} \\ x_{-\ell,-k}y_{-\ell',-k'} \\ x_{-k,k+\ell}y_{-k',k'+\ell'} \end{pmatrix}_{6_{[[k+k'],[\ell+\ell']]}} \oplus \begin{pmatrix} x_{\ell,-k-\ell}y_{k',\ell'} \\ x_{k,\ell}y_{-k'-\ell',k'} \\ x_{-k-\ell,k}y_{\ell',-k'-\ell'} \\ x_{k+\ell,-\ell}y_{-\ell',-k'} \\ x_{-\ell,-k}y_{-k',k'+\ell'} \\ x_{-k,k+\ell}y_{k'+\ell',-\ell'} \end{pmatrix}_{6_{[[k-k'-\ell'],[\ell+k']]}}$$

$$\oplus \begin{pmatrix} x_{\ell,-k-\ell}y_{-k'-\ell',k'} \\ x_{k,\ell}y_{\ell',-k'-\ell'} \\ x_{-k-\ell,k}y_{k',\ell'} \\ x_{k+\ell,-\ell}y_{-k',k'+\ell'} \\ x_{-\ell,-k}y_{k'+\ell',-\ell'} \\ x_{-k,k+\ell}y_{-\ell',-k'} \end{pmatrix}_{6_{[[k+\ell'],[\ell-k'-\ell']]}} \oplus \begin{pmatrix} x_{\ell,-k-\ell}y_{-k',k'+\ell'} \\ x_{k,\ell}y_{-\ell',-k'} \\ x_{-k-\ell,k}y_{k'+\ell',-\ell'} \\ x_{k+\ell,-\ell}y_{-k'-\ell',k'} \\ x_{-\ell,-k}y_{k',\ell'} \\ x_{-k,k+\ell}y_{\ell',-k'-\ell'} \end{pmatrix}_{6_{[[k-\ell'],[\ell-k']]}}$$

13.1 $\Delta(6N^2)$ with $N/3 \neq$ Integer   129

$$\oplus \begin{pmatrix} x_{\ell,-k-\ell}y_{-\ell',-k'} \\ x_{k,\ell}y_{k'+\ell',-\ell'} \\ x_{-k-\ell,k}y_{-k',k'+\ell'} \\ x_{k+\ell,-\ell}y_{k',\ell'} \\ x_{-\ell,-k}y_{\ell',-k'-\ell'} \\ x_{-k,k+\ell}y_{-k'-\ell',k'} \end{pmatrix}_{6_{[[k+k'+\ell'],[\ell-\ell']]}} \oplus \begin{pmatrix} x_{\ell,-k-\ell}y_{k'+\ell',-\ell'} \\ x_{k,\ell}y_{-k',k'+\ell'} \\ x_{-k-\ell,k}y_{-\ell',-k'} \\ x_{k+\ell,-\ell}y_{\ell',-k'-\ell'} \\ x_{-\ell,-k}y_{-k'-\ell',k'} \\ x_{-k,k+\ell}y_{k',\ell'} \end{pmatrix}_{6_{[[k-k'],[\ell+k'+\ell']]}}$$

(13.28)

Similarly, products of the sextets **6** and the triplets $\mathbf{3}_{1k}$ and $\mathbf{3}_{2k}$ are:

$$\begin{pmatrix} x_{\ell,-k-\ell} \\ x_{k,\ell} \\ x_{-k-\ell,k} \\ x_{k+\ell,-\ell} \\ x_{-\ell,-k} \\ x_{-k,k+\ell} \end{pmatrix}_{6_{[[k],[\ell]]}} \otimes \begin{pmatrix} y_{k',0} \\ y_{-k',k'} \\ y_{0,-k'} \end{pmatrix}_{\mathbf{3}_{1k'}}$$

$$= \begin{pmatrix} x_{\ell,-k-\ell}y_{-k',k'} \\ x_{k,\ell}y_{0,-k'} \\ x_{-k-\ell,k}y_{k',0} \\ x_{k+\ell,-\ell}y_{-k',k'} \\ x_{-\ell,-k}y_{k',0} \\ x_{-k,k+\ell}y_{0,-k'} \end{pmatrix}_{6_{[[k],[\ell-k']]}} \oplus \begin{pmatrix} x_{\ell,-k-\ell}y_{k',0} \\ x_{k,\ell}y_{-k',k'} \\ x_{-k-\ell,k}y_{0,-k'} \\ x_{k+\ell,-\ell}y_{0,-k'} \\ x_{-\ell,-k}y_{-k',k'} \\ x_{-k,k+\ell}y_{k',0} \end{pmatrix}_{6_{[[k-k'],[\ell+k']]}}$$

$$\oplus \begin{pmatrix} x_{\ell,-k-\ell}y_{0,-k'} \\ x_{k,\ell}y_{k',0} \\ x_{-k-\ell,k}y_{-k',k'} \\ x_{k+\ell,-\ell}y_{k',0} \\ x_{-\ell,-k}y_{0,-k'} \\ x_{-k,k+\ell}y_{-k',k'} \end{pmatrix}_{6_{[[k+k'],[\ell]]}} \quad , \quad (13.29)$$

$$\begin{pmatrix} x_{\ell,-k-\ell} \\ x_{k,\ell} \\ x_{-k-\ell,k} \\ x_{k+\ell,-\ell} \\ x_{-\ell,-k} \\ x_{-k,k+\ell} \end{pmatrix}_{6_{[[k],[\ell]]}} \otimes \begin{pmatrix} y_{k',0} \\ y_{-k',k'} \\ y_{0,-k'} \end{pmatrix}_{\mathbf{3}_{2k'}}$$

$$= \begin{pmatrix} x_{\ell,-k-\ell}y_{-k',k'} \\ x_{k,\ell}y_{0,-k'} \\ x_{-k-\ell,k}y_{k',0} \\ -x_{k+\ell,-\ell}y_{-k',k'} \\ -x_{-\ell,-k}y_{k',0} \\ -x_{-k,k+\ell}y_{0,-k'} \end{pmatrix}_{6_{[[k],[\ell-k']]}} \oplus \begin{pmatrix} x_{\ell,-k-\ell}y_{k',0} \\ x_{k,\ell}y_{-k',k'} \\ x_{-k-\ell,k}y_{0,-k'} \\ -x_{k+\ell,-\ell}y_{0,-k'} \\ -x_{-\ell,-k}y_{-k',k'} \\ -x_{-k,k+\ell}y_{k',0} \end{pmatrix}_{6_{[[k-k'],[\ell+k']]}}$$

$$\oplus \begin{pmatrix} x_{\ell,-k-\ell}y_{0,-k'} \\ x_{k,\ell}y_{k',0} \\ x_{-k-\ell,k}y_{-k',k'} \\ -x_{k+\ell,-\ell}y_{k',0} \\ -x_{-\ell,-k}y_{0,-k'} \\ -x_{-k,k+\ell}y_{-k',k'} \end{pmatrix}_{6_{[[k+k'],[\ell]]}}, \tag{13.30}$$

$$\begin{pmatrix} x_{k,0} \\ x_{-k,k} \\ x_{0,-k} \end{pmatrix}_{3_{1k}} \otimes \begin{pmatrix} y_{k',0} \\ y_{-k',k'} \\ y_{0,-k'} \end{pmatrix}_{3_{1k'}} = \begin{pmatrix} x_{k,0}y_{k',0} \\ x_{-k,k}y_{-k',k'} \\ x_{0,-k}y_{0,-k'} \end{pmatrix}_{3_{1(k+k')}} \oplus \begin{pmatrix} x_{0,-k}y_{-k',k'} \\ x_{k,0}y_{0,-k'} \\ x_{-k,k}y_{k',0} \\ x_{k,0}y_{-k',k'} \\ x_{0,-k}y_{k',0} \\ x_{-k,k}y_{0,-k'} \end{pmatrix}_{6_{[[k],[-k']]}}, \tag{13.31}$$

$$\begin{pmatrix} x_{k,0} \\ x_{-k,k} \\ x_{0,-k} \end{pmatrix}_{3_{1k}} \otimes \begin{pmatrix} y_{k',0} \\ y_{-k',k'} \\ y_{0,-k'} \end{pmatrix}_{3_{2k'}} = \begin{pmatrix} x_{k,0}y_{k',0} \\ x_{-k,k}y_{-k',k'} \\ x_{0,-k}y_{0,-k'} \end{pmatrix}_{3_{2(k+k')}} \oplus \begin{pmatrix} x_{0,-k}y_{-k',k'} \\ x_{k,0}y_{0,-k'} \\ x_{-k,k}y_{k',0} \\ -x_{k,0}y_{-k',k'} \\ -x_{0,-k}y_{k',0} \\ -x_{-k,k}y_{0,-k'} \end{pmatrix}_{6_{[[k],[-k']]}}, \tag{13.32}$$

$$\begin{pmatrix} x_{k,0} \\ x_{-k,k} \\ x_{0,-k} \end{pmatrix}_{3_{2k}} \otimes \begin{pmatrix} y_{k',0} \\ y_{-k',k'} \\ y_{0,-k'} \end{pmatrix}_{3_{2k'}} = \begin{pmatrix} x_{k,0}y_{k',0} \\ x_{-k,k}y_{-k',k'} \\ x_{0,-k}y_{0,-k'} \end{pmatrix}_{3_{1(k+k')}} \oplus \begin{pmatrix} x_{0,-k}y_{-k',k'} \\ x_{k,0}y_{0,-k'} \\ x_{-k,k}y_{k',0} \\ x_{k,0}y_{-k',k'} \\ x_{0,-k}y_{k',0} \\ x_{-k,k}y_{0,-k'} \end{pmatrix}_{6_{[[k],[-k']]}}, \tag{13.33}$$

$$\begin{pmatrix} x_{\ell,-k-\ell} \\ x_{k,\ell} \\ x_{-k-\ell,k} \\ x_{k+\ell,-\ell} \\ x_{-\ell,-k} \\ x_{-k,k+\ell} \end{pmatrix}_{6_{[[k],[\ell]]}} \otimes \begin{pmatrix} y_1 \\ y_2 \end{pmatrix}_2 = \begin{pmatrix} x_{\ell,-k-\ell}y_1 \\ \omega x_{k,\ell}y_1 \\ \omega^2 x_{-k-\ell,k}y_1 \\ x_{k+\ell,-\ell}y_2 \\ \omega x_{-\ell,-k}y_2 \\ \omega^2 x_{-k,k+\ell}y_2 \end{pmatrix}_{6_{[[k],[\ell]]}} \oplus \begin{pmatrix} x_{\ell,-k-\ell}y_2 \\ \omega^2 x_{k,\ell}y_2 \\ \omega x_{-k-\ell,k}y_2 \\ x_{k+\ell,-\ell}y_1 \\ \omega^2 x_{-\ell,-k}y_1 \\ \omega x_{-k,k+\ell}y_1 \end{pmatrix}_{6_{[[k],[\ell]]}}, \tag{13.34}$$

$$\begin{pmatrix} x_{k,0} \\ x_{-k,k} \\ x_{0,-k} \end{pmatrix}_{3_{1k}} \otimes \begin{pmatrix} y_1 \\ y_2 \end{pmatrix}_2 = \begin{pmatrix} x_{k,0}(y_1 + \omega^2 y_2) \\ \omega x_{-k,k}(y_1 + y_2) \\ x_{0,-k}(\omega^2 y_1 + y_2) \end{pmatrix}_{3_{1k}} \oplus \begin{pmatrix} x_{k,0}(y_1 - \omega^2 y_2) \\ \omega x_{-k,k}(y_1 - y_2) \\ x_{0,-k}(\omega^2 y_1 - y_2) \end{pmatrix}_{3_{2k}}, \tag{13.35}$$

13.2 $\Delta(6N^2)$ with $N/3$ Integer

$$\begin{pmatrix} x_{k,0} \\ x_{-k,k} \\ x_{0,-k} \end{pmatrix}_{3_{2k}} \otimes \begin{pmatrix} y_1 \\ y_2 \end{pmatrix}_2 = \begin{pmatrix} x_{k,0}(y_1 + \omega^2 y_2) \\ \omega x_{-k,k}(y_1 + y_2) \\ x_{0,-k}(\omega^2 y_1 + y_2) \end{pmatrix}_{3_{2k}} \oplus \begin{pmatrix} x_{k,0}(y_1 - \omega^2 y_2) \\ \omega x_{-k,k}(y_1 - y_2) \\ x_{0,-k}(\omega^2 y_1 - y_2) \end{pmatrix}_{3_{1k}}, \quad (13.36)$$

$$\begin{pmatrix} x_1 \\ x_2 \end{pmatrix}_2 \otimes \begin{pmatrix} y_1 \\ y_2 \end{pmatrix}_2 = (x_1 y_2 + x_2 y_1)_{1_0} \oplus (x_1 y_2 - x_2 y_1)_{1_1} \oplus \begin{pmatrix} x_2 y_2 \\ x_1 y_1 \end{pmatrix}_2. \quad (13.37)$$

Tensor products with trivial singlets remain the same representation, while products with non-trivial singlets are:

$$\begin{pmatrix} x_1 \\ x_2 \\ x_3 \\ x_4 \\ x_5 \\ x_6 \end{pmatrix}_{6_{[[k],[\ell]]}} \otimes (y)_{1_1} = \begin{pmatrix} x_1 y \\ x_2 y \\ x_3 y \\ -x_4 y \\ -x_5 y \\ -x_6 y \end{pmatrix}_{6_{[[k],[\ell]]}}, \qquad \begin{pmatrix} x_1 \\ x_2 \\ x_3 \end{pmatrix}_{3_{1k}} \otimes (y)_{1_1} = \begin{pmatrix} x_1 y \\ x_2 y \\ x_3 y \end{pmatrix}_{3_{2k}},$$

(13.38)

$$\begin{pmatrix} x_1 \\ x_2 \\ x_3 \end{pmatrix}_{3_{2k}} \otimes (y)_{1_1} = \begin{pmatrix} x_1 y \\ x_2 y \\ x_3 y \end{pmatrix}_{3_{1k}}, \qquad \begin{pmatrix} x_1 \\ x_2 \end{pmatrix}_2 \otimes (y)_{1_1} = \begin{pmatrix} x_1 y \\ -x_2 y \end{pmatrix}_2. \quad (13.39)$$

13.2 $\Delta(6N^2)$ with $N/3$ Integer

13.2.1 Conjugacy Classes

The conjugacy classes of $\Delta(3N^2)$ are useful to obtain those of $\Delta(6N^2)$, except for the elements involving c.

Conjugacy classes containing the elements $a^\ell a'^m$ are almost the same as those of $\Delta(6N^2)$ when $N/3 \neq$ integer. One difference is that, when $N/3$ is an integer, there are classes with one element. Since

$$ba^\ell a'^{-\ell} b^{-1} = a^{-2\ell} a'^{-\ell}, \qquad ca^\ell a'^{-\ell} c^{-1} = a^\ell a'^{-\ell}, \quad (13.40)$$

and $\ell = -2\ell$ if $\ell = N/3$ or $2N/3$, there are classes $\{a^{N/3} a'^{2N/3}\}$ and $\{a^{2N/3} a'^{N/3}\}$.

Now consider the elements $ba^\ell a'^m$ and $b^2 a^\ell a'^m$. In $\Delta(3N^2)$ with $N/3$ integer, the elements $ba^{p-n-3m} a'^n$ with $m = 0, 1, \ldots, (N-3)/3$, and $n = 0, 1, \ldots, N-1$, belong to the same conjugacy class. Furthermore, the elements $b^2 a^{p-n-3m} a'^n$ with $m = 0, 1, \ldots, (N-3)/3$, and $n = 0, 1, \ldots, N-1$, belong to the same conjugacy class. Since

$$cba^{p-n-3m} a'^n c^{-1} = b^2 a^{-n} a'^{-p+n+3m}, \quad (13.41)$$

these elements belong to the same conjugacy class for $\Delta(6N^2)$.

Similarly to what was done for $\Delta(6N^2)$ when $N/3 \neq$ integer, we can study the conjugacy classes of elements involving c. It turns out that these conjugacy classes are the same for the group $\Delta(6N^2)$ when $N/3 \neq$ integer and when $N/3$ is an integer.

Here we summarize the conjugacy classes of $\Delta(6N^2)$ when $N/3$ is an integer:

$C_1:$ $\{e\}$,

$C_1^{(p)}:$ $\{a^p a'^{-p}\}$, $p = N/3, 2N/3$,

$C_3^{(k)}:$ $\{a^k a'^{-k}, a^{-2k} a'^{-k}, a^k a'^{2k}\}$, $k = 1, \ldots, N-1$, $k \neq N/3, 2N/3$,

$C_6^{(\ell,m)}:$ $\{a^\ell a'^m, a^{m-\ell} a'^{-\ell}, a^{-m} a'^{\ell-m}, a^{-m} a'^{-\ell}, a^{m-\ell} a'^m, a^\ell a'^{\ell-m}\}$,
$\ell, m = 0, \ldots, N-1$,

$C_{2N^2/3}^{(q)}:$ $\{ba^{q-n-3m} a'^n, b^2 a^{-n} a'^{n+3m-q} \mid n = 0, 1, \ldots, N-1,$
$m = 0, 1, \ldots, \dfrac{N-3}{3}\},$
$q = 0, 1, 2,$

$C_{3N}^{(\ell)}:$ $\{ca^{\ell+n} a'^n, b^2 ca^{-\ell} a'^{-\ell-n}, bca^{-n} a'^\ell \mid n = 0, 1, \ldots, N-1\}$,
$\ell = 0, \ldots, N-1$,

(13.42)

where ℓ, m in $C_6^{(\ell,m)}$ do not satisfy (13.6). The order h of each element in the given conjugacy classes is found to be

$C_1:$ $h = 1$,

$C_1^{(p)}:$ $h = 3$,

$C_3^{(k)}:$ $h = N/\gcd(N, k)$,

$C_6^{(\ell,m)}:$ $h = N/\gcd(N, \ell, m)$,   (13.43)

$C_{2N^2/3}^{(q)}:$ $h = 3$,

$C_{3N}^{(\ell)}:$ $h = 2N/\gcd(N, \ell)$.

The numbers of each of the conjugacy classes $C_1^{(p)}$, $C_3^{(k)}$, $C_6^{(\ell,m)}$, $C_{2N^2/3}^{(q)}$, and $C_{3N}^{(\ell)}$ are 2, $N-3$, $(N^2 - 3N + 6)/6$, 3, and N, respectively. The total number of conjugacy classes is thus equal to $(N^2 + 9N + 24)/6$. The relations (2.18) and (2.19) for $\Delta(6N^2)$ with $N/3$ integer are

$$m_1 + 2^2 m_2 + 3^2 m_3 + \cdots = 6N^2, \quad (13.44)$$

$$m_1 + m_2 + m_3 + \cdots = \frac{N^2 + 9N + 24}{6}. \quad (13.45)$$

The solution is $(m_1, m_2, m_3, m_6) = (2, 4, 2(N-1), [N(N-3)]/6)$.

13.2.2 Characters and Representations

There are two singlets $\mathbf{1}_k$. Their characters are the same as those of the group $\Delta(6N^2)$ with $N/3$ integer. That is, because $c^2 = e$, the characters of the two singlets have two possible values $\chi_{1_k}(c) = (-1)^k$ with $k = 0, 1$, and they correspond to two singlets $\mathbf{1}_k$. Note that $\chi_{1_k}(a) = \chi_{1_k}(a') = \chi_{1_k}(b) = 1$. The group $\Delta(6N^2)$ with $N/3$ integer has four doublets $\mathbf{2}_k$ with $k = 1, 2, 3, 4$. One of them, namely $\mathbf{2}_1$, is the same as the doublet (13.19) of $\Delta(6N^2)$ with $N/3 \neq$ integer. The other three doublet representations will be obtained following the discussion of the sextet representations.

We thus consider the sextet representations, which are the same as those of $\Delta(6N^2)$ with $N/3$ integer, viz., (13.20) and (13.21), i.e.,

$$b = \begin{pmatrix} b_1 & 0 \\ 0 & b_2 \end{pmatrix}, \quad c = \begin{pmatrix} 0 & 1 \\ 1 & 0 \end{pmatrix}, \quad a = \begin{pmatrix} a_1 & 0 \\ 0 & a_2 \end{pmatrix}, \quad a' = \begin{pmatrix} a'_1 & 0 \\ 0 & a'_2 \end{pmatrix}, \tag{13.46}$$

where

$$b_1 = \begin{pmatrix} 0 & 1 & 0 \\ 0 & 0 & 1 \\ 1 & 0 & 0 \end{pmatrix}, \quad b_2 = \begin{pmatrix} 0 & 0 & 1 \\ 1 & 0 & 0 \\ 0 & 1 & 0 \end{pmatrix}, \tag{13.47}$$

$$a_1 = a'^{-1}_2 = \begin{pmatrix} \rho^l & 0 & 0 \\ 0 & \rho^k & 0 \\ 0 & 0 & \rho^{-l-k} \end{pmatrix}, \quad a_2 = a'^{-1}_1 = \begin{pmatrix} \rho^{l+k} & 0 & 0 \\ 0 & \rho^{-l} & 0 \\ 0 & 0 & \rho^{-k} \end{pmatrix}. \tag{13.48}$$

In certain cases, the above representation becomes reducible. For $N/3$ an integer, $\mathbf{6}_{[[k],[\ell]]}$ is reducible if $(k, \ell) = (N/3, N/3), (2N/3, 2N/3)$, $k + \ell = 0 \pmod{N}$, $k = 0$, $\ell \neq 0$, and $\ell = 0$, $k \neq 0$. Thus, the number of irreducible sextet representations is $[N^2 - 2 - N - 2(N - 1)]/6$.

Among the above reducible sextet representations, the irreducible triplets are obtained in the same way as (13.25) for $\Delta(6N^2)$ with $N/3 \neq$ integer, i.e.,

$$b(\mathbf{3}_{1k}) = \begin{pmatrix} 0 & 1 & 0 \\ 0 & 0 & 1 \\ 1 & 0 & 0 \end{pmatrix}, \quad c(\mathbf{3}_{1k}) = \begin{pmatrix} 0 & 0 & 1 \\ 0 & 1 & 0 \\ 1 & 0 & 0 \end{pmatrix},$$

$$a(\mathbf{3}_{1k}) = \begin{pmatrix} \rho^k & 0 & 0 \\ 0 & \rho^{-k} & 0 \\ 0 & 0 & 1 \end{pmatrix}, \quad a'(\mathbf{3}_{1k}) = \begin{pmatrix} 1 & 0 & 0 \\ 0 & \rho^k & 0 \\ 0 & 0 & \rho^{-k} \end{pmatrix}, \tag{13.49}$$

$$b(\mathbf{3}_{2k}) = b(\mathbf{3}_{1k}), \quad c(\mathbf{3}_{2k}) = -c(\mathbf{3}_{1k}),$$
$$a(\mathbf{3}_{2k}) = a(\mathbf{3}_{1k}), \quad a'(\mathbf{3}_{2k}) = a'(\mathbf{3}_{1k}). \tag{13.50}$$

In addition, one can obtain the three doublets by diagonalizing the above reducible sextet representations:

$$b(\mathbf{2}_2) = b(\mathbf{2}_1), \quad c(\mathbf{2}_2) = c(\mathbf{2}_1), \quad a(\mathbf{2}_2) = a'(\mathbf{2}_2) = \begin{pmatrix} \omega^2 & 0 \\ 0 & \omega \end{pmatrix}, \quad (13.51)$$

$$b(\mathbf{2}_3) = b(\mathbf{2}_1), \quad c(\mathbf{2}_3) = c(\mathbf{2}_1), \quad a(\mathbf{2}_3) = a'(\mathbf{2}_3) = \begin{pmatrix} \omega & 0 \\ 0 & \omega^2 \end{pmatrix}, \quad (13.52)$$

$$b(\mathbf{2}_4) = 1, \quad c(\mathbf{2}_4) = c(\mathbf{2}_1), \quad a(\mathbf{2}_4) = a'(\mathbf{2}_4) = \begin{pmatrix} \omega & 0 \\ 0 & \omega^2 \end{pmatrix}. \quad (13.53)$$

The characters are shown in Table 13.2.

13.2.3 Tensor Products

When $3N$ is an integer, the number of doublets is increased. Since products with other representations are the same as those for $3N \neq$ integer, it suffices to examine products with additional doublets:

$$\begin{pmatrix} x_{\ell,-k-\ell} \\ x_{k,\ell} \\ x_{-k-\ell,k} \\ x_{k+\ell,-\ell} \\ x_{-\ell,-k} \\ x_{-k,k+\ell} \end{pmatrix}_{\mathbf{6}_{[[k],[\ell]]}} \otimes \begin{pmatrix} y_1 \\ y_2 \end{pmatrix}_{\mathbf{2}_2} = \begin{pmatrix} x_{\ell,-k-\ell} y_1 \\ \omega x_{k,\ell} y_1 \\ \omega^2 x_{-k-\ell,k} y_1 \\ x_{k+\ell,-\ell} y_2 \\ \omega x_{-\ell,-k} y_2 \\ \omega^2 x_{-k,k+\ell} y_2 \end{pmatrix}_{\mathbf{6}_{[[k+2N/3],[\ell+2N/3]]}}$$

$$\oplus \begin{pmatrix} x_{\ell,-k-\ell} y_2 \\ \omega^2 x_{k,\ell} y_2 \\ \omega x_{-k-\ell,k} y_2 \\ x_{k+\ell,-\ell} y_1 \\ \omega^2 x_{-\ell,-k} y_1 \\ \omega x_{-k,k+\ell} y_1 \end{pmatrix}_{\mathbf{6}_{[[k+N/3],[\ell+N/3]]}}, \quad (13.54)$$

13.2 $\Delta(6N^2)$ with $N/3$ Integer

Table 13.2 Characters of $\Delta(6N^2)$ for $N/3$ integer

	h	χ_{1_r}	χ_{2_1}	χ_{2_2}	χ_{2_3}	χ_{2_4}	$\chi_{3_{1_k}}$	$\chi_{3_{2_k}}$	$\chi_{6_{[(k),(\ell)]}}$
C_1	1	1	2	2	2	2	3	3	6
$C_1^{(m)}$	3	1	2	2	2	2	$\rho^{-2mk} + 2\rho^{mk}$	$\rho^{-2mk} + 2\rho^{mk}$	$2\rho^{m(k-\ell)} + 2\rho^{-m(2k+\ell)} + 2\rho^{m(k+2\ell)}$
$C_3^{(m)}$	$N/\gcd(N,m)$	1	2	2	2	2	$\rho^{-2mk} + 2\rho^{mk}$	$\rho^{-2mk} + 2\rho^{mk}$	$2\rho^{m(k-\ell)} + 2\rho^{-m(2k+\ell)} + 2\rho^{m(k+2\ell)}$
$C_6^{(m,n)}$	$N/\gcd(N,m,n)$	1	2	$\omega^{m+n} + \omega^{2m+2n}$	$\omega^{m+n} + \omega^{2m+2n}$	$\omega^{m+n} + \omega^{2m+2n}$	$\rho^{mk} + \rho^{-nk} + \rho^{(-m+n)k}$	$\rho^{mk} + \rho^{-nk} + \rho^{(-m+n)k}$	$\rho^{mk+n\ell} + \rho^{(-m+n)k-m\ell} + \rho^{-nk+(m-n)\ell} + \rho^{-nk-m\ell} + \rho^{mk+(m-n)\ell} + \rho^{(-m+n)k+n\ell}$
$C_{2N^2/3}^{(\tau)}$	3	1	-1	$\omega^{2+\tau} + \omega^{(4+2\tau)}$	$\omega^{1+\tau} + \omega^{2+2\tau}$	$\omega^{\tau} + \omega^{2\tau}$	0	0	0
$C_{3N}^{(m)}$	$2N/\gcd(N,m)$	$(-1)^r$	0	0	0	0	ρ^{-mk}	$-\rho^{-mk}$	0

$$\begin{pmatrix} x_{\ell,-k-\ell} \\ x_{k,\ell} \\ x_{-k-\ell,k} \\ x_{k+\ell,-\ell} \\ x_{-\ell,-k} \\ x_{-k,k+\ell} \end{pmatrix}_{\mathbf{6}_{[[k],[\ell]]}} \otimes \begin{pmatrix} y_1 \\ y_2 \end{pmatrix}_{\mathbf{2}_3} = \begin{pmatrix} x_{\ell,-k-\ell} y_1 \\ \omega x_{k,\ell} y_1 \\ \omega^2 x_{-k-\ell,k} y_1 \\ x_{k+\ell,-\ell} y_2 \\ \omega x_{-\ell,-k} y_2 \\ \omega^2 x_{-k,k+\ell} y_2 \end{pmatrix}_{\mathbf{6}_{[[k+N/3],[\ell+N/3]]}}$$

$$\oplus \begin{pmatrix} x_{\ell,-k-\ell} y_2 \\ \omega^2 x_{k,\ell} y_2 \\ \omega x_{-k-\ell,k} y_2 \\ x_{k+\ell,-\ell} y_1 \\ \omega^2 x_{-\ell,-k} y_1 \\ \omega x_{-k,k+\ell} y_1 \end{pmatrix}_{\mathbf{6}_{[[k+2N/3],[\ell+2N/3]]}}, \quad (13.55)$$

$$\begin{pmatrix} x_{\ell,-k-\ell} \\ x_{k,\ell} \\ x_{-k-\ell,k} \\ x_{k+\ell,-\ell} \\ x_{-\ell,-k} \\ x_{-k,k+\ell} \end{pmatrix}_{\mathbf{6}_{[[k],[\ell]]}} \otimes \begin{pmatrix} y_1 \\ y_2 \end{pmatrix}_{\mathbf{2}_4} = \begin{pmatrix} x_{\ell,-k-\ell} y_1 \\ x_{k,\ell} y_1 \\ x_{-k-\ell,k} y_1 \\ x_{k+\ell,-\ell} y_2 \\ x_{-\ell,-k} y_2 \\ x_{-k,k+\ell} y_2 \end{pmatrix}_{\mathbf{6}_{[[k+N/3],[\ell+N/3]]}}$$

$$\oplus \begin{pmatrix} x_{\ell,-k-\ell} y_2 \\ x_{k,\ell} y_2 \\ x_{-k-\ell,k} y_2 \\ x_{k+\ell,-\ell} y_1 \\ x_{-\ell,-k} y_1 \\ x_{-k,k+\ell} y_1 \end{pmatrix}_{\mathbf{6}_{[[k+2N/3],[\ell+2N/3]]}}, \quad (13.56)$$

$$\begin{pmatrix} x_{k,0} \\ x_{-k,k} \\ x_{0,-k} \end{pmatrix}_{\mathbf{3}_{1k}} \otimes \begin{pmatrix} y_1 \\ y_2 \end{pmatrix}_{\mathbf{2}_2} = \begin{pmatrix} x_{k,0} y_1 \\ \omega x_{-k,k} y_1 \\ \omega^2 x_{0,-k} y_1 \\ x_{0,-k} y_2 \\ \omega x_{-k,k} y_2 \\ \omega^2 x_{k,0} y_2 \end{pmatrix}_{\mathbf{6}_{[[-k+2N/3],[k+2N/3]]}}, \quad (13.57)$$

$$\begin{pmatrix} x_{k,0} \\ x_{-k,k} \\ x_{0,-k} \end{pmatrix}_{\mathbf{3}_{2k}} \otimes \begin{pmatrix} y_1 \\ y_2 \end{pmatrix}_{\mathbf{2}_2} = \begin{pmatrix} x_{k,0} y_1 \\ \omega x_{-k,k} y_1 \\ \omega^2 x_{0,-k} y_1 \\ -x_{0,-k} y_2 \\ -\omega x_{-k,k} y_2 \\ -\omega^2 x_{k,0} y_2 \end{pmatrix}_{\mathbf{6}_{[[-k+2N/3],[k+2N/3]]}}, \quad (13.58)$$

13.2 $\Delta(6N^2)$ with $N/3$ Integer

$$\begin{pmatrix} x_{k,0} \\ x_{-k,k} \\ x_{0,-k} \end{pmatrix}_{3_{1k}} \otimes \begin{pmatrix} y_1 \\ y_2 \end{pmatrix}_{2_3} = \begin{pmatrix} x_{k,0}y_1 \\ \omega x_{-k,k}y_1 \\ \omega^2 x_{0,-k}y_1 \\ x_{0,-k}y_2 \\ \omega x_{-k,k}y_2 \\ \omega^2 x_{k,0}y_2 \end{pmatrix}_{6_{[[-k+N/3],[k+N/3]]}}, \quad (13.59)$$

$$\begin{pmatrix} x_{k,0} \\ x_{-k,k} \\ x_{0,-k} \end{pmatrix}_{3_{2k}} \otimes \begin{pmatrix} y_1 \\ y_2 \end{pmatrix}_{2_3} = \begin{pmatrix} x_{k,0}y_1 \\ \omega x_{-k,k}y_1 \\ \omega^2 x_{0,-k}y_1 \\ -x_{0,-k}y_2 \\ -\omega x_{-k,k}y_2 \\ -\omega^2 x_{k,0}y_2 \end{pmatrix}_{6_{[[-k+N/3],[k+N/3]]}}, \quad (13.60)$$

$$\begin{pmatrix} x_{k,0} \\ x_{-k,k} \\ x_{0,-k} \end{pmatrix}_{3_{1k}} \otimes \begin{pmatrix} y_1 \\ y_2 \end{pmatrix}_{2_4} = \begin{pmatrix} x_{k,0}y_1 \\ x_{-k,k}y_1 \\ x_{0,-k}y_1 \\ x_{0,-k}y_2 \\ x_{-k,k}y_2 \\ x_{k,0}y_2 \end{pmatrix}_{6_{[[-k+N/3],[k+N/3]]}}, \quad (13.61)$$

$$\begin{pmatrix} x_{k,0} \\ x_{-k,k} \\ x_{0,-k} \end{pmatrix}_{3_{2k}} \otimes \begin{pmatrix} y_1 \\ y_2 \end{pmatrix}_{2_4} = \begin{pmatrix} x_{k,0}y_1 \\ x_{-k,k}y_1 \\ x_{0,-k}y_1 \\ -x_{0,-k}y_2 \\ -x_{-k,k}y_2 \\ -x_{k,0}y_2 \end{pmatrix}_{6_{[[-k+N/3],[k+N/3]]}}. \quad (13.62)$$

The tensor products among doublets are as follows:

$$\begin{pmatrix} x_1 \\ x_2 \end{pmatrix}_{2_k} \otimes \begin{pmatrix} y_1 \\ y_2 \end{pmatrix}_{2_k} = (x_1 y_2 + x_2 y_1)_{1_0} \oplus (x_1 y_2 - x_2 y_1)_{1_1} \oplus \begin{pmatrix} x_2 y_2 \\ x_1 y_1 \end{pmatrix}_{2_k}, \quad (13.63)$$

for $k = 1, 2, 3, 4$,

$$\begin{pmatrix} x_1 \\ x_2 \end{pmatrix}_{2_1} \otimes \begin{pmatrix} y_1 \\ y_2 \end{pmatrix}_{2_2} = \begin{pmatrix} x_2 y_2 \\ x_1 y_1 \end{pmatrix}_{2_3} \oplus \begin{pmatrix} x_1 y_2 \\ x_2 y_1 \end{pmatrix}_{2_4}, \quad (13.64)$$

$$\begin{pmatrix} x_1 \\ x_2 \end{pmatrix}_{2_1} \otimes \begin{pmatrix} y_1 \\ y_2 \end{pmatrix}_{2_3} = \begin{pmatrix} x_2 y_2 \\ x_1 y_1 \end{pmatrix}_{2_2} \oplus \begin{pmatrix} x_2 y_1 \\ x_1 y_2 \end{pmatrix}_{2_4}, \quad (13.65)$$

$$\begin{pmatrix} x_1 \\ x_2 \end{pmatrix}_{2_1} \otimes \begin{pmatrix} y_1 \\ y_2 \end{pmatrix}_{2_4} = \begin{pmatrix} x_1 y_2 \\ x_2 y_1 \end{pmatrix}_{2_2} \oplus \begin{pmatrix} x_1 y_1 \\ x_2 y_2 \end{pmatrix}_{2_3}, \quad (13.66)$$

$$\begin{pmatrix}x_1\\x_2\end{pmatrix}_{2_2}\otimes\begin{pmatrix}y_1\\y_2\end{pmatrix}_{2_3}=\begin{pmatrix}x_2y_2\\x_1y_1\end{pmatrix}_{2_1}\oplus\begin{pmatrix}x_1y_2\\x_2y_1\end{pmatrix}_{2_4}, \qquad (13.67)$$

$$\begin{pmatrix}x_1\\x_2\end{pmatrix}_{2_2}\otimes\begin{pmatrix}y_1\\y_2\end{pmatrix}_{2_4}=\begin{pmatrix}x_1y_1\\x_2y_2\end{pmatrix}_{2_1}\oplus\begin{pmatrix}x_1y_2\\x_2y_1\end{pmatrix}_{2_3}, \qquad (13.68)$$

$$\begin{pmatrix}x_1\\x_2\end{pmatrix}_{2_3}\otimes\begin{pmatrix}y_1\\y_2\end{pmatrix}_{2_4}=\begin{pmatrix}x_1y_2\\x_2y_1\end{pmatrix}_{2_1}\oplus\begin{pmatrix}x_1y_1\\x_2y_2\end{pmatrix}_{2_2}. \qquad (13.69)$$

Tensor products of non-trivial singlets with additional doublets are the same as those of **2** for $N/3 \neq$ integer.

13.3 $\Delta(54)$

Here we consider a simple example of $\Delta(6N^2)$. The group $\Delta(6)$ is nothing but S_3 and $\Delta(24)$ is isomorphic to S_4. Thus, the simplest non-trivial example is $\Delta(54)$.

13.3.1 Conjugacy Classes

All elements of $\Delta(54)$ can be written in the form $b^k c^\ell a^m a'^n$, where $k, m, n = 0, 1, 2$, and $\ell = 0, 1$. Half of them are the elements of $\Delta(27)$, whose conjugacy classes are shown in (10.37). Since $cac^{-1} = a'^{-1}$ and $ca'c^{-1} = a^{-1}$, the conjugacy classes $C_1^{(1)}$ and $C_1^{(2)}$ of $\Delta(27)$ still correspond to the conjugacy classes of $\Delta(54)$. However, the conjugacy classes $C_3^{(0,1)}$ and $C_3^{(0,2)}$ of $\Delta(27)$ are combined into one class. Similarly, since $cba^k a'^\ell c^{-1} = b^2 ca^k a'^\ell c^{-1}$, the conjugacy classes $C_3^{(1,p)}$ and $C_3^{(2,p')}$ of $\Delta(27)$ for $p + p' = 0 \pmod 3$ are combined into one class of $\Delta(54)$.

Let us consider the conjugacy classes of elements involving c. For example, we obtain

$$a^k a'^\ell (ca^m) a^{-k} a'^{-\ell} = ca^{m+p} a'^p, \qquad (13.70)$$

where $p = -k - \ell$. Thus, the element ca^m is conjugate to $ca^{m+p} a'^p$ with $p = 0, 1, 2$. Furthermore, it is found that

$$b(ca^{m+p} a'^p) b^{-1} = b^2 ca^{-m} a'^{-m-p}, \qquad (13.71)$$

$$b(ca^{-m} a'^{-m-p}) b^{-1} = bca^{-p} a'^m. \qquad (13.72)$$

These elements thus belong to the same conjugacy class.

13.3 $\Delta(54)$

Using the above results, the elements of $\Delta(54)$ are classified into the following conjugacy classes:

$$
\begin{aligned}
&C_1: &&\{e\}, &&h=1,\\
&C_1^{(1)}: &&\{aa'^2\}, &&h=3,\\
&C_1^{(2)}: &&\{a^2a'\}, &&h=3,\\
&C_6^{(0,1)}: &&\{a', a, a^2a'^2, a'^2, a^2, aa'\}, &&h=3,\\
&C_6^{(0)}: &&\{b, ba^2a', baa'^2, b^2, b^2a^2a', b^2aa'^2\}, &&h=3,\\
&C_6^{(1)}: &&\{ba, ba', ba^2a'^2, b^2a^2, b^2a'^2, b^2aa'\}, &&h=3,\\
&C_6^{(2)}: &&\{ba^2, baa', ba'^2, b^2a, b^2a^2a'^2, b^2a'\}, &&h=3,\\
&C_9^{(0)}: &&\{c^p a'^p, b^2 ca'^{-p}, bca^{-p} \mid p=0,1,2\}, &&h=2,\\
&C_9^{(1)}: &&\{ca^{1+p}a'^p, b^2 ca^2a'^{-1-p}, bca^{-p}a' \mid p=0,1,2\}, &&h=6,\\
&C_9^{(2)}: &&\{ca^{2+p}a'^p, b^2 caa'^{-2-p}, bca^{-p}a'^2 \mid p=0,1,2\}, &&h=6.
\end{aligned}
$$
(13.73)

The total number of conjugacy classes is equal to ten. The relations (2.18) and (2.19) for $\Delta(54)$ lead to

$$m_1 + 2^2 m_2 + 3^2 m_3 + \cdots = 54, \tag{13.74}$$

$$m_1 + m_2 + m_3 + \cdots = 10. \tag{13.75}$$

The solution is $(m_1, m_2, m_3) = (2, 4, 4)$, whence there are two singlets, four doublets, and four triplets.

13.3.2 Characters and Representations

We start by discussing the two singlets. It is straightforward to show that $\chi_{1_k}(a) = \chi_{1_k}(a') = \chi_{1_k}(b) = 1$ for the two singlets from the above conjugacy class structure. In addition, since $c^2 = e$, the two values are $\chi_{1_k}(c) = (-1)^k$ with $k = 0, 1$. These correspond to the two singlets.

Next, we consider the triplets. For example, the generators a, a', b, and c can be represented by

$$a = \begin{pmatrix} \omega^k & 0 & 0 \\ 0 & \omega^{2k} & 0 \\ 0 & 0 & 1 \end{pmatrix}, \quad a' = \begin{pmatrix} 1 & 0 & 0 \\ 0 & \omega^k & 0 \\ 0 & 0 & \omega^{2k} \end{pmatrix},$$

$$b = \begin{pmatrix} 0 & 1 & 0 \\ 0 & 0 & 1 \\ 1 & 0 & 0 \end{pmatrix}, \quad c = \begin{pmatrix} 0 & 0 & 1 \\ 0 & 1 & 0 \\ 1 & 0 & 0 \end{pmatrix},$$
(13.76)

Table 13.3 Characters of $\Delta(54)$

	χ_{1_0}	χ_{1_1}	χ_{2_1}	χ_{2_2}	χ_{2_3}	χ_{2_4}	$\chi_{3_{11}}$	$\chi_{3_{12}}$	$\chi_{3_{21}}$	$\chi_{3_{22}}$
C_1	1	1	2	2	2	2	3	3	3	3
$C_1^{(1)}$	1	1	2	2	2	2	3ω	$3\omega^2$	3ω	$3\omega^2$
$C_1^{(2)}$	1	1	2	2	2	2	$3\omega^2$	3ω	$3\omega^2$	3ω
$C_6^{(0,1)}$	1	1	2	-1	-1	-1	0	0	0	0
$C_6^{(0)}$	1	1	-1	-1	-1	2	0	0	0	0
$C_6^{(1)}$	1	1	-1	2	-1	-1	0	0	0	0
$C_6^{(2)}$	1	1	-1	-1	2	-1	0	0	0	0
$C_9^{(0)}$	1	-1	0	0	0	0	1	1	-1	-1
$C_9^{(1)}$	1	-1	0	0	0	0	ω^2	ω	$-\omega^2$	$-\omega$
$C_9^{(2)}$	1	-1	0	0	0	0	ω	ω^2	$-\omega$	$-\omega^2$

on $\mathbf{3}_{1k}$ for $k=1,2$. Obviously, the $\Delta(54)$ algebra is still satisfied when c is replaced by $-c$. That is, the generators a, a', b, and c are represented by

$$a = \begin{pmatrix} \omega^k & 0 & 0 \\ 0 & \omega^{2k} & 0 \\ 0 & 0 & 1 \end{pmatrix}, \quad a' = \begin{pmatrix} 1 & 0 & 0 \\ 0 & \omega^k & 0 \\ 0 & 0 & \omega^{2k} \end{pmatrix},$$

$$b = \begin{pmatrix} 0 & 1 & 0 \\ 0 & 0 & 1 \\ 1 & 0 & 0 \end{pmatrix}, \quad c = \begin{pmatrix} 0 & 0 & -1 \\ 0 & -1 & 0 \\ -1 & 0 & 0 \end{pmatrix},$$

(13.77)

on $\mathbf{3}_{2k}$ for $k=1,2$. The characters χ_3 for $\mathbf{3}_{1k}$ and $\mathbf{3}_{2k}$ are shown in Table 13.3.

Now, consider the doublets. There are four doublet representations of $\Delta(54)$. The generators a, a', b, and c are represented by

$$a = a' = \begin{pmatrix} 1 & 0 \\ 0 & 1 \end{pmatrix}, \quad b = \begin{pmatrix} \omega & 0 \\ 0 & \omega^2 \end{pmatrix}, \quad c = \begin{pmatrix} 0 & 1 \\ 1 & 0 \end{pmatrix}, \quad \text{on } \mathbf{2}_1, \quad (13.78)$$

$$a = a' = \begin{pmatrix} \omega^2 & 0 \\ 0 & \omega \end{pmatrix}, \quad b = \begin{pmatrix} \omega & 0 \\ 0 & \omega^2 \end{pmatrix}, \quad c = \begin{pmatrix} 0 & 1 \\ 1 & 0 \end{pmatrix}, \quad \text{on } \mathbf{2}_2, \quad (13.79)$$

$$a = a' = \begin{pmatrix} \omega & 0 \\ 0 & \omega^2 \end{pmatrix}, \quad b = \begin{pmatrix} \omega & 0 \\ 0 & \omega^2 \end{pmatrix}, \quad c = \begin{pmatrix} 0 & 1 \\ 1 & 0 \end{pmatrix}, \quad \text{on } \mathbf{2}_3, \quad (13.80)$$

$$a = a' = \begin{pmatrix} \omega & 0 \\ 0 & \omega^2 \end{pmatrix}, \quad b = \begin{pmatrix} 1 & 0 \\ 0 & 1 \end{pmatrix}, \quad c = \begin{pmatrix} 0 & 1 \\ 1 & 0 \end{pmatrix}, \quad \text{on } \mathbf{2}_4. \quad (13.81)$$

The characters χ_2 for $\mathbf{2}_{1,2,3,4}$ are shown in Table 13.3.

13.3.3 Tensor Products

The tensor products between triplets are as follows:

$$\begin{pmatrix}x_1\\x_2\\x_3\end{pmatrix}_{3_{11}} \otimes \begin{pmatrix}y_1\\y_2\\y_3\end{pmatrix}_{3_{11}} = \begin{pmatrix}x_1y_1\\x_2y_2\\x_3y_3\end{pmatrix}_{3_{12}} \oplus \begin{pmatrix}x_2y_3+x_3y_2\\x_3y_1+x_1y_3\\x_1y_2+x_2y_1\end{pmatrix}_{3_{12}} \oplus \begin{pmatrix}x_2y_3-x_3y_2\\x_3y_1-x_1y_3\\x_1y_2-x_2y_1\end{pmatrix}_{3_{22}},$$
(13.82)

$$\begin{pmatrix}x_1\\x_2\\x_3\end{pmatrix}_{3_{12}} \otimes \begin{pmatrix}y_1\\y_2\\y_3\end{pmatrix}_{3_{12}} = \begin{pmatrix}x_1y_1\\x_2y_2\\x_3y_3\end{pmatrix}_{3_{11}} \oplus \begin{pmatrix}x_2y_3+x_3y_2\\x_3y_1+x_1y_3\\x_1y_2+x_2y_1\end{pmatrix}_{3_{11}} \oplus \begin{pmatrix}x_2y_3-x_3y_2\\x_3y_1-x_1y_3\\x_1y_2-x_2y_1\end{pmatrix}_{3_{21}},$$
(13.83)

$$\begin{pmatrix}x_1\\x_2\\x_3\end{pmatrix}_{3_{21}} \otimes \begin{pmatrix}y_1\\y_2\\y_3\end{pmatrix}_{3_{21}} = \begin{pmatrix}x_1y_1\\x_2y_2\\x_3y_3\end{pmatrix}_{3_{12}} \oplus \begin{pmatrix}x_2y_3+x_3y_2\\x_3y_1+x_1y_3\\x_1y_2+x_2y_1\end{pmatrix}_{3_{12}} \oplus \begin{pmatrix}x_2y_3-x_3y_2\\x_3y_1-x_1y_3\\x_1y_2-x_2y_1\end{pmatrix}_{3_{22}},$$
(13.84)

$$\begin{pmatrix}x_1\\x_2\\x_3\end{pmatrix}_{3_{22}} \otimes \begin{pmatrix}y_1\\y_2\\y_3\end{pmatrix}_{3_{22}} = \begin{pmatrix}x_1y_1\\x_2y_2\\x_3y_3\end{pmatrix}_{3_{11}} \oplus \begin{pmatrix}x_2y_3+x_3y_2\\x_3y_1+x_1y_3\\x_1y_2+x_2y_1\end{pmatrix}_{3_{11}} \oplus \begin{pmatrix}x_2y_3-x_3y_2\\x_3y_1-x_1y_3\\x_1y_2-x_2y_1\end{pmatrix}_{3_{21}},$$
(13.85)

$$\begin{pmatrix}x_1\\x_2\\x_3\end{pmatrix}_{3_{11}} \otimes \begin{pmatrix}y_1\\y_2\\y_3\end{pmatrix}_{3_{12}} = (x_1y_1+x_2y_2+x_3y_3)_{1_0} \oplus \begin{pmatrix}x_1y_1+\omega^2 x_2y_2+\omega x_3y_3\\ \omega x_1y_1+\omega^2 x_2y_2+x_3y_3\end{pmatrix}_{2_1}$$

$$\oplus \begin{pmatrix}x_1y_2+\omega^2 x_2y_3+\omega x_3y_1\\ \omega x_1y_3+\omega^2 x_2y_1+x_3y_2\end{pmatrix}_{2_2}$$

$$\oplus \begin{pmatrix}x_1y_3+\omega^2 x_2y_1+\omega x_3y_2\\ \omega x_1y_2+\omega^2 x_2y_3+x_3y_1\end{pmatrix}_{2_3}$$

$$\oplus \begin{pmatrix}x_1y_3+x_2y_1+x_3y_2\\ x_1y_2+x_2y_3+x_3y_1\end{pmatrix}_{2_4},$$
(13.86)

$$\begin{pmatrix}x_1\\x_2\\x_3\end{pmatrix}_{3_{11}} \otimes \begin{pmatrix}y_1\\y_2\\y_3\end{pmatrix}_{3_{21}} = \begin{pmatrix}x_1y_1\\x_2y_2\\x_3y_3\end{pmatrix}_{3_{22}} \oplus \begin{pmatrix}x_3y_2-x_2y_3\\x_1y_3-x_3y_1\\x_2y_1-x_1y_2\end{pmatrix}_{3_{12}} \oplus \begin{pmatrix}x_3y_2+x_2y_3\\x_1y_3+x_3y_1\\x_2y_1+x_1y_2\end{pmatrix}_{3_{22}},$$
(13.87)

$$\begin{pmatrix}x_1\\x_2\\x_3\end{pmatrix}_{3_{11}} \otimes \begin{pmatrix}y_1\\y_2\\y_3\end{pmatrix}_{3_{22}} = (x_1y_1+x_2y_2+x_3y_3)_{1_1} \oplus \begin{pmatrix}x_1y_1+\omega^2 x_2y_2+\omega x_3y_3\\-\omega x_1y_1-\omega^2 x_2y_2-x_3y_3\end{pmatrix}_{2_1}$$

$$\oplus \begin{pmatrix}x_1y_2+\omega^2 x_2y_3+\omega x_3y_1\\-\omega x_1y_3-\omega^2 x_2y_1-x_3y_2\end{pmatrix}_{2_2}$$

$$\oplus \begin{pmatrix}x_1y_3+\omega^2 x_2y_1+\omega x_3y_2\\-\omega x_1y_2-\omega^2 x_2y_3-x_3y_1\end{pmatrix}_{2_3}$$

$$\oplus \begin{pmatrix}x_1y_3+x_2y_1+x_3y_2\\-x_1y_2-x_2y_3-x_3y_1\end{pmatrix}_{2_4}, \quad (13.88)$$

$$\begin{pmatrix}x_1\\x_2\\x_3\end{pmatrix}_{3_{12}} \otimes \begin{pmatrix}y_1\\y_2\\y_3\end{pmatrix}_{3_{21}} = (x_1y_1+x_2y_2+x_3y_3)_{1_1} \oplus \begin{pmatrix}x_1y_1+\omega^2 x_2y_2+\omega x_3y_3\\-\omega x_1y_1-\omega^2 x_2y_2-x_3y_3\end{pmatrix}_{2_1}$$

$$\oplus \begin{pmatrix}x_1y_3+\omega^2 x_2y_1+\omega x_3y_2\\-\omega x_1y_2-\omega^2 x_2y_3-x_3y_1\end{pmatrix}_{2_2}$$

$$\oplus \begin{pmatrix}x_1y_2+\omega^2 x_2y_3+\omega x_3y_1\\-\omega x_1y_3-\omega^2 x_2y_1-x_3y_2\end{pmatrix}_{2_3}$$

$$\oplus \begin{pmatrix}x_1y_2+x_2y_3+x_3y_1\\-x_1y_3-x_2y_1-x_3y_2\end{pmatrix}_{2_4}. \quad (13.89)$$

$$\begin{pmatrix}x_1\\x_2\\x_3\end{pmatrix}_{3_{12}} \otimes \begin{pmatrix}y_1\\y_2\\y_3\end{pmatrix}_{3_{22}} = \begin{pmatrix}x_1y_1\\x_2y_2\\x_3y_3\end{pmatrix}_{3_{21}} \oplus \begin{pmatrix}x_3y_2-x_2y_3\\x_1y_3-x_3y_1\\x_2y_1-x_1y_2\end{pmatrix}_{3_{11}} \oplus \begin{pmatrix}x_3y_2+x_2y_3\\x_1y_3+x_3y_1\\x_2y_1+x_1y_2\end{pmatrix}_{3_{21}}, \quad (13.90)$$

$$\begin{pmatrix}x_1\\x_2\\x_3\end{pmatrix}_{3_{21}} \otimes \begin{pmatrix}y_1\\y_2\\y_3\end{pmatrix}_{3_{22}} = (x_1y_1+x_2y_2+x_3y_3)_{1_0} \oplus \begin{pmatrix}x_1y_1+\omega^2 x_2y_2+\omega x_3y_3\\ \omega x_1y_1+\omega^2 x_2y_2+x_3y_3\end{pmatrix}_{2_1}$$

$$\oplus \begin{pmatrix}x_1y_2+\omega^2 x_2y_3+\omega x_3y_1\\ \omega x_1y_3+\omega^2 x_2y_1+x_3y_2\end{pmatrix}_{2_2}$$

$$\oplus \begin{pmatrix}x_3y_2+\omega^2 x_1y_3+\omega x_2y_1\\ x_1y_2+\omega x_2y_3+\omega^2 x_3y_1\end{pmatrix}_{2_3}$$

$$\oplus \begin{pmatrix}x_1y_3+x_2y_1+x_3y_2\\ x_1y_2+x_2y_3+x_3y_1\end{pmatrix}_{2_4}, \quad (13.91)$$

13.3 Δ(54)

The tensor products between doublets are:

$$\begin{pmatrix} x_1 \\ x_2 \end{pmatrix}_{2_k} \otimes \begin{pmatrix} y_1 \\ y_2 \end{pmatrix}_{2_k} = (x_1 y_2 + x_2 y_1)_{1_0} \oplus (x_1 y_2 - x_2 y_1)_{1_1} \oplus \begin{pmatrix} x_2 y_2 \\ x_1 y_1 \end{pmatrix}_{2_k}, \quad (13.92)$$

for $k = 1, 2, 3, 4$,

$$\begin{pmatrix} x_1 \\ x_2 \end{pmatrix}_{2_1} \otimes \begin{pmatrix} y_1 \\ y_2 \end{pmatrix}_{2_2} = \begin{pmatrix} x_2 y_2 \\ x_1 y_1 \end{pmatrix}_{2_3} \oplus \begin{pmatrix} x_1 y_2 \\ x_2 y_1 \end{pmatrix}_{2_4}, \quad (13.93)$$

$$\begin{pmatrix} x_1 \\ x_2 \end{pmatrix}_{2_1} \otimes \begin{pmatrix} y_1 \\ y_2 \end{pmatrix}_{2_3} = \begin{pmatrix} x_2 y_2 \\ x_1 y_1 \end{pmatrix}_{2_2} \oplus \begin{pmatrix} x_2 y_1 \\ x_1 y_2 \end{pmatrix}_{2_4}, \quad (13.94)$$

$$\begin{pmatrix} x_1 \\ x_2 \end{pmatrix}_{2_1} \otimes \begin{pmatrix} y_1 \\ y_2 \end{pmatrix}_{2_4} = \begin{pmatrix} x_1 y_2 \\ x_2 y_1 \end{pmatrix}_{2_2} \oplus \begin{pmatrix} x_1 y_1 \\ x_2 y_2 \end{pmatrix}_{2_3}, \quad (13.95)$$

$$\begin{pmatrix} x_1 \\ x_2 \end{pmatrix}_{2_2} \otimes \begin{pmatrix} y_1 \\ y_2 \end{pmatrix}_{2_3} = \begin{pmatrix} x_2 y_2 \\ x_1 y_1 \end{pmatrix}_{2_1} \oplus \begin{pmatrix} x_1 y_2 \\ x_2 y_1 \end{pmatrix}_{2_4}, \quad (13.96)$$

$$\begin{pmatrix} x_1 \\ x_2 \end{pmatrix}_{2_2} \otimes \begin{pmatrix} y_1 \\ y_2 \end{pmatrix}_{2_4} = \begin{pmatrix} x_1 y_1 \\ x_2 y_2 \end{pmatrix}_{2_1} \oplus \begin{pmatrix} x_1 y_2 \\ x_2 y_1 \end{pmatrix}_{2_3}, \quad (13.97)$$

$$\begin{pmatrix} x_1 \\ x_2 \end{pmatrix}_{2_3} \otimes \begin{pmatrix} y_1 \\ y_2 \end{pmatrix}_{2_4} = \begin{pmatrix} x_1 y_2 \\ x_2 y_1 \end{pmatrix}_{2_1} \oplus \begin{pmatrix} x_1 y_1 \\ x_2 y_2 \end{pmatrix}_{2_2}. \quad (13.98)$$

The tensor products between doublets and triplets are:

$$\begin{pmatrix} x_1 \\ x_2 \end{pmatrix}_{2_1} \otimes \begin{pmatrix} y_1 \\ y_2 \\ y_3 \end{pmatrix}_{3_{1k}} = \begin{pmatrix} x_1 y_1 + \omega^2 x_2 y_1 \\ \omega x_1 y_2 + \omega x_2 y_2 \\ \omega^2 x_1 y_3 + x_2 y_3 \end{pmatrix}_{3_{1k}} \oplus \begin{pmatrix} x_1 y_1 - \omega^2 x_2 y_1 \\ \omega x_1 y_2 - \omega x_2 y_2 \\ \omega^2 x_1 y_3 - x_2 y_3 \end{pmatrix}_{3_{2k}}, \quad (13.99)$$

$$\begin{pmatrix} x_1 \\ x_2 \end{pmatrix}_{2_1} \otimes \begin{pmatrix} y_1 \\ y_2 \\ y_3 \end{pmatrix}_{3_{2k}} = \begin{pmatrix} x_1 y_1 + \omega^2 x_2 y_1 \\ \omega x_1 y_2 + \omega x_2 y_2 \\ \omega^2 x_1 y_3 + x_2 y_3 \end{pmatrix}_{3_{2k}} \oplus \begin{pmatrix} x_1 y_1 - \omega^2 x_2 y_1 \\ \omega x_1 y_2 - \omega x_2 y_2 \\ \omega^2 x_1 y_3 - x_2 y_3 \end{pmatrix}_{3_{1k}}, \quad (13.100)$$

$$\begin{pmatrix} x_1 \\ x_2 \end{pmatrix}_{2_2} \otimes \begin{pmatrix} y_1 \\ y_2 \\ y_3 \end{pmatrix}_{3_{11}} = \begin{pmatrix} \omega x_1 y_2 + x_2 y_3 \\ \omega^2 x_1 y_3 + \omega^2 x_2 y_1 \\ x_1 y_1 + \omega x_2 y_2 \end{pmatrix}_{3_{11}} \oplus \begin{pmatrix} \omega x_1 y_2 - x_2 y_3 \\ \omega^2 x_1 y_3 - \omega^2 x_2 y_1 \\ x_1 y_1 - \omega x_2 y_2 \end{pmatrix}_{3_{21}}, \quad (13.101)$$

$$\begin{pmatrix}x_1\\x_2\end{pmatrix}_{2_2} \otimes \begin{pmatrix}y_1\\y_2\\y_3\end{pmatrix}_{3_{21}} = \begin{pmatrix}\omega x_1 y_2 + x_2 y_3\\ \omega^2 x_1 y_3 + \omega^2 x_2 y_1\\ x_1 y_1 + \omega x_2 y_2\end{pmatrix}_{3_{21}} \oplus \begin{pmatrix}\omega x_1 y_2 - x_2 y_3\\ \omega^2 x_1 y_3 - \omega^2 x_2 y_1\\ x_1 y_1 - \omega x_2 y_2\end{pmatrix}_{3_{11}},$$
(13.102)

$$\begin{pmatrix}x_1\\x_2\end{pmatrix}_{2_2} \otimes \begin{pmatrix}y_1\\y_2\\y_3\end{pmatrix}_{3_{12}} = \begin{pmatrix}\omega x_1 y_3 + x_2 y_2\\ \omega^2 x_1 y_1 + \omega^2 x_2 y_3\\ x_1 y_2 + \omega x_2 y_1\end{pmatrix}_{3_{12}} \oplus \begin{pmatrix}\omega x_1 y_3 - x_2 y_2\\ \omega^2 x_1 y_1 - \omega^2 x_2 y_3\\ x_1 y_2 - \omega x_2 y_1\end{pmatrix}_{3_{22}},$$
(13.103)

$$\begin{pmatrix}x_1\\x_2\end{pmatrix}_{2_2} \otimes \begin{pmatrix}y_1\\y_2\\y_3\end{pmatrix}_{3_{22}} = \begin{pmatrix}\omega x_1 y_3 + x_2 y_2\\ \omega^2 x_1 y_1 + \omega^2 x_2 y_3\\ x_1 y_2 + \omega x_2 y_1\end{pmatrix}_{3_{22}} \oplus \begin{pmatrix}\omega x_1 y_3 - x_2 y_2\\ \omega^2 x_1 y_1 - \omega^2 x_2 y_3\\ x_1 y_2 - \omega x_2 y_1\end{pmatrix}_{3_{12}},$$
(13.104)

$$\begin{pmatrix}x_1\\x_2\end{pmatrix}_{2_3} \otimes \begin{pmatrix}y_1\\y_2\\y_3\end{pmatrix}_{3_{11}} = \begin{pmatrix}\omega x_1 y_3 + x_2 y_2\\ \omega^2 x_1 y_1 + \omega^2 x_2 y_3\\ x_1 y_2 + \omega x_2 y_1\end{pmatrix}_{3_{11}} \oplus \begin{pmatrix}\omega x_1 y_3 - x_2 y_2\\ \omega^2 x_1 y_1 - \omega^2 x_2 y_3\\ x_1 y_2 - \omega x_2 y_1\end{pmatrix}_{3_{21}},$$
(13.105)

$$\begin{pmatrix}x_1\\x_2\end{pmatrix}_{2_3} \otimes \begin{pmatrix}y_1\\y_2\\y_3\end{pmatrix}_{3_{21}} = \begin{pmatrix}\omega x_1 y_3 + x_2 y_2\\ \omega^2 x_1 y_1 + \omega^2 x_2 y_3\\ x_1 y_2 + \omega x_2 y_1\end{pmatrix}_{3_{21}} \oplus \begin{pmatrix}\omega x_1 y_3 - x_2 y_2\\ \omega^2 x_1 y_1 - \omega^2 x_2 y_3\\ x_1 y_2 - \omega x_2 y_1\end{pmatrix}_{3_{11}},$$
(13.106)

$$\begin{pmatrix}x_1\\x_2\end{pmatrix}_{2_3} \otimes \begin{pmatrix}y_1\\y_2\\y_3\end{pmatrix}_{3_{12}} = \begin{pmatrix}\omega x_1 y_2 + x_2 y_3\\ \omega^2 x_1 y_3 + \omega^2 x_2 y_1\\ x_1 y_1 + \omega x_2 y_2\end{pmatrix}_{3_{12}} \oplus \begin{pmatrix}\omega x_1 y_2 - x_2 y_3\\ \omega^2 x_1 y_3 - \omega^2 x_2 y_1\\ x_1 y_1 - \omega x_2 y_2\end{pmatrix}_{3_{22}},$$
(13.107)

$$\begin{pmatrix}x_1\\x_2\end{pmatrix}_{2_3} \otimes \begin{pmatrix}y_1\\y_2\\y_3\end{pmatrix}_{3_{22}} = \begin{pmatrix}\omega x_1 y_2 + x_2 y_3\\ \omega^2 x_1 y_3 + \omega^2 x_2 y_1\\ x_1 y_1 + \omega x_2 y_2\end{pmatrix}_{3_{22}} \oplus \begin{pmatrix}\omega x_1 y_2 - x_2 y_3\\ \omega^2 x_1 y_3 - \omega^2 x_2 y_1\\ x_1 y_1 - \omega x_2 y_2\end{pmatrix}_{3_{12}},$$
(13.108)

$$\begin{pmatrix}x_1\\x_2\end{pmatrix}_{2_4} \otimes \begin{pmatrix}y_1\\y_2\\y_3\end{pmatrix}_{3_{11}} = \begin{pmatrix}x_1 y_3 + x_2 y_2\\ x_1 y_1 + x_2 y_3\\ x_1 y_2 + x_2 y_1\end{pmatrix}_{3_{11}} \oplus \begin{pmatrix}x_1 y_3 - x_2 y_2\\ x_1 y_1 - x_2 y_3\\ x_1 y_2 - x_2 y_1\end{pmatrix}_{3_{21}}, \quad (13.109)$$

$$\begin{pmatrix}x_1\\x_2\end{pmatrix}_{2_4} \otimes \begin{pmatrix}y_1\\y_2\\y_3\end{pmatrix}_{3_{21}} = \begin{pmatrix}x_1 y_3 + x_2 y_2\\ x_1 y_1 + x_2 y_3\\ x_1 y_2 + x_2 y_1\end{pmatrix}_{3_{21}} \oplus \begin{pmatrix}x_1 y_3 - x_2 y_2\\ x_1 y_1 - x_2 y_3\\ x_1 y_2 - x_2 y_1\end{pmatrix}_{3_{11}}, \quad (13.110)$$

$$\begin{pmatrix}x_1\\x_2\end{pmatrix}_{2_4} \otimes \begin{pmatrix}y_1\\y_2\\y_3\end{pmatrix}_{3_{12}} = \begin{pmatrix}x_1 y_2 + x_2 y_3\\ x_1 y_3 + x_2 y_1\\ x_1 y_1 + x_2 y_2\end{pmatrix}_{3_{12}} \oplus \begin{pmatrix}x_1 y_2 - x_2 y_3\\ x_1 y_3 - x_2 y_1\\ x_1 y_1 - x_2 y_2\end{pmatrix}_{3_{22}}, \quad (13.111)$$

$$\begin{pmatrix} x_1 \\ x_2 \end{pmatrix}_{2_4} \otimes \begin{pmatrix} y_1 \\ y_2 \\ y_3 \end{pmatrix}_{3_{22}} = \begin{pmatrix} x_1 y_2 + x_2 y_3 \\ x_1 y_3 + x_2 y_1 \\ x_1 y_1 + x_2 y_2 \end{pmatrix}_{3_{22}} \oplus \begin{pmatrix} x_1 y_2 - x_2 y_3 \\ x_1 y_3 - x_2 y_1 \\ x_1 y_1 - x_2 y_2 \end{pmatrix}_{3_{12}}. \qquad (13.112)$$

Finally, the tensor products of the non-trivial singlet $\mathbf{1}_1$ with other representations are:

$$\mathbf{2}_k \otimes \mathbf{1}_1 = \mathbf{2}_k, \qquad \mathbf{3}_{1k} \otimes \mathbf{1}_1 = \mathbf{3}_{2k}, \qquad \mathbf{3}_{2k} \otimes \mathbf{1}_1 = \mathbf{3}_{1k}. \qquad (13.113)$$

References

1. Escobar, J.A., Luhn, C.: J. Math. Phys. **50**, 013524 (2009). arXiv:0809.0639 [hep-th]

Chapter 14
Subgroups and Decompositions of Multiplets

In particle physics, a symmetry is often broken to a subgroup to describe low energy phenomena. Therefore, it is very important to study the breaking patterns of discrete groups and decompositions of multiplets. In this chapter, we discuss decompositions of multiplets for the groups studied in the previous chapters.

Suppose that a finite group G has order N and that M is a divisor of N. Then, Lagrange's theorem implies that a finite group H with order M is a candidate for a subgroup of G (see Appendix A).

An irreducible representation r_G of G can be decomposed into irreducible representations $r_{H,m}$ of its subgroup H as $r_G = \sum_m r_{H,m}$. If the trivial singlet of H is included in such a decomposition $\sum_m r_{H,m}$, and a scalar field with such a trivial singlet develops its vacuum expectation value (VEV), the group G breaks to H. On the other hand, if a scalar field in a multiplet r_G develops its VEV and it does not correspond to the trivial singlet of H, the group G breaks not to H, but to another group.

In the following sections, we consider decompositions of multiplets of G into multiplets of subgroups. For a finite group G, there are several chains of subgroups, viz.,

$$G \to G_1 \to \cdots \to G_k \to Z_N \to \{e\},$$

$$G \to G'_1 \to \cdots \to G'_m \to Z_M \to \{e\},$$

and so on. It should be obvious that the smallest non-trivial subgroup in these chains will be an Abelian group such as Z_N or Z_M. Since we concentrate on subgroups discussed explicitly in the previous chapters, we consider the largest subgroup, i.e., G_1 or G'_1, in each chain of subgroups.

14.1 S_3

We begin with S_3 because it is the smallest non-Abelian discrete group, with order equal to $2 \times 3 = 6$. There are then two candidates for subgroups. One is a group of order two and the other a group of order three. The former corresponds to Z_2 and the latter to Z_3. As discussed in Sect. 3.1, S_3 consists of $\{e, a, b, ab, ba, bab\}$, where $a^2 = e$ and $(ab)^3 = e$. Now the subgroup Z_2 consists of, e.g., $\{e, a\}$, or another combination such as $\{e, b\}$ or $\{e, bab\}$, which also correspond to Z_2. The subgroup Z_3 consists of $\{e, ab, ba = (ab)^2\}$.

The group S_3 has two singlets **1** and **1'**, and one doublet **2**. Both subgroups Z_2 and Z_3 are Abelian. Thus, decompositions of multiplets under Z_2 and Z_3 are rather simple. We shall examine these decompositions in what follows. The breaking pattern of S_3 is summarized in Table 14.1.

14.1.1 $S_3 \to Z_3$

The elements $\{e, ab, ba\}$ of S_3 constitute the Z_3 subgroup, which is a normal subgroup. There is no other choice to obtain a Z_3 subgroup. There are three singlet representations $\mathbf{1}_k$ with $k = 0, 1, 2$ for Z_3, that is, $ab = \omega^k$ on $\mathbf{1}_k$. Recall that $\chi_1(ab) = \chi_{1'}(ab) = 1$ for both **1** and **1'** of S_3. Thus, both **1** and **1'** of S_3 correspond to $\mathbf{1}_0$ of Z_3. On the other hand, the doublet **2** of S_3 decomposes into two singlets of Z_3. Since $\chi_2(ab) = -1$, the S_3 doublet **2** decomposes into $\mathbf{1}_1$ and $\mathbf{1}_2$ of Z_3.

In order to understand this decomposition explicitly, we take the two-dimensional representation of the group element ab as given in (2.28), viz.,

$$ab = \begin{pmatrix} -1/2 & -\sqrt{3}/2 \\ \sqrt{3}/2 & -1/2 \end{pmatrix}. \tag{14.1}$$

Then the doublet (x_1, x_2) decomposes into two non-trivial singlets, viz.,

$$\mathbf{1}_1 : x_1 - ix_2, \qquad \mathbf{1}_2 : x_1 + ix_2. \tag{14.2}$$

14.1.2 $S_3 \to Z_2$

The subgroup Z_2 of S_3 consists of, e.g., $\{e, a\}$. It has two singlet representations $\mathbf{1}_k$, for $k = 0, 1$, that is, $a = (-1)^k$ on $\mathbf{1}_k$. Recall that $\chi_1(a) = 1$ and $\chi_{1'}(a) = -1$ for **1** and **1'** of S_3. Thus, **1** and **1'** of S_3 correspond to $\mathbf{1}_0$ and $\mathbf{1}_1$ of Z_2, respectively. On the other hand, the doublet **2** of S_3 decomposes into two singlets of Z_2. Since $\chi_2(a) = -1$, the S_3 doublet **2** decomposes into $\mathbf{1}_0$ and $\mathbf{1}_1$ of Z_2. Indeed, the element a is represented on **2** in (2.28) by

$$a = \begin{pmatrix} 1 & 0 \\ 0 & -1 \end{pmatrix}. \tag{14.3}$$

Then for the doublet (x_1, x_2), the elements x_1 and x_2 correspond to $x_1 = \mathbf{1}_0$ and $x_2 = \mathbf{1}_1$, respectively.

14.2 S_4

Table 14.1 Breaking pattern of S_3

S_3	Z_3	S_3	Z_2
1	1_0	1	1_0
1'	1_0	1'	1_1
2	$1_1 + 1_2$	2	$1_0 + 1_1$

In addition to $\{e, a\}$, there are other Z_2 subgroups, viz., $\{e, b\}$ and $\{e, aba\}$. In both cases, the same results are obtained when we choose a proper basis. These are examples of Abelian subgroups.

For non-Abelian subgroups, the same situation arises. That is, different elements of a finite group G can generate the same subgroup. We exemplify this by considering D_6. All elements of D_6 can be written in the form $a^m b^k$ for $m = 0, 1, \ldots, 5$, and $k = 0, 1$, where $a^6 = e$ and $bab = a^{-1}$. Denoting $\tilde{a} = a^2$, we find that the elements $\tilde{a}^m b^k$ for $m = 0, 1, 2$, and $k = 0, 1$, correspond to the subgroup $D_3 \simeq S_3$. On the other hand, denoting $\tilde{b} = ab$, we find that the elements $\tilde{a}^m \tilde{b}^k$ for $m = 0, 1, 2$, and $k = 0, 1$, correspond to another D_3 subgroup. The decompositions of D_6 multiplets into D_3 multiplets are the same for both D_3 subgroups when we move to a proper basis.

14.2 S_4

As mentioned in Chap. 13, the group S_4 is isomorphic to $\Delta(24)$ and $(Z_2 \times Z_2) \rtimes S_3$. It is convenient to use the terminology of $(Z_2 \times Z_2) \rtimes S_3$. That is, all the elements are expressed in the form $b^k c^\ell a^m a'^n$ with $k = 0, 1, 2$, and $\ell, m, n = 0, 1$ (see Chap. 13). The generators a, a', b, and c are related to the notation of Sect. 3.2 in the following way:

$$b = c_1, \qquad c = f_1, \qquad a = a_4, \qquad a' = a_2. \tag{14.4}$$

They satisfy the algebraic relations

$$b^3 = c^2 = (bc)^2 = a^2 = a'^2 = e, \qquad aa' = a'a,$$
$$bab^{-1} = a^{-1}a'^{-1}, \qquad ba'b^{-1} = a, \tag{14.5}$$
$$cac^{-1} = a'^{-1}, \qquad ca'a^{-1} = a^{-1}.$$

Furthermore, their representations on **1**, **1'**, **2**, **3**, and **3'** are shown in Table 14.2. As subgroups, S_4 contains non-Abelian groups S_3, A_4, and $\Sigma(8)$, the latter being $(Z_2 \times Z_2) \rtimes Z_2$. Thus, the decompositions of S_4 are non-trivial compared with those of S_3. The breaking pattern of S_4 is summarized in Table 14.3.

Table 14.2 Representations of S_4 elements

	1	1′	2	3	3′
b	1	1	$\begin{pmatrix} \omega & 0 \\ 0 & \omega^2 \end{pmatrix}$	$\begin{pmatrix} 0 & 1 & 0 \\ 0 & 0 & 1 \\ 1 & 0 & 0 \end{pmatrix}$	$\begin{pmatrix} 0 & 1 & 0 \\ 0 & 0 & 1 \\ 1 & 0 & 0 \end{pmatrix}$
c	1	−1	$\begin{pmatrix} 0 & 1 \\ 1 & 0 \end{pmatrix}$	$\begin{pmatrix} 0 & 0 & 1 \\ 0 & 1 & 0 \\ 1 & 0 & 0 \end{pmatrix}$	$\begin{pmatrix} 0 & 0 & -1 \\ 0 & -1 & 0 \\ -1 & 0 & 0 \end{pmatrix}$
a	1	1	$\begin{pmatrix} 1 & 0 \\ 0 & 1 \end{pmatrix}$	$\begin{pmatrix} -1 & 0 & 0 \\ 0 & -1 & 0 \\ 0 & 0 & 1 \end{pmatrix}$	$\begin{pmatrix} -1 & 0 & 0 \\ 0 & -1 & 0 \\ 0 & 0 & 1 \end{pmatrix}$
a'	1	1	$\begin{pmatrix} 1 & 0 \\ 0 & 1 \end{pmatrix}$	$\begin{pmatrix} 1 & 0 & 0 \\ 0 & -1 & 0 \\ 0 & 0 & -1 \end{pmatrix}$	$\begin{pmatrix} 1 & 0 & 0 \\ 0 & -1 & 0 \\ 0 & 0 & -1 \end{pmatrix}$

Table 14.3 Breaking pattern of S_4

S_4	S_3	S_4	A_4	S_4	$\Sigma(8)$
1	1	1	1	1	1_{+0}
1′	1′	1′	1	1′	1_{-0}
2	2	2	1′ + 1″	2	$1_{+0} + 1_{-0}$
3	1 + 2	3	3	3	$1_{+1} + 2$
3′	1′ + 2	3′	3	3′	$1_{-1} + 2$

Table 14.4 Representations of S_3 elements

	1	1′	2
b	1	1	$\begin{pmatrix} \omega & 0 \\ 0 & \omega^2 \end{pmatrix}$
c	1	−1	$\begin{pmatrix} 0 & 1 \\ 1 & 0 \end{pmatrix}$

14.2.1 $S_4 \to S_3$

The subgroup S_3 elements are $\{a_1, b_1, d_1, d_1, e_1, f_1\}$. Alternatively, they can be denoted by $b^k c^\ell$ with $k = 0, 1, 2$, and $\ell = 0, 1$, i.e., $\{e, b, b^2, c, bc, b^2c\}$. Among them, Table 14.4 shows the representations of the generators b and c on **1**, **1′**, and **2** of S_3. Then the singlets **1** and **1′** and the doublet **2** remain the same representation of S_3, i.e., **1**, **1′**, and **2** for each. Triplets **3** and **3′** are decomposed to **1 + 2** and **1′ + 2**. The

14.2 S_4

Table 14.5 Representations of A_4 elements

		1	1'	1''	3
	b	1	ω	ω^2	$\begin{pmatrix} 0 & 1 & 0 \\ 0 & 0 & 1 \\ 1 & 0 & 0 \end{pmatrix}$
	a	1	1	1	$\begin{pmatrix} -1 & 0 & 0 \\ 0 & -1 & 0 \\ 0 & 0 & 1 \end{pmatrix}$
	a'	1	1	1	$\begin{pmatrix} 1 & 0 & 0 \\ 0 & -1 & 0 \\ 0 & 0 & -1 \end{pmatrix}$

components of **3** (x_1, x_2, x_3) decompose to **1** and **2** according to

$$\mathbf{1}: x_1 + x_2 + x_3, \qquad \mathbf{2}: (x_1 + \omega^2 x_2 + \omega x_3, \omega x_1 + x_2 + \omega^2 x_3), \qquad (14.6)$$

and the components of **3'** decompose to **1'** and **2** according to

$$\mathbf{1'}: x_1 + x_2 + x_3, \qquad \mathbf{2}: (x_1 + \omega^2 x_2 + \omega x_3, -\omega x_1 - x_2 - \omega^2 x_3). \qquad (14.7)$$

14.2.2 $S_4 \to A_4$

The A_4 subgroup consists of $b^k a^m a'^n$ with $k = 0, 1, 2$, and $m, n = 0, 1$. Recall that A_4 is isomorphic to $\Delta(12)$. Table 14.5 shows the representations of the generators b, a, and a' on **1**, **1'**, **1''**, and **3** of A_4. Then the representations **1**, **1'**, **2**, **3**, and **3'** of S_4 decompose to **1**, **1**, **1'** + **1''**, **3**, and **3**, respectively.

14.2.3 $S_4 \to \Sigma(8)$

The subgroup $\Sigma(8)$, i.e., $(Z_2 \times Z_2) \rtimes Z_2$, consists of elements $c^\ell a^m a'^n$ with $\ell, m, n = 0, 1$. Table 14.6 shows the representations of the generators c, a, and a' on $\mathbf{1}_{+0}$, $\mathbf{1}_{+1}$, $\mathbf{1}_{-0}$, $\mathbf{1}_{-1}$, and $\mathbf{2}_{1,0}$ of $\Sigma(8)$.

The representations **1**, **1'**, **2**, **3**, and **3'** of S_4 decompose to $\mathbf{1}_{+0}$, $\mathbf{1}_{-0}$, $\mathbf{1}_{+0} + \mathbf{1}_{-0}$, $\mathbf{1}_{+1} + \mathbf{2}$, and $\mathbf{1}_{-1} + \mathbf{2}$, respectively. The components of **3** (x_1, x_2, x_3) decompose to $\mathbf{1}_{+1}$ and **2** according to

$$\mathbf{1}_{+1}: x_2, \qquad \mathbf{2}: (x_3, x_1), \qquad (14.8)$$

and the components of **3'** decompose to $\mathbf{1}_{-1}$ and **2** according to

$$\mathbf{1}_{-1}: x_2, \qquad \mathbf{2}: (x_3, -x_1). \qquad (14.9)$$

Table 14.6 Representations of $\Sigma(8)$ elements

	$\mathbf{1}_{+0}$	$\mathbf{1}_{+1}$	$\mathbf{1}_{-0}$	$\mathbf{1}_{-1}$	$\mathbf{2}_{1,0}$
c	1	1	-1	-1	$\begin{pmatrix} 0 & 1 \\ 1 & 0 \end{pmatrix}$
a	1	-1	1	-1	$\begin{pmatrix} 1 & 0 \\ 0 & -1 \end{pmatrix}$
a'	1	-1	1	-1	$\begin{pmatrix} -1 & 0 \\ 0 & 1 \end{pmatrix}$

Table 14.7 Representations of $\tilde{a}$, $\tilde{a}'$, and $\tilde{b}$ in $\Delta(12)$

	$\mathbf{1}_k$	$\mathbf{3}$
a	1	$\begin{pmatrix} -1 & 0 & 0 \\ 0 & 1 & 0 \\ 0 & 0 & -1 \end{pmatrix}$
a'	1	$\begin{pmatrix} -1 & 0 & 0 \\ 0 & -1 & 0 \\ 0 & 0 & 1 \end{pmatrix}$
b	ω^k	$\begin{pmatrix} 0 & 1 & 0 \\ 0 & 0 & 1 \\ 1 & 0 & 0 \end{pmatrix}$

Table 14.8 Breaking pattern of A_4

A_4	Z_3	A_4	$Z_2 \times Z_2$
$\mathbf{1}_k$	$\mathbf{1}_k$	$\mathbf{1}_k$	$\mathbf{1}_{0,0}$
$\mathbf{3}$	$\mathbf{1}_0 + \mathbf{1}_1 + \mathbf{1}_2$	$\mathbf{3}$	$\mathbf{1}_{1,1} + \mathbf{1}_{0,1} + \mathbf{1}_{1,0}$

14.3 A_4

The group A_4 is isomorphic to $\Delta(12)$. Here, we apply the generic results of $\Delta(3N^3)$ to the group A_4. All elements of $\Delta(12)$ can be written in the form $b^k a^m a'^n$ with $k = 0, 1, 2$, and $m, n = 0, 1$. Table 14.7 shows the representations of generators a, a', and b on each representation. Regarding subgroups, A_4 contains Abelian groups Z_3 and $Z_2 \times Z_2$. The breaking pattern of A_4 is summarized in Table 14.8.

14.3.1 $A_4 \to Z_3$

The group Z_3 consists of $\{e, b, b^2\}$. The representations $\mathbf{1}_k$ and $\mathbf{3}$ of $\Delta(12)$ decompose to $\mathbf{1}_k$ and $\mathbf{1}_0 + \mathbf{1}_1 + \mathbf{1}_2$, respectively. Decomposition of the triplet (x_1, x_2, x_3) is obtained by $\mathbf{1}_0 : x_1 + x_2 + x_3$, $\mathbf{1}_1 : x_1 + \omega^2 x_2 + \omega x_3$, and $\mathbf{1}_2 : x_1 + \omega x_2 + \omega^2 x_3$.

14.4 A_5

Table 14.9 Breaking pattern of A_5

A_5	A_4	A_5	D_5	A_5	D_3
1	1	1	1_+	1	1_+
3	3	3	$1_- + 2_1$	3	$1_- + 2$
3'	3	3'	$1_- + 2_2$	3'	$1_- + 2$
4	$1+3$	4	$2_1 + 2_2$	4	$1_+ + 1_- + 2$
5	$1' + 1'' + 3$	5	$1_+ + 2_1 + 2_2$	5	$1_+ + 2 + 2$

14.3.2 $A_4 \to Z_2 \times Z_2$

The subgroup $Z_2 \times Z_2$ consists of $\{e, a, a', aa'\}$. The representations 1_k and 3 of A_4 decompose to $1_{0,0}$ and $1_{1,1} + 1_{0,1} + 1_{1,0}$, respectively.

14.4 A_5

All elements of A_5 can be expressed as products of $s = a$ and $t = bab$, as shown in Sect. 4.2. Regarding subgroups, A_5 contains the non-Abelian groups A_4, D_5, and S_3. The breaking pattern of A_5 is summarized in Table 14.9.

14.4.1 $A_5 \to A_4$

The subgroup A_4 has elements $\{e, b, \tilde{a}, b\tilde{a}b^2, b^2\tilde{a}b, b\tilde{a}, \tilde{a}b, \tilde{a}b\tilde{a}, b^2\tilde{a}, b^2\tilde{a}b\tilde{a}b\}$, where $\tilde{a} = ab^2aba$. We denote $\tilde{t} = b$ and $\tilde{s} = \tilde{a}$. These satisfy the relations

$$\tilde{s}^2 = \tilde{t}^3 = (\tilde{s}\tilde{t})^3 = e, \tag{14.10}$$

and correspond to the generators s and t of the group A_4 in Sect. 4.1. The representations $\mathbf{1}$, $\mathbf{3}$, $\mathbf{3'}$, $\mathbf{4}$, and $\mathbf{5}$ of A_5 decompose to $\mathbf{1}$, $\mathbf{3}$, $\mathbf{3}$, $\mathbf{1+3}$, and $\mathbf{1' + 1'' + 3}$, respectively.

14.4.2 $A_5 \to D_5$

The subgroup D_5 consists of the elements $a^k \tilde{a}^m$ with $k = 0, 1$, and $m = 0, 1, 2, 3, 4$, where $\tilde{a} \equiv bab^2a$. These satisfy $a^2 = \tilde{a}^5 = e$ and $a\tilde{a}a = \tilde{a}^4$. In order to identify the D_5 basis used in Chap. 6, we define $\tilde{b} = abab^2a$. Table 14.10 shows the representations of these generators $\tilde{a}$ and $\tilde{b}$ on 1_+, 1_-, 2_1, and 2_2 of D_5. The representations $\mathbf{1}$, $\mathbf{3}$, $\mathbf{3'}$, $\mathbf{4}$, and $\mathbf{5}$ of A_5 then decompose to 1_+, $1_- + 2_1$, $1_- + 2_2$, $2_1 + 2_2$, and $1_+ + 2_1 + 2_2$, respectively.

Table 14.10 Representations of elements of D_5

	1_+	1_-	2_1	2_2
$\tilde{a}$	1	1	$\begin{pmatrix} \exp 2\pi i/5 & 0 \\ 0 & -\exp 2\pi i/5 \end{pmatrix}$	$\begin{pmatrix} \exp 4\pi i/5 & 0 \\ 0 & -\exp 4\pi i/5 \end{pmatrix}$
$\tilde{b}$	1	-1	$\begin{pmatrix} 0 & 1 \\ 1 & 0 \end{pmatrix}$	$\begin{pmatrix} 0 & 1 \\ 1 & 0 \end{pmatrix}$

Table 14.11 Representations of T'

	1	$1'$	$1''$	2	$2'$	$2''$	3
s	1	1	1	$-\frac{i}{\sqrt{3}}\begin{pmatrix} 1 & \sqrt{2} \\ \sqrt{2} & -1 \end{pmatrix}$	$-\frac{i}{\sqrt{3}}\begin{pmatrix} 1 & \sqrt{2} \\ \sqrt{2} & -1 \end{pmatrix}$	$-\frac{i}{\sqrt{3}}\begin{pmatrix} 1 & \sqrt{2} \\ \sqrt{2} & -1 \end{pmatrix}$	$\frac{1}{3}\begin{pmatrix} -1 & 2 & 2 \\ 2 & -1 & 2 \\ 2 & 2 & -1 \end{pmatrix}$
r	1	1	1	$\begin{pmatrix} -1 & 0 \\ 0 & -1 \end{pmatrix}$	$\begin{pmatrix} -1 & 0 \\ 0 & -1 \end{pmatrix}$	$\begin{pmatrix} -1 & 0 \\ 0 & -1 \end{pmatrix}$	$\begin{pmatrix} 1 & 0 & 0 \\ 0 & 1 & 0 \\ 0 & 0 & 1 \end{pmatrix}$
t	1	ω	ω^2	$\begin{pmatrix} \omega & 0 \\ 0 & \omega^2 \end{pmatrix}$	$\begin{pmatrix} \omega^2 & 0 \\ 0 & 1 \end{pmatrix}$	$\begin{pmatrix} 1 & 0 \\ 0 & \omega \end{pmatrix}$	$\begin{pmatrix} 1 & 0 & 0 \\ 0 & \omega & 0 \\ 0 & 0 & \omega^2 \end{pmatrix}$

14.4.3 $A_5 \to S_3 \simeq D_3$

Recall that the group S_3 is isomorphic to D_3. The subgroup D_3 consists of elements $b^k \tilde{a}^m$ with $k = 0, 1, 2$, and $m = 0, 1$, where we define $\tilde{a} = ab^2 ab^2 ab$. These generators satisfy $\tilde{a}^2 = e$ and $\tilde{a}b\tilde{a} = b^2$. The representations **1**, **3**, **3'**, **4**, and **5** of A_5 then decompose to 1_+, $1_- + 2$, $1_- + 2$, $1_+ + 1_- + 2$, and $1_+ + 2 + 2$, respectively.

14.5 T'

All elements of T' can be expressed in terms of the generators s, t, and r, which satisfy the algebraic relations $s^2 = r$, $r^2 = t^3 = (st)^3 = e$, and $rt = tr$. Table 14.11 shows the different representations of s, t, and r. Regarding subgroups, T' contains Z_6, Z_4, and Q_4. The breaking pattern of T' is summarized in Table 14.12.

14.5.1 $T' \to Z_6$

The subgroup Z_6 consists of elements a^m, with $m = 0, \ldots, 5$, where $a = rt$ and $a^6 = e$. The group Z_6 has six singlet representations 1_n with $n = 0, \ldots, 5$. On the

14.6 General D_N

Table 14.12 Breaking pattern of T'

T'	Z_6	T'	Z_4	T'	Q_4
1	1_0	1	1_0	1	1_{++}
1'	1_2	1'	1_0	1'	1_{++}
1''	1_4	1''	1_0	1''	1_{++}
2	$1_1 + 1_5$	2	$1_1 + 1_3$	2	2
2'	$1_3 + 1_5$	2'	$1_1 + 1_3$	2'	2
2''	$1_3 + 1_5$	2''	$1_1 + 1_3$	2''	2
3	$1_0 + 1_2 + 1_4$	3	$1_0 + 1_2 + 1_2$	3	$1_{+-} + 1_{-+} + 1_{--}$

singlet 1_n, the generator a is represented by $a = e^{2\pi i n/6}$. The representations **1, 1', 1'', 2, 2', 2''**, and **3** of T' thus decompose to $1_0, 1_2, 1_4, 1_5 + 1_1, 1_5 + 1_3, 1_3 + 1_5$, and $1_0 + 1_2 + 1_4$, respectively.

14.5.2 $T' \to Z_4$

The subgroup Z_4 consists of elements $\{e, s, s^2, s^3\}$. Z_4 has four singlet representations 1_m with $m = 0, 1, 2, 3$. On the singlet 1_m, the generator s is represented by $s = e^{\pi i m/2}$. All the doublets **2, 2'**, and **2''** of T' decompose to two singlets 1_1 and 1_3 of Z_4 according to

$$1_1 : \frac{1+\sqrt{3}}{\sqrt{2}} x_1 + x_2, \qquad 1_3 : -\frac{-1+\sqrt{3}}{\sqrt{2}} i x_1 + x_2,$$

where (x_1, x_2) correspond to the doublets. In addition, the triplet $3 : (x_1, x_2, x_3)$ decomposes to singlets $1_0 + 1_2 + 1_2$ according to $1_0 : (x_1 + x_2 + x_3)$, $1_2 : (-x_1 + x_3)$, and $1_2 : (-x_1 + x_2)$.

14.5.3 $T' \to Q_4$

We consider the subgroup Q_4, which consists of elements $s^m b^k$ with $m = 0, 1, 2, 3$, and $k = 0, 1$. The generator b is defined by $b = tst^2$. The representations **1, 1', 1'', 2, 2', 2''**, and **3** of T' decompose to $1_{++}, 1_{++}, 1_{++}, 2, 2, 2$, and $1_{+-} + 1_{-+} + 1_{--}$, respectively.

14.6 General D_N

Since the group D_N is isomorphic to $Z_N \rtimes Z_2$, D_M and Z_N appear as subgroups of D_N in addition to Z_2, where M is a divisor of N. Recall that all elements of D_N can

Table 14.13 Breaking pattern of D_N for even N

D_N	Z_2	D_N	Z_N
$\mathbf{1}_{++}$	$\mathbf{1}_0$	$\mathbf{1}_{++}$	$\mathbf{1}_0$
$\mathbf{1}_{+-}$	$\mathbf{1}_0$	$\mathbf{1}_{+-}$	$\mathbf{1}_{N/2}$
$\mathbf{1}_{-+}$	$\mathbf{1}_1$	$\mathbf{1}_{-+}$	$\mathbf{1}_{N/2}$
$\mathbf{1}_{--}$	$\mathbf{1}_1$	$\mathbf{1}_{--}$	$\mathbf{1}_0$
$\mathbf{2}_k$	$\mathbf{1}_0 + \mathbf{1}_1$	$\mathbf{2}_k$	$\mathbf{1}_k + \mathbf{1}_{N-k}$

D_N	D_M (M is even)	D_N	D_M (M is odd)
$\mathbf{1}_{++}$	$\mathbf{1}_{++}$	$\mathbf{1}_{++}$	$\mathbf{1}_+$
$\mathbf{1}_{+-}$	$\mathbf{1}_{++}$ (M/N is even)	$\mathbf{1}_{+-}$	$\mathbf{1}_+$
	$\mathbf{1}_{+-}$ (M/N is odd)	$\mathbf{1}_{-+}$	$\mathbf{1}_-$
$\mathbf{1}_{-+}$	$\mathbf{1}_{--}$ (M/N is even)	$\mathbf{1}_{--}$	$\mathbf{1}_-$
	$\mathbf{1}_{-+}$ (M/N is odd)	$\mathbf{2}_k$	$\mathbf{2}_{k'}$ ($k = k' + Mn$)
$\mathbf{1}_{--}$	$\mathbf{1}_{--}$		$\tilde{\mathbf{2}}_{M-k'}$ ($k = M - k'$)
$\mathbf{2}_k$	$\mathbf{2}_{k'}$ ($k = k' + Mn$)		$\mathbf{1}_+ + \mathbf{1}_-$ ($k = Mn$)
	$\tilde{\mathbf{2}}_{M/2-k'}$ ($k = M/2 - k' + Mn$)		
	$\mathbf{1}_{+-} + \mathbf{1}_{-+}$ ($k = M(2n+1)/2$)		
	$\mathbf{1}_{++} + \mathbf{1}_{--}$ ($k = Mn$)		

be written in the form $a^m b^k$ with $m = 0, \ldots, N-1$, and $k = 0, 1$. There are singlets and doublets $\mathbf{2}_k$, where $k = 1, \ldots, N/2 - 1$, for N even and $k = 1, \ldots, (N-1)/2$, for N odd. On the doublet $\mathbf{2}_k$, the generators a and b are represented by

$$a = \begin{pmatrix} \rho^k & 0 \\ 0 & \rho^{-k} \end{pmatrix}, \quad b = \begin{pmatrix} 0 & 1 \\ 1 & 0 \end{pmatrix}, \tag{14.11}$$

where $\rho = e^{2\pi i/N}$. For N even, there are four singlets $\mathbf{1}_{\pm\pm}$. The generator b is represented by $b = 1$ on $\mathbf{1}_{+\pm}$, while $b = -1$ on $\mathbf{1}_{-\pm}$. The generator a is represented by $a = 1$ on $\mathbf{1}_{++}$ and $\mathbf{1}_{--}$, while $a = -1$ on $\mathbf{1}_{+-}$ and $\mathbf{1}_{+-}$. For N odd, there are two singlets $\mathbf{1}_\pm$. The generator b is represented by $b = 1$ on $\mathbf{1}_+$ and $b = -1$ on $\mathbf{1}_-$, while $a = 1$ on both singlets. The general breaking patterns of D_N are summarized in Table 14.13 for even N and Table 14.14 for odd N.

14.6.1 $D_N \to Z_2$

The two elements e and b generate the Z_2 subgroup. Obviously, there are two singlet representations $\mathbf{1}_0$ and $\mathbf{1}_1$, where the subscript denotes the Z_2 charge. That is, we have $b = 1$ on $\mathbf{1}_0$ and $b = -1$ on $\mathbf{1}_1$.

14.6 General D_N

Table 14.14 Breaking pattern of D_N for odd N

D_N	Z_2	D_N	Z_N	D_N	D_M
$\mathbf{1}_+$	$\mathbf{1}_0$	$\mathbf{1}_+$	$\mathbf{1}_0$	$\mathbf{1}_+$	$\mathbf{1}_+$
$\mathbf{1}_-$	$\mathbf{1}_0$	$\mathbf{1}_-$	$\mathbf{1}_0$	$\mathbf{1}_-$	$\mathbf{1}_-$
$\mathbf{2}_k$	$\mathbf{1}_0 + \mathbf{1}_1$	$\mathbf{2}_k$	$\mathbf{1}_k + \mathbf{1}_{N-k}$	$\mathbf{2}_k$	$\mathbf{2}_{k'}\ (N = k' + Mn)$
					$\tilde{\mathbf{2}}_{M-k'}\ (k = Mn - k')$
					$\mathbf{1}_+ + \mathbf{1}_-\ (k = Mn)$

When N is even, the singlets $\mathbf{1}_{++}$ and $\mathbf{1}_{+-}$ of D_N become $\mathbf{1}_0$ of Z_2 and the singlets $\mathbf{1}_{-+}$ and $\mathbf{1}_{--}$ of D_N become $\mathbf{1}_1$ of Z_2. The doublets $\mathbf{2}_k$ of D_N, viz., (x_1, x_2), decompose to two singlets according to $\mathbf{1}_0 : x_1 + x_2$ and $\mathbf{1}_1 : x_1 - x_2$.

When N is odd, the singlet $\mathbf{1}_+$ of D_N becomes $\mathbf{1}_0$ of Z_2 and the singlet $\mathbf{1}_-$ of D_N becomes $\mathbf{1}_1$ of Z_2. The decompositions of doublets $\mathbf{2}_k$ are the same as for N even.

14.6.2 $D_N \to Z_N$

The subgroup Z_N consists of the elements $\{e, a, \ldots, a^{N-1}\}$. Obviously, it is a normal subgroup of D_N and there are N types of irreducible singlet representation $\mathbf{1}_0, \mathbf{1}_1, \ldots, \mathbf{1}_{N-1}$. On $\mathbf{1}_k$, the generator a is represented by $a = \rho^k$.

When N is even, the singlets $\mathbf{1}_{++}$ and $\mathbf{1}_{--}$ of D_N become $\mathbf{1}_0$ of Z_N and the singlets $\mathbf{1}_{+-}$ and $\mathbf{1}_{-+}$ of D_N become $\mathbf{1}_{N/2}$ of Z_N. The doublets $\mathbf{2}_k$ and (x_1, x_2) decompose to two singlets according to $\mathbf{1}_k : x_1$ and $\mathbf{1}_{N-k} : x_2$.

When N is odd, both $\mathbf{1}_+$ and $\mathbf{1}_-$ of D_N become $\mathbf{1}_0$ of Z_N. The decompositions of doublets $\mathbf{2}_k$ are the same as for N even.

14.6.3 $D_N \to D_M$

The above decompositions of D_N are rather straightforward, because the subgroups are Abelian. Here we consider the subgroup D_M, where M is a divisor of N. The decompositions of D_N to D_M are expected to be non-trivial. We denote $\tilde{a} = a^\ell$, where $\ell = N/M$ and ℓ is therefore an integer. The subgroup D_M consists of elements $\tilde{a}^m b^k$ with $m = 0, \ldots, M-1$, and $k = 0, 1$. There are three relevant combinations of (N, M), i.e., $(N, M) =$ (even, even), (even, odd), and (odd, odd).

We start with the combination $(N, M) =$ (even, even). Recall that ab of D_N is represented by $ab = 1$ on $\mathbf{1}_{\pm+}$ and $ab = -1$ on $\mathbf{1}_{\pm-}$. Thus, the representations of $a^\ell b$ depend on whether ℓ is even or odd. When ℓ is odd, ab and $a^\ell b$ are represented in the same way on each of the above singlets. On the other hand, when ℓ is even, we always have the singlet representations with $a^\ell = 1$. The doublets $\mathbf{2}_k$ of D_N correspond to the doublets $\mathbf{2}_{k'}$ of D_M when $k = k'$ (mod M). In addition, when $k =$

$-k'$ (mod M), the doublets $\mathbf{2}_k$ (x_1, x_2) of D_N correspond to the doublets $\mathbf{2}_{M-k'}$ (x_2, x_1) of D_M. That is, the components are swapped over and we denote it by $\tilde{\mathbf{2}}_{M-k'}$. Furthermore, the other doublets $\mathbf{2}_k$ of D_N decompose to two singlets of D_M according to $\mathbf{1}_{+-} + \mathbf{1}_{-+}$ with $\mathbf{1}_{+-} : x_1 + x_2$ and $\mathbf{1}_{-+} : x_1 - x_2$ for $k = (M/2)$ (mod M) and $\mathbf{1}_{++} + \mathbf{1}_{--}$ with $\mathbf{1}_{++} : x_1 + x_2$ and $\mathbf{1}_{--} : x_1 - x_2$ for $k = 0$ (mod M).

Next we consider the case $(N, M) = $ (even, odd). In this case, the singlets $\mathbf{1}_{++}$, $\mathbf{1}_{+-}$, $\mathbf{1}_{-+}$, and $\mathbf{1}_{--}$ of D_N become $\mathbf{1}_+$, $\mathbf{1}_+$, $\mathbf{1}_-$, and $\mathbf{1}_-$ of D_M, respectively. The doublets $\mathbf{2}_k$ of D_N correspond to the doublets $\mathbf{2}_{k'}$ of D_M when $k = k'$ (mod M). In addition, when $k = -k'$ (mod M), the doublets $\mathbf{2}_k$ (x_1, x_2) of D_N correspond to the doublets $\mathbf{2}_{M-k'}$ (x_2, x_1) of D_M. Furthermore, when $k = 0$ (mod M), the other doublets $\mathbf{2}_k$ of D_N decompose to two singlets of D_M according to $\mathbf{1}_+ + \mathbf{1}_-$, where $\mathbf{1}_+ : x_1 + x_2$ and $\mathbf{1}_- : x_1 - x_2$.

We now consider the case $(N, M) = $ (odd, odd). In this case, the singlets $\mathbf{1}_+$ and $\mathbf{1}_-$ of D_N become $\mathbf{1}_+$ and $\mathbf{1}_-$ of D_M. The doublets $\mathbf{2}_k$ of D_N correspond to the doublets $\mathbf{2}_{k'}$ of D_M when $k = k'$ (mod M). In addition, when $k = -k'$ (mod M), the doublets $\mathbf{2}_k$ (x_1, x_2) of D_N correspond to the doublets $\mathbf{2}_{M-k'}$ (x_2, x_1) of D_M. Furthermore, when $k = 0$ (mod M), the other doublets $\mathbf{2}_k$ of D_N decompose to two singlets of D_M according to $\mathbf{1}_+ + \mathbf{1}_-$, where $\mathbf{1}_+ : x_1 + x_2$ and $\mathbf{1}_- : x_1 - x_2$.

14.7 D_4

Here we study D_4, which is the second smallest discrete symmetry. All elements of D_4 can be written in the form $a^m b^k$ with $m = 0, 1, 2, 3$, and $k = 0, 1$. Since the order of D_4 is 8, it contains order 2 and 4 subgroups. There are two types of order 4 group, corresponding to $Z_2 \times Z_2$ and Z_4. All subgroups are Abelian so the decompositions are rather simple.

14.7.1 $D_4 \to Z_4$

The subgroup Z_4 consists of the elements $\{e, a, a^2, a^3\}$. Obviously, it is a normal subgroup of D_4 and there are four types of irreducible singlet representation $\mathbf{1}_m$ with $m = 0, 1, 2, 3$, where a is represented by $a = e^{\pi i m/2}$. From the characters of the group D_4, it is found that singlet representations $\mathbf{1}_{++}$ and $\mathbf{1}_{--}$ of D_4 correspond to $\mathbf{1}_0$ of Z_4, while $\mathbf{1}_{+-}$ and $\mathbf{1}_{-+}$ of D_4 correspond to $\mathbf{1}_2$ of Z_4. For the D_4 doublet $\mathbf{2}$, it is convenient to use the diagonal basis for the matrix a, so that

$$a = \begin{pmatrix} i & 0 \\ 0 & -i \end{pmatrix}. \tag{14.12}$$

Then we can read off that the doublet $\mathbf{2} : (x_1, x_2)$ decomposes to two singlets according to $\mathbf{1}_1 : x_1$ and $\mathbf{1}_3 : x_2$.

14.8 General Q_N

Table 14.15 Representations of Q_N for $N = 4n$

($N = 4n$)	$\mathbf{1}_{++}$	$\mathbf{1}_{+-}$	$\mathbf{1}_{-+}$	$\mathbf{1}_{--}$	$\mathbf{2}_{k=\text{odd}}$	$\mathbf{2}_{k=\text{even}}$
a	1	-1	-1	1	$\begin{pmatrix} \rho^k & 0 \\ 0 & \rho^{-k} \end{pmatrix}$	$\begin{pmatrix} \rho^k & 0 \\ 0 & \rho^{-k} \end{pmatrix}$
b	1	1	-1	-1	$\begin{pmatrix} 0 & i \\ i & 0 \end{pmatrix}$	$\begin{pmatrix} 0 & 1 \\ 1 & 0 \end{pmatrix}$

14.7.2 $D_4 \to Z_2 \times Z_2$

We denote $\tilde{a} = a^2$. Then the subgroup $Z_2 \times Z_2$ consists of elements $\{e, \tilde{a}, b, \tilde{a}b\}$, where $\tilde{a}b = b\tilde{a}$ and $\tilde{a}^2 = b^2 = e$. Their representations are clearly quite simple, that is, $\mathbf{1}_{\pm\pm}$, whose $Z_2 \times Z_2$ charges are determined by $\tilde{a} = \pm 1$ and $b = \pm 1$. We use the notation that the first (second) subscript of $\mathbf{1}_{\pm\pm}$ denotes the Z_2 charge for $\tilde{a}$ (b). Then the singlets $\mathbf{1}_{++}$ and $\mathbf{1}_{+-}$ of D_4 correspond to $\mathbf{1}_{++}$ of $Z_2 \times Z_2$, while $\mathbf{1}_{-+}$ and $\mathbf{1}_{--}$ of D_4 correspond to $\mathbf{1}_{+-}$ of $Z_2 \times Z_2$. The doublet $\mathbf{2}$ of D_4 decomposes to $\mathbf{1}_{-+}$ and $\mathbf{1}_{--}$ of Z_2.

In addition to the above, there is another choice of $Z_2 \times Z_2$ subgroup which consists of the elements $\{e, a^2, ab, a^3b\}$. In this case, we obtain the same decomposition of D_4.

14.7.3 $D_4 \to Z_2$

Furthermore, both Z_4 and $Z_2 \times Z_2$ include the subgroup Z_2. The decomposition of D_4 to Z_2 is rather straightforward.

14.8 General Q_N

Recall that all elements of Q_N can be written in the form $a^m b^k$ with $m = 0, \ldots, N-1$, and $k = 0, 1$, where $a^N = e$ and $b^2 = a^{N/2}$. Similarly to D_N with N even, there are four singlets $\mathbf{1}_{\pm\pm}$ and doublets $\mathbf{2}_k$ with $k = 1, \ldots, N/2 - 1$. Tables 14.15 and 14.16 show the representations of a and b on these representations for $N = 4n$ and $N = 4n + 2$. In general, the group Q_N includes Z_4, Z_N, and Q_M as subgroups. The breaking patterns of Q_N are summarized in Table 14.17 for $N = 4n$ and Table 14.18 for $N = 4n + 2$.

Table 14.16 Representations of Q_N for $N = 4n + 2$

($N = 4n+2$)	$\mathbf{1}_{++}$	$\mathbf{1}_{+-}$	$\mathbf{1}_{-+}$	$\mathbf{1}_{--}$	$\mathbf{2}_{k=odd}$	$\mathbf{2}_{k=even}$
a	1	-1	-1	1	$\begin{pmatrix} \rho^k & 0 \\ 0 & \rho^{-k} \end{pmatrix}$	$\begin{pmatrix} \rho^k & 0 \\ 0 & \rho^{-k} \end{pmatrix}$
b	1	i	$-i$	-1	$\begin{pmatrix} 0 & i \\ i & 0 \end{pmatrix}$	$\begin{pmatrix} 0 & 1 \\ 1 & 0 \end{pmatrix}$

Table 14.17 Breaking pattern of Q_N for $N = 4n$. All parameters n, m, k, k', n' are integers

Q_N	Z_4	Q_N	Z_N
$\mathbf{1}_{++}$	$\mathbf{1}_0$	$\mathbf{1}_{++}$	$\mathbf{1}_0$
$\mathbf{1}_{+-}$	$\mathbf{1}_0$	$\mathbf{1}_{+-}$	$\mathbf{1}_{N/2}$
$\mathbf{1}_{-+}$	$\mathbf{1}_2$	$\mathbf{1}_{-+}$	$\mathbf{1}_2$
$\mathbf{1}_{--}$	$\mathbf{1}_2$	$\mathbf{1}_{--}$	$\mathbf{1}_0$
$\mathbf{2}_k$	$\mathbf{1}_1 + \mathbf{1}_3$	$\mathbf{2}_k$	$\mathbf{1}_k + \mathbf{1}_{N-k}$

Q_N	Q_M ($M = 4m$)	Q_N	Q_M ($M = 4m+2$)
$\mathbf{1}_{++}$	$\mathbf{1}_{++}$	$\mathbf{1}_{++}$	$\mathbf{1}_{++}$
$\mathbf{1}_{+-}$	$\mathbf{1}_{++}$ (M/N is even)	$\mathbf{1}_{+-}$	$\mathbf{1}_{++}$
	$\mathbf{1}_{+-}$ (M/N is odd)	$\mathbf{1}_{-+}$	$\mathbf{1}_{--}$
$\mathbf{1}_{-+}$	$\mathbf{1}_{--}$ (M/N is even)	$\mathbf{1}_{--}$	$\mathbf{1}_{--}$
	$\mathbf{1}_{-+}$ (M/N is odd)	$\mathbf{2}_k$	$\mathbf{2}_{k'}$ ($k = k' + Mn'$)
$\mathbf{1}_{--}$	$\mathbf{1}_{--}$		$\tilde{\mathbf{2}}_{M-k'}$ ($k = Mn' - k'$)
$\mathbf{2}_k$	$\mathbf{2}_{k'}$ ($k = k' + Mn'$)		$\mathbf{1}_{+-} + \mathbf{1}_{-+}$ ($k = M(2n'+1)/2$)
	$\tilde{\mathbf{2}}_{M-k'}$ ($k = Mn' - k'$)		$\mathbf{1}_{++} + \mathbf{1}_{--}$ ($k = Mn'$)
	$\mathbf{1}_{+-} + \mathbf{1}_{-+}$ ($k = M(2n'+1)/2$)		
	$\mathbf{1}_{++} + \mathbf{1}_{--}$ ($k = Mn'$)		

14.8.1 $Q_N \to Z_4$

First we consider the subgroup Z_4, which consists of the elements $\{e, b, b^2, b^3\}$. Obviously, there are four singlet representations $\mathbf{1}_m$ for Z_4, and the generator b is represented by $b = e^{\pi i m/2}$ on $\mathbf{1}_m$.

When $N = 4n$, $\mathbf{1}_{++}$ and $\mathbf{1}_{+-}$ of Q_N correspond to $\mathbf{1}_0$ of Z_4, while $\mathbf{1}_{-+}$ and $\mathbf{1}_{--}$ of Q_N correspond to $\mathbf{1}_2$ of Z_4. The doublets $\mathbf{2}_k$ of Q_N, viz., (x_1, x_2), decompose to two singlets $\mathbf{1}_1 : (x_1 - ix_2)$ and $\mathbf{1}_3 : (x_1 + ix_2)$.

When $N = 4n + 2$, $\mathbf{1}_{++}, \mathbf{1}_{+-}, \mathbf{1}_{-+},$ and $\mathbf{1}_{--}$ of Q_N correspond to $\mathbf{1}_0, \mathbf{1}_1, \mathbf{1}_2,$ and $\mathbf{1}_3$ of Z_4, respectively. The decompositions of doublets $\mathbf{2}_k$ are the same as for $N = 4n$.

14.8 General Q_N

Table 14.18 Breaking pattern of Q_N for $N = 4n + 2$ and $M = 4m + 2$. All parameters n, m, k, k', n' are integers

Q_N	Z_4	Q_N	Z_N	Q_N	Q_M
1_{++}	1_0	1_{++}	1_0	1_{++}	1_{++}
1_{+-}	1_0	1_{+-}	$1_{N/2}$	1_{+-}	1_{++}
1_{-+}	1_2	1_{-+}	$1_{N/2}$	1_{-+}	1_{--}
1_{--}	1_2	1_{--}	1_0	1_{--}	1_{--}
2_k	$1_1 + 1_3$	2_k	$1_k + 1_{N-k}$	2_k	$2_{k'}$ ($N = k' + Mn'$)
					$\tilde{2}_{M-k'}$ ($N = Mn' - k'$)
					$1_{+-} + 1_{-+}$ ($N = M(2n'+1)/2$)
					$1_{++} + 1_{--}$ ($N = Mn'$)

14.8.2 $Q_N \to Z_N$

We now consider the subgroup Z_N, which consists of the elements $\{e, a, \ldots, a^{N-1}\}$. Obviously, it is normal subgroup of Q_N and there are N types of irreducible singlet representation $1_0, 1_1, \ldots, 1_{N-1}$. On the singlet 1_m of Z_N, the generator a is represented by $a = \rho^m$. The singlets 1_{++} and 1_{--} of Q_N correspond to 1_0 of Z_N and the singlets 1_{+-} and 1_{-+} of Q_N correspond to $1_{N/2}$ of Z_N. The doublets 2_k and (x_1, x_2) of Q_N decompose to two singlets $1_k : x_1$ and $1_{N-k} : x_2$.

14.8.3 $Q_N \to Q_M$

We consider the subgroup Q_M, where M is a divisor of N. We define $\tilde{a} = a^\ell$ with $\ell = N/M$, where ℓ is thus an integer. The subgroup Q_M consists of all elements $\tilde{a}^m b^k$ with $m = 0, \ldots, M-1$, and $k = 0, 1$. There are three relevant combinations (N, M), i.e., $(N, M) = (4n, 4m)$, $(4n, 4m+2)$, and $(4n+2, 4m+2)$.

We start with the combination $(N, M) = (4n, 4m)$, where $\ell = N/M$ can be even or odd. Recall that ab of Q_N is represented by $ab = 1$ on $1_{\pm+}$ and $ab = -1$ on $1_{\pm-}$. Thus, the representations of $a^\ell b$ depend on whether ℓ is even or odd. When ℓ is odd, ab and $a^\ell b$ are represented in the same way as on each of the above singlets. On the other hand, when ℓ is even, we always have the singlet representations with $a^\ell = 1$. The doublets 2_k of Q_N correspond to the doublets $2_{k'}$ of Q_M when $k = k'$ (mod M). In addition, when $k = -k'$ (mod M), the doublets 2_k (x_1, x_2) of Q_N correspond to the doublets $2_{M-k'}$ (x_2, x_1) of Q_M. Furthermore, the other doublets 2_k of Q_N decompose to two singlets of Q_M according to $1_{+-} + 1_{-+}$ with $1_{+-} :$ $x_1 + x_2$ and $1_{-+} : x_1 - x_2$ for $k = (M/2)$ (mod M) and $1_{++} + 1_{--}$ with $1_{++} :$ $x_1 + x_2$ and $1_{--} : x_1 - x_2$ for $k = 0$ (mod M).

Next we consider the case $(N, M) = (4n, 4m+2)$, where ℓ must be even. Similarly to the above case with ℓ even, the singlets $1_{++}, 1_{+-}, 1_{-+}$, and 1_{--} of Q_N correspond to $1_{++}, 1_{++}, 1_{--}$, and 1_{--} of Q_M. The results for decompositions of doublets are also the same as for the above case with $(N, M) = (4n, 4m)$ and $\ell = N/M$ even.

Next, we consider the case $(N, M) = (4n + 2, 4m + 2)$, where ℓ must be odd. In this case, the results for decompositions are the same as for the case with $(N, M) = (4n, 4m)$ and $\ell = N/M$ odd.

14.9 Q_4

All elements of Q_4 can be expressed in the form $a^m b^k$ with $m = 0, 1, 2, 3$, and $k = 0, 1$. Since the order of Q_4 is equal to 8, it contains order 2 and 4 subgroups. There are several order 4 subgroups which correspond to Z_4 groups.

14.9.1 $Q_4 \to Z_4$

For example, the elements $\{e, a, a^2, a^3\}$ comprise one Z_4 subgroup. It is clearly a normal subgroup of Q_4 and there are four types of irreducible singlet representation $\mathbf{1}_m$ with $m = 0, 1, 2, 3$, where a is represented by $a = e^{\pi i m/2}$. From the characters of the group Q_4, it is found that $\mathbf{1}_{++}$ and $\mathbf{1}_{--}$ of Q_4 correspond to $\mathbf{1}_0$ of Z_4, while $\mathbf{1}_{-+}$ and $\mathbf{1}_{+-}$ of Q_4 correspond to $\mathbf{1}_2$ of Z_4. For the doublets of Q_4, it is convenient to use the diagonal basis for the matrix a so that

$$a = \begin{pmatrix} i & 0 \\ 0 & -i \end{pmatrix}. \tag{14.13}$$

Then we find that the doublet $\mathbf{2}$ (x_1, x_2) decomposes to two singlets $\mathbf{1}_1 : x_1$ and $\mathbf{1}_3 : x_2$.

Other Z_4 subgroups can be found, namely, $\{e, b, b^2, b^3\}$ and $\{e, ab, (ab)^2, (ab)^3\}$. For these Z_4 subgroups, we obtain the same results when we choose a proper basis. Furthermore, Z_2 subgroups can be found from the above Z_4 groups. The decomposition of Z_4 to Z_2 is rather straightforward.

14.10 QD_{2N}

Since the group QD_{2N} is isomorphic to $Z_N \rtimes Z_2$, D_M and Z_N appear as subgroups of D_N in addition to Z_2, where M is a divisor of N. Recall that all elements of QD_{2N} can be expressed in the form $a^m b^k$ with $m = 0, \ldots, N - 1$, and $k = 0, 1$. There are four singlets and $(N/2 - 1)$ doublets $\mathbf{2}_k$, where $k = 1, \ldots, N/2 - 1$. On the doublet $\mathbf{2}_k$, the generators a and b are represented by

$$a = \begin{pmatrix} \rho^k & 0 \\ 0 & \rho^{k(N/2-1)} \end{pmatrix}, \quad b = \begin{pmatrix} 0 & 1 \\ 1 & 0 \end{pmatrix}, \tag{14.14}$$

14.10 QD_{2N}

Table 14.19 Breaking pattern of QD_{2N}

QD_{2N}	Z_2	QD_{2N}	Z_N	QD_{2N}	$D_{N/2}$
1_{++}	1_0	1_{++}	1_0	1_{++}	1_{++}
1_{-+}	1_0	1_{-+}	$1_{N/2}$	1_{-+}	1_{+-}
1_{+-}	1_1	1_{+-}	1_0	1_{+-}	1_{--}
1_{--}	1_1	1_{--}	$1_{N/2}$	1_{--}	1_{-+}
2_k	$1_0 + 1_1$	2_k	$1_k + 1_{k(N/2-1)}$	2_k	$2_{k'}$ $(k = k' + N/4)$
					$1_{+-} + 1_{-+}$ $(k = N/4)$

where $\rho = e^{2\pi i/N}$. There are four singlets $1_{ss'}$ with $s, s' = \pm$. The generator a is represented by $a = 1$ on $1_{+s'}$, while $a = -1$ on $1_{-s'}$. The generator b is represented by $b = 1$ on 1_{s+}, while $b = -1$ on 1_{s-}. The general breaking pattern of QD_{2N} is summarized in Table 14.19.

14.10.1 $QD_{2N} \to Z_2$

The two elements e and b generate the Z_2 subgroup. Obviously, there are two singlet representations 1_0 and 1_1, where the subscript denotes the Z_2 charge. That is, we have $b = 1$ on 1_0 and $b = -1$ on 1_1.

The singlets 1_{++} and 1_{-+} of QD_{2N} become 1_0 of Z_2, while the singlets 1_{+-} and 1_{--} of QD_{2N} become 1_1 of Z_2. The doublets 2_k (x_1, x_2) of QD_{2N} decompose to two singlets $1_0 : x_1 + x_2$ and $1_1 : x_1 - x_2$.

14.10.2 $QD_{2N} \to Z_N$

The subgroup Z_N consists of the elements $\{e, a, \ldots, a^{N-1}\}$. Obviously it is a normal subgroup of D_N and there are N types of irreducible singlet representation $1_0, 1_1, \ldots, 1_{N-1}$. On the 1_k, the generator a is represented by $a = \rho^k$.

The singlets 1_{++} and 1_{+-} of QD_{2N} become 1_0 of Z_N, while the singlets 1_{-+} and 1_{--} of QD_{2N} become $1_{N/2}$ of Z_N. The doublets 2_k and (x_1, x_2) decompose to two singlets $1_k : x_1$ and $1_{k(N/2-1)} : x_2$.

14.10.3 $QD_{2N} \to D_{N/2}$

The above decompositions of QD_{2N} are rather straightforward because the subgroups are Abelian. The decomposition from QD_{2N} to $D_{N/2}$ is expected to be non-trivial. We define $\tilde{a} = a^2$. The subgroup $D_{N/2}$ consists of all elements of the form $\tilde{a}^m b^k$ with $m = 0, \ldots, N/2 - 1$, and $k = 0, 1$.

Table 14.20 Representations of $\Sigma(2N^2)$

	$\mathbf{1}_{+n}$	$\mathbf{1}_{-n}$	$\mathbf{2}_{p,q}$
a	ρ^n	ρ^n	$\begin{pmatrix} \rho^q & 0 \\ 0 & \rho^p \end{pmatrix}$
a'	ρ^n	ρ^n	$\begin{pmatrix} \rho^p & 0 \\ 0 & \rho^q \end{pmatrix}$
b	1	-1	$\begin{pmatrix} 0 & 1 \\ 1 & 0 \end{pmatrix}$

The singlet representations $\mathbf{1}_{st}$ decompose to $\mathbf{1}_{tu}$ of $D_{N/2}$ with $u = st$. The doublets $\mathbf{2}_k$ of QD_{2N} correspond to the doublets $\mathbf{2}_{k'}$ of $D_{N/2}$ when $k = k'$ (mod $N/2$). In addition, the doublet $\mathbf{2}_{N/4}$ of D_N decomposes to two singlets of $D_{N/2}$ according to $\mathbf{1}_{+-} + \mathbf{1}_{-+}$ with $\mathbf{1}_{+-}: x_1 + x_2$ and $\mathbf{1}_{-+}: x_1 - x_2$.

14.11 General $\Sigma(2N^2)$

Recall that all elements of the group $\Sigma(2N^2)$ can be written in the form $b^k a^m a'^n$ with $k = 0, 1$, and $m, n = 0, 1, \ldots, N - 1$. The generators a, a', and b satisfy $a^N = a'^N = b^2 = e$, $aa' = a'a$, and $bab = a'$, that is, a, a', and b correspond to Z_N, Z'_N, and Z_2 of $(Z_N \times Z'_N) \rtimes Z_2$, respectively. Table 14.20 shows the different representations of these generators. The number of doublets $\mathbf{2}_{p,q}$ is equal to $N(N-1)/2$ with the relation $p > q$.

In general, the group $\Sigma(2N^2)$ contains Z_{2N}, $Z_N \times Z_N$, D_N, Q_N, and $\Sigma(2M^2)$ as subgroups. The breaking pattern of $\Sigma(2N^2)$ is summarized in Table 14.21.

14.11.1 $\Sigma(2N^2) \to Z_{2N}$

The group $\Sigma(2N^2)$ always includes a subgroup Z_{2N}. We consider the elements of Z_{2N} as $(ba)^m$ with $m = 0, \ldots, 2N-1$. There are $2N$ singlet representations $\mathbf{1}_m$ for Z_{2N} and the generator ba is represented by $b = \rho^m$ on $\mathbf{1}_m$, where $\rho = e^{\pi i/N}$. The representations $\mathbf{1}_{+n}$, $\mathbf{1}_{-n}$, and $\mathbf{2}_{\ell,m}$ of $\Sigma(2N^2)$ then decompose to $\mathbf{1}_{2n}$, $\mathbf{1}_{2n+N}$, and $\mathbf{1}_{\ell+m} + \mathbf{1}_{\ell+m+N}$, respectively. The components of doublets (x_ℓ, x_m) correspond to $\mathbf{1}_{\ell+m}: (\rho^\ell x_\ell + \rho^m x_m)$ and $\mathbf{1}_{\ell+m+N}: (\rho^\ell x_\ell - \rho^m x_m)$.

14.11.2 $\Sigma(2N^2) \to Z_N \times Z_N$

The subgroup $Z_N \times Z_N$ consists of the elements $a^m a'^n$ with $m, n = 0, \ldots, N - 1$. Obviously, it is a normal subgroup of $\Sigma(2N^2)$. There are N^2 singlet representations

14.11 General $\Sigma(2N^2)$

Table 14.21 Breaking pattern of $\Sigma(2N^2)$. All parameters n, m, k, k', n' are integers

$\Sigma(2N^2)$	Z_{2N}	$\Sigma(2N^2)$	$Z_N \times Z_N$
$\mathbf{1}_{+n}$	$\mathbf{1}_{2n}$	$\mathbf{1}_{+n}$	$\mathbf{1}_{n,n}$
$\mathbf{1}_{-n}$	$\mathbf{1}_{2n+N}$	$\mathbf{1}_{-n}$	$\mathbf{1}_{n,n}$
$\mathbf{2}_{\ell,m}$	$\mathbf{1}_{\ell+m} + \mathbf{1}_{\ell+m+N}$	$\mathbf{2}_{\ell,m}$	$\mathbf{1}_{\ell,m} + \mathbf{1}_{m,\ell}$

$\Sigma(2N^2)$	D_N (N is even)	$\Sigma(2N^2)$	D_N (N is odd)
$\mathbf{1}_{+n}$	$\mathbf{1}_{++}$	$\mathbf{1}_{+n}$	$\mathbf{1}_{+}$
$\mathbf{1}_{-n}$	$\mathbf{1}_{--}$	$\mathbf{1}_{-n}$	$\mathbf{1}_{-}$
$\mathbf{2}_{\ell,m}$	$\mathbf{2}_{k'}$ ($\ell = m + k'$)	$\mathbf{2}_{\ell,m}$	$\mathbf{2}_{k'}$ ($\ell = m + k'$)
	$\tilde{\mathbf{2}}_{k'}$ ($\ell = m - k'$)		$\tilde{\mathbf{2}}_{N-k'}$ ($\ell = m - k'$)
	$\mathbf{1}_{+-} + \mathbf{1}_{-+}$ ($\ell = m + N/2$)		

$\Sigma(2N^2)$	Q_N	$\Sigma(2N^2)$	$\Sigma(2M^2)$
$\mathbf{1}_{+n}$	$\mathbf{1}_{++}$ (n is even)	$\mathbf{1}_{+n}$	$\mathbf{1}_{+n}$
	$\mathbf{1}_{--}$ (n is odd)	$\mathbf{1}_{-n}$	$\mathbf{1}_{-n}$
$\mathbf{1}_{-n}$	$\mathbf{1}_{--}$ (n is even)	$\mathbf{2}_{\ell,m}$	$\mathbf{2}_{\ell',m'}$ ($\ell = \ell' + Mn, m = m' + Mn'$)
	$\mathbf{1}_{++}$ (n is odd)		
$\mathbf{2}_{\ell,m}$	$\mathbf{2}_{k'}$ ($\ell = m + k'$)		
	$\mathbf{1}_{+-} + \mathbf{1}_{-+}$ ($\ell = m + N/2$)		

$\mathbf{1}_{m,n}$ and the generators a and a' are represented by $a = \rho^m$ and $a' = \rho^n$ on $\mathbf{1}_{m,n}$. The representations $\mathbf{1}_{+n}$, $\mathbf{1}_{-n}$, and $\mathbf{2}_{\ell,m}$ of $\Sigma(2N^2)$ then decompose to $\mathbf{1}_{n,n}$, $\mathbf{1}_{n,n}$, and $\mathbf{1}_{\ell,m} + \mathbf{1}_{m,\ell}$, respectively.

14.11.3 $\Sigma(2N^2) \to D_N$

We now consider D_N as a subgroup of $\Sigma(2N^2)$. We define $\tilde{a} = a^{-1}a'$. Then the subgroup D_N consists of the elements $\tilde{a}^m b^k$ with $k = 0, 1$, and $m = 0, \ldots, N-1$. Table 14.22 shows the representations of the generators $\tilde{a}$ and b on each representation of $\Sigma(2N^2)$.

First we consider the case where N is even. The doublets $\mathbf{2}_{p,q}$ of $\Sigma(2N^2)$ are still doublets of D_N, except when $p - q = N/2$. On the other hand, when $p - q = N/2$, the doublets decompose to two singlets of D_N. The representations $\mathbf{1}_{+n}, \mathbf{1}_{-n}, \mathbf{2}_{q+k',q}, \mathbf{2}_{q-k',q}$, and $\mathbf{2}_{q+N/2,q}$ of $\Sigma(2N^2)$ then decompose to $\mathbf{1}_{++}, \mathbf{1}_{--}, \mathbf{2}_{k'}, \tilde{\mathbf{2}}_{k'}$, and $\mathbf{1}_{+-} + \mathbf{1}_{-+}$, respectively.

Table 14.22 Representations of $\tilde{a}$ and b in $\Sigma(2N^2)$

	$\mathbf{1}_{+n}$	$\mathbf{1}_{-n}$	$\mathbf{2}_{p,q}$
$\tilde{a}$	1	1	$\begin{pmatrix} \rho^{p-q} & 0 \\ 0 & \rho^{-(p-q)} \end{pmatrix}$
b	1	-1	$\begin{pmatrix} 0 & 1 \\ 1 & 0 \end{pmatrix}$

Table 14.23 Representations of $\tilde{a}$ and $\tilde{b}$ in $\Sigma(2N^2)$

	$\mathbf{1}_{+n}$	$\mathbf{1}_{-n}$	$\mathbf{2}_{p,q}$
$\tilde{a}$	1	1	$\begin{pmatrix} \rho^{p-q} & 0 \\ 0 & \rho^{-(p-q)} \end{pmatrix}$
$\tilde{b}$	1	-1	$\begin{pmatrix} 0 & (-1)^q \\ (-1)^p & 0 \end{pmatrix}$

Next, we consider the case where N is odd. In this case, the representations $\mathbf{1}_{+n}$, $\mathbf{1}_{-n}$, $\mathbf{2}_{q+k',q}$, and $\mathbf{2}_{q-k',q}$ of $\Sigma(2N^2)$ decompose to $\mathbf{1}_{+}$, $\mathbf{1}_{-}$, $\mathbf{2}_{k'}$, and $\tilde{\mathbf{2}}_{N-k'}$, respectively.

14.11.4 $\Sigma(2N^2) \to Q_N$

We consider Q_N as a subgroup of $\Sigma(2N^2)$ with N even. We define $\tilde{a} = a^{-1}a'$ and $\tilde{b} = ba'^{N/2}$. Then the subgroup Q_N consists of all elements of the form $\tilde{a}^m \tilde{b}^k$ with $m = 0, \ldots, N-1$, and $k = 0, 1$. Table 14.23 shows the various representations of these generators $\tilde{a}$ and $\tilde{b}$ for each representation of $\Sigma(2N^2)$. The singlets $\mathbf{1}_{+n}$ and $\mathbf{1}_{-n}$ of $\Sigma(2N^2)$ then become singlets of Q_N, namely, $\mathbf{1}_{++}$ and $\mathbf{1}_{--}$ for even n and $\mathbf{1}_{--}$ and $\mathbf{1}_{++}$ for odd n. The decompositions of doublets are obtained in a similar way to the decomposition for $\Sigma(2N^2) \to D_N$, whence $\mathbf{2}_{q+k',q}$ and $\mathbf{2}_{q+N/2}$ go to $\mathbf{2}_{k'}$ and $\mathbf{1}_{+-} + \mathbf{1}_{-+}$, respectively.

14.11.5 $\Sigma(2N^2) \to \Sigma(2M^2)$

We consider the subgroup $\Sigma(2M^2)$, where M is a divisor of N. We denote $\tilde{a} = a^{\ell}$ and $\tilde{a}' = a'^{\ell}$ with $\ell = N/M$, whence ℓ is an integer. The subgroup $\Sigma(2M^2)$ consists of all elements of the form $b^k \tilde{a}^m \tilde{a}'^n$ with $k = 0, 1$, and $m, n = 0, \ldots, M-1$. Table 14.24 shows the representations of $\tilde{a}$, $\tilde{a}'$, and $\tilde{b}$ on each representation of $\Sigma(2N^2)$. The representations $\mathbf{1}_{+n}$, $\mathbf{1}_{-n}$, and $\mathbf{2}_{p+Mn,q+Mn'}$ of $\Sigma(2N^2)$ then correspond to the representations $\mathbf{1}_{+n}$, $\mathbf{1}_{-n}$, and $\mathbf{2}_{p,q}$ of $\Sigma(2M^2)$, where n, n' are integers.

Table 14.24 Representations of $\tilde{a}$, $\tilde{a}'$, and $\tilde{b}$ in $\Sigma(2N^2)$

	$\mathbf{1}_{+n}$	$\mathbf{1}_{-n}$	$\mathbf{2}_{p,q}$
$\tilde{a}$	$\rho^{n\ell}$	$\rho^{n\ell}$	$\begin{pmatrix} \rho^{q\ell} & 0 \\ 0 & \rho^{p\ell} \end{pmatrix}$
$\tilde{a}'$	$\rho^{n\ell}$	$\rho^{n\ell}$	$\begin{pmatrix} \rho^{p\ell} & 0 \\ 0 & \rho^{q\ell} \end{pmatrix}$
b	1	-1	$\begin{pmatrix} 0 & 1 \\ 1 & 0 \end{pmatrix}$

14.12 $\Sigma(32)$

The group $\Sigma(32)$ contains subgroups D_4, Q_4, and $\Sigma(8)$, as well as Abelian groups. In addition, it is useful to construct a discrete group as a subgroup of known groups. Here, we show one example, namely, $(Z_4 \times Z_2) \rtimes Z_2$ as a subgroup of $\Sigma(32) \simeq (Z_4 \times Z_4) \rtimes Z_2$.

All elements of the group $\Sigma(32)$ can be written in the form $b^k a^m a'^n$ with $k = 0, 1$, and $m, n = 0, 1, 2, 3$. The generators, a, a', and b satisfy $a^4 = a'^4 = b^2 = e$, $aa' = a'a$, and $bab = a'$. Here we define $\tilde{a} = aa'$ and $\tilde{a}' = a^2$, where $\tilde{a}^4 = e$ and $\tilde{a}'^2 = e$. Then the elements $b^k \tilde{a}^m \tilde{a}'^n$ with $k, n = 0, 1$, and $m = 0, 1, 2, 3$, generate a closed subalgebra, i.e., $(Z_4 \times Z_2) \rtimes Z_2$ with ten conjugacy classes. It has eight singlets $\mathbf{1}_{\pm 0}$, $\mathbf{1}_{\pm 1}$, $\mathbf{1}_{\pm 2}$, and $\mathbf{1}_{\pm 3}$ and two doublets, $\mathbf{2}_1$ and $\mathbf{2}_2$. These conjugacy classes and characters are shown in Table 14.25. From this table, we can find decompositions of the $\Sigma(32)$ representations $\mathbf{1}_{\pm 0, \pm 1, \pm 2, \pm 3}$, $\mathbf{2}_{1,0}$, $\mathbf{2}_{3,2}$, $\mathbf{2}_{3,0}$, $\mathbf{2}_{2,1}$, $\mathbf{2}_{2,0}$, and $\mathbf{2}_{3,1}$ to representations of $(Z_4 \times Z_2) \rtimes Z_2$, viz., $\mathbf{1}_{\pm 0, \pm 1, \pm 0, \pm 1}$, $\mathbf{2}_1$, $\mathbf{2}_2$, $\mathbf{1}_{+3} + \mathbf{1}_{-3}$, and $\mathbf{1}_{+2} + \mathbf{1}_{-2}$, respectively.

Table 14.25 Conjugacy classes and characters of $(Z_4 \times Z_2) \rtimes Z_2$

		h	$\chi_{\pm 0}$	$\chi_{\pm 1}$	$\chi_{\pm 2}$	$\chi_{\pm 3}$	χ_{2_1}	χ_{2_2}
C_1:	$\{e\}$,	1	1	1	1	1	2	2
$C_1^{(1)}$:	$\{\tilde{a}\tilde{a}'\}$,	4	1	-1	1	-1	$2i$	$-2i$
$C_1^{(2)}$:	$\{\tilde{a}^2\tilde{a}'^2\}$,	2	1	1	1	1	-2	-2
$C_1^{(3)}$:	$\{\tilde{a}^3\tilde{a}'^3\}$,	4	1	-1	1	-1	$-2i$	$2i$
$C_2^{(0)}$:	$\{b, b\tilde{a}^2\tilde{a}'^2\}$,	2	± 1	± 1	± 1	± 1	0	0
$C_2^{(0)}$:	$\{b\tilde{a}\tilde{a}', b\tilde{a}^3\tilde{a}'^3\}$,	4	± 1	∓ 1	± 1	∓ 1	0	0
$C_2^{(0)}$:	$\{b\tilde{a}^2, b\tilde{a}'^2\}$,	4	± 1	∓ 1	∓ 1	± 1	0	0
$C_2^{(0)}$:	$\{b\tilde{a}\tilde{a}'^3, b\tilde{a}^3\tilde{a}'\}$,	2	± 1	± 1	∓ 1	∓ 1	0	0
$C_2^{(2,0)}$:	$\{\tilde{a}^2, \tilde{a}'^2\}$,	2	1	-1	-1	1	0	0
$C_2^{(3,1)}$:	$\{\tilde{a}\tilde{a}'^3, \tilde{a}^3\tilde{a}'\}$,	2	1	1	-1	-1	0	0

Table 14.26 Representations of a, a', and b in $\Delta(3N^2)$ for $N/3 \neq$ integer

	$\mathbf{1}_k$	$\mathbf{3}_{[k][\ell]}$
a	1	$\begin{pmatrix} \rho^\ell & 0 & 0 \\ 0 & \rho^k & 0 \\ 0 & 0 & \rho^{-k-\ell} \end{pmatrix}$
a'	1	$\begin{pmatrix} \rho^{-k-\ell} & 0 & 0 \\ 0 & \rho^\ell & 0 \\ 0 & 0 & \rho^k \end{pmatrix}$
b	ω^k	$\begin{pmatrix} 0 & 1 & 0 \\ 0 & 0 & 1 \\ 1 & 0 & 0 \end{pmatrix}$

Table 14.27 Representations of a, a', and b in $\Delta(3N^2)$ for $N/3$ an integer

	$\mathbf{1}_{k,\ell}$	$\mathbf{3}_{[k][\ell]}$
a	ω^ℓ	$\begin{pmatrix} \rho^\ell & 0 & 0 \\ 0 & \rho^k & 0 \\ 0 & 0 & \rho^{-k-\ell} \end{pmatrix}$
a'	ω^ℓ	$\begin{pmatrix} \rho^{-k-\ell} & 0 & 0 \\ 0 & \rho^\ell & 0 \\ 0 & 0 & \rho^k \end{pmatrix}$
b	ω^k	$\begin{pmatrix} 0 & 1 & 0 \\ 0 & 0 & 1 \\ 1 & 0 & 0 \end{pmatrix}$

14.13 General $\Delta(3N^2)$

All elements of $\Delta(3N^2)$ can be written in the form $b^k a^m a'^n$ with $k = 0, 1, 2,$ and $m, n = 0, \ldots, N - 1$, where the generators, b, a, and a' correspond to Z_3, Z_N, and Z'_N of $(Z_N \times Z'_N) \rtimes Z_3$, respectively. Table 14.26 shows the different representations of the generators b, a, and a' on each representation of $\Delta(3N^2)$ for $N/3 \neq$ integer. Table 14.27 shows the same when $N/3$ is an integer.

In general, the group $\Delta(3N^2)$ includes subgroups Z_3, $Z_N \times Z_N$, and $\Delta(3M^2)$, where M is a divisor of N. In addition, when N has certain special values, it includes T_N as a subgroup, as shown in Chap. 11. We consider the above decompositions in what follows. A summary of the breaking patterns is shown in Table 14.28 for $N/3$ integer and Table 14.29 for $N/3 \neq$ integer.

14.13 General $\Delta(3N^2)$

Table 14.28 Breaking pattern of $\Delta(3N^2)$ for $N/3$ an integer

$\Delta(3N^2)$	Z_3	$\Delta(3N^2)$	$Z_N \times Z_N$
$\mathbf{1}_{k,\ell}$		$\mathbf{1}_{k,\ell}$	$\mathbf{1}_{N\ell/3, N\ell/3}$
$\mathbf{3}_{[k][\ell]}$	$\mathbf{1}_0 + \mathbf{1}_1 + \mathbf{1}_2$	$\mathbf{3}_{[k][\ell]}$	$\mathbf{1}_{\ell,-k-\ell} + \mathbf{1}_{k,\ell} + \mathbf{1}_{-k-\ell,k}$

$\Delta(3N^2)$		$\Delta(3M^2)$ $(M/3 = \text{integer})$	
$\mathbf{1}_{k,\ell}$		$\mathbf{1}_{k, N\ell/M}$	
$\mathbf{3}_{[k][\ell]}$		$\mathbf{3}_{[k'][\ell']}$ $(k = k' + Mn, \ell = \ell' + Mn')$	

$\Delta(3N^2)$		$\Delta(3M^2)$ $(M/3 \neq \text{integer})$	
$\mathbf{1}_{k,\ell}$		$\mathbf{1}_k$	
$\mathbf{3}_{[k][\ell]}$		$\mathbf{3}_{[k'][\ell']}$ $(k = k' + Mn, \ell = \ell' + Mn')$	

Table 14.29 Breaking pattern of $\Delta(3N^2)$ for $N/3 \neq$ integer

$\Delta(3N^2)$	Z_3	$\Delta(3N^2)$	$Z_N \times Z_N$
$\mathbf{1}_k$	$\mathbf{1}_k$	$\mathbf{1}_k$	$\mathbf{1}_{0,0}$
$\mathbf{3}_{[k][\ell]}$	$\mathbf{1}_0 + \mathbf{1}_1 + \mathbf{1}_2$	$\mathbf{3}_{[k][\ell]}$	$\mathbf{1}_{\ell,-k-\ell} + \mathbf{1}_{k,\ell} + \mathbf{1}_{-k-\ell,k}$

$\Delta(3N^2)$	T_N	$\Delta(3N^2)$	$\Delta(3M^2)$
$\mathbf{1}_k$	$\mathbf{1}_k$	$\mathbf{1}_k$	$\mathbf{1}_k$
$\mathbf{3}_{[k][\ell]}$	$\mathbf{3}_{[l-mk]}$	$\mathbf{3}_{[k][\ell]}$	$\mathbf{3}_{[k'][\ell']}$ $(k = k' + Mn, \ell = \ell' + Mn')$

14.13.1 $\Delta(3N^2) \to Z_3$

The subgroup Z_3 consists of elements $\{e, b, b^2\}$. There are three singlet representations $\mathbf{1}_m$ with $m = 0, 1, 2$, for Z_3, while the generator b is represented by $b = \omega^m$ on $\mathbf{1}_m$. When $N/3 \neq$ integer, the representations $\mathbf{1}_k$ and $\mathbf{3}_{[k][\ell]}$ of $\Delta(3N^2)$ decompose to $\mathbf{1}_k$ and $\mathbf{1}_0 + \mathbf{1}_1 + \mathbf{1}_2$, respectively. On the other hand, when $N/3$ is an integer the representations $\mathbf{1}_{k,\ell}$ and $\mathbf{3}_{[k][\ell]}$ of $\Delta(3N^2)$ decompose to $\mathbf{1}_k$ and $\mathbf{1}_0 + \mathbf{1}_1 + \mathbf{1}_2$. In both cases, the triplet components (x_1, x_2, x_3) of $\Delta(3N^2)$ are decomposed to singlets of Z_3 according to $\mathbf{1}_0 : x_1 + x_2 + x_3$, $\mathbf{1}_1 : x_1 + \omega^2 x_2 + \omega x_3$, and $\mathbf{1}_2 : x_1 + \omega x_2 + \omega^2 x_3$.

14.13.2 $\Delta(3N^2) \to Z_N \times Z_N$

The subgroup $Z_N \times Z_N$ consists of elements $\{a^m a'^n\}$ with $m, n = 0, 1, \ldots, N-1$. There are N^2 singlet representations $\mathbf{1}_{m,n}$ and the generators a and a' are represented by $a = \rho^m$ and $a' = \rho^n$ on $\mathbf{1}_{m,n}$. When $N/3 \neq$ integer, the representations $\mathbf{1}_k$

and $3_{[k][\ell]}$ of $\Delta(3N^2)$ decompose to $1_{0,0}$ and $1_{\ell,-k-\ell} + 1_{k,\ell} + 1_{-k-\ell,k}$, respectively. In addition, when $N/3$ is an integer, the representations $1_{k,\ell}$ and $3_{[k][\ell]}$ decompose to $1_{N\ell/3,N\ell/3}$ and $1_{\ell,-k-\ell} + 1_{k,\ell} + 1_{-k-\ell,k}$, respectively.

14.13.3 $\Delta(3N^2) \to T_N$

When N is any prime number except 3 or any power of such a prime number, $\Delta(3N^2)$ has T_N as a subgroup. In the basis of $\Delta(3N^2)$ and T_N used in the text, the subgroup T_N consists of all elements of the form $\{\tilde{a}^n b^k\}$, where $\tilde{a} = a^{-m^2} a'^m$, $n = 0, 1, \ldots, N-1$, $k = 0, 1, 2$, and m is defined in Chap. 11 on T_N. For each representation $3_{[k][\ell]}$, we have

$$\tilde{a} = \begin{pmatrix} \rho^{\ell-mk} & 0 & 0 \\ 0 & \rho^{m(\ell-mk)} & 0 \\ 0 & 0 & \rho^{m^2(\ell-mk)} \end{pmatrix}. \quad (14.15)$$

This can be compared with the representations $3_{[k]}$, i.e., the triplets of T_N:

$$\tilde{a} = \begin{pmatrix} \rho^k & 0 & 0 \\ 0 & \rho^{km} & 0 \\ 0 & 0 & \rho^{km^2} \end{pmatrix}. \quad (14.16)$$

Thus the triplets $3_{[k][\ell]}$ of $\Delta(3N^2)$ decompose to $3_{[\ell-mk]}$ of T_N. On the other hand, if $\ell - mk = 0 \pmod{N}$, the triplets decompose to singlets $1_0 + 1_1 + 1_2$ of T_N and their components (x_1, x_2, x_3) correspond to

$$1_0 : x_1 + x_2 + x_3, \quad 1_1 : x_1 + \omega^2 x_2 + \omega x_3, \quad 1_2 : x_1 + \omega x_2 + \omega^2 x_3. \quad (14.17)$$

For instance, when $N = 7$, there are 16 triplets in $\Delta(147)$ and two triplets in T_7. Triplets $3_{[0][1]}, 3_{[0][2]}, 3_{[0][4]}, 3_{[1][3]}, 3_{[1][4]}, 3_{[2][6]}$, and $3_{[4][5]}$ decompose to 3_1 of T_7. In the same way, triplets $3_{[0][3]}, 3_{[0][5]}, 3_{[0][6]}, 3_{[1][1]}, 3_{[2][2]}, 3_{[3][5]}$, and $3_{[4][4]}$ decompose to 3_2 of T_7. The remaining triplets, i.e., $3_{[1][2]}$ and $3_{[3][6]}$, decompose to $1_0 + 1_1 + 1_2$ of T_7.

As far as singlets are concerned, since the representations are $a = 1$, $a' = 1$, $\tilde{a} = 1$, and $b = \omega^n$ in both $\Delta(3N^2)$ and T_N, their singlets correspond.

14.13.4 $\Delta(3N^2) \to \Delta(3M^2)$

We consider the subgroup $\Delta(3M^2)$, where M is a divisor of N. We define $\tilde{a} = a^p$ and $\tilde{a}' = a'^p$, with $p = N/M$, whence p is an integer. The subgroup $\Delta(3M^2)$ consists of all elements of the form $b^k \tilde{a}^m \tilde{a}'^n$ with $k = 0, 1, 2$, and $m, n = 0, \ldots, M-1$.

14.13 General $\Delta(3N^2)$

Table 14.30 Representations of $\tilde{a}$, $\tilde{a}'$, and $\tilde{b}$ in $\Delta(3N^2)$ for $N/3 =$ integer

	$\mathbf{1}_{k,\ell}$	$\mathbf{3}_{[k][\ell]}$
$\tilde{a}$	$\omega^{p\ell}$	$\begin{pmatrix} \rho^{p\ell} & 0 & 0 \\ 0 & \rho^{pk} & 0 \\ 0 & 0 & \rho^{-p(k+\ell)} \end{pmatrix}$
$\tilde{a}'$	$\omega^{p\ell}$	$\begin{pmatrix} \rho^{-p(k+\ell)} & 0 & 0 \\ 0 & \rho^{p\ell} & 0 \\ 0 & 0 & \rho^{pk} \end{pmatrix}$
$\tilde{b}$	ω^k	$\begin{pmatrix} 0 & 1 & 0 \\ 0 & 0 & 1 \\ 1 & 0 & 0 \end{pmatrix}$

Table 14.31 Representations of $\tilde{a}$, $\tilde{a}'$, and $\tilde{b}$ in $\Delta(3N^2)$ for $N/3 \neq$ integer

	$\mathbf{1}_k$	$\mathbf{3}_{[k][\ell]}$
$\tilde{a}$	1	$\begin{pmatrix} \rho^{p\ell} & 0 & 0 \\ 0 & \rho^{pk} & 0 \\ 0 & 0 & \rho^{-p(k+\ell)} \end{pmatrix}$
$\tilde{a}'$	1	$\begin{pmatrix} \rho^{-p(k+\ell)} & 0 & 0 \\ 0 & \rho^{p\ell} & 0 \\ 0 & 0 & \rho^{pk} \end{pmatrix}$
$\tilde{b}$	ω^k	$\begin{pmatrix} 0 & 1 & 0 \\ 0 & 0 & 1 \\ 1 & 0 & 0 \end{pmatrix}$

Table 14.30 shows the various representations of $\tilde{a}$, $\tilde{a}'$, and $\tilde{b}$ on each representation of $\Delta(3N^2)$ for $N/3$ integer. In addition, Table 14.31 shows the representations of $\tilde{a}$, $\tilde{a}'$, and $\tilde{b}$ on each representation of $\Delta(3N^2)$ for $N/3 \neq$ integer. There are three types of combination (N, M), i.e., (1) both $N/3$ and $M/3$ are integers, (2) $N/3$ is an integer, but $M/3$ is not an integer, and (3) neither $N/3$ nor $M/3$ is an integer.

When both $N/3$ and $M/3$ are integers, the representations $\mathbf{1}_{k,\ell}$ and $\mathbf{3}_{[k+Mn][\ell+Mn']}$ of $\Delta(3N^2)$ decompose to representations $\mathbf{1}_{k,p\ell}$ and $\mathbf{3}_{[k][\ell]}$ of $\Delta(3M^2)$, where n and n' are integers.

Next we consider the case where $N/3$ is an integer and $M/3 \neq$ integer, where $p = N/M$ must be $3n$. In this case, the representations $\mathbf{1}_{k,\ell}$ and $\mathbf{3}_{[k+Mn][\ell+Mn']}$ of $\Delta(3N^2)$ decompose to representations $\mathbf{1}_k$ and $\mathbf{3}_{[k][\ell]}$ of $\Delta(3M^2)$, respectively.

The last case is when neither $N/3$ nor $M/3$ is an integer. In this case, the representations $\mathbf{1}_k$ and $\mathbf{3}_{[k+Mn][\ell+Mn']}$ of $\Delta(3N^2)$ decompose to representations $\mathbf{1}_k$ and $\mathbf{3}_{[k][\ell]}$ of $\Delta(3M^2)$.

Table 14.32 Representations of a, a', and b in $\Delta(27)$

	$\mathbf{1}_{k,\ell}$	$\mathbf{3}_{[k][\ell]}$
a	ω^ℓ	$\begin{pmatrix} \omega^\ell & 0 & 0 \\ 0 & \omega^k & 0 \\ 0 & 0 & \omega^{-k-\ell} \end{pmatrix}$
a'	ω^ℓ	$\begin{pmatrix} \omega^{-k-\ell} & 0 & 0 \\ 0 & \omega^\ell & 0 \\ 0 & 0 & \omega^k \end{pmatrix}$
b	ω^k	$\begin{pmatrix} 0 & 1 & 0 \\ 0 & 0 & 1 \\ 1 & 0 & 0 \end{pmatrix}$

Table 14.33 Breaking pattern of $\Delta(27)$

$\Delta(27)$	Z_3	$\Delta(27)$	$Z_3 \times Z_3$
$\mathbf{1}_{k,\ell}$	$\mathbf{1}_k$	$\mathbf{1}_{k,\ell}$	$\mathbf{1}_{\ell,\ell}$
$\mathbf{3}_{[k][\ell]}$	$\mathbf{1}_0 + \mathbf{1}_1 + \mathbf{1}_2$	$\mathbf{3}_{[k][\ell]}$	$\mathbf{1}_{\ell,-k-\ell} + \mathbf{1}_{k,\ell} + \mathbf{1}_{-k-\ell,k}$

14.14 $\Delta(27)$

All elements of the group $\Delta(27)$ can be expressed in the form $b^k a^m a'^n$ with $k = 0, 1, 2$, and $m, n = 0, 1, 2$, where the generators b, a, and a' correspond to the Z_3 groups of $(Z_3 \times Z_3) \rtimes Z_3$. Table 14.32 shows the representations of the generators b, a, and a' for each representation of $\Delta(27)$. As subgroups, the group $\Delta(27)$ contains Abelian subgroups Z_3 and $Z_3 \times Z_3$. The breaking pattern of $\Delta(27)$ is summarized in Table 14.33.

14.14.1 $\Delta(27) \to Z_3$

The subgroup Z_3 consists of the elements $\{e, b, b^2\}$. There are three singlet representations $\mathbf{1}_m$ with $m = 0, 1, 2$, for Z_3 and the generator b is represented by $b = \omega^m$ on $\mathbf{1}_m$. When $N/3 \neq$ integer, the representations $\mathbf{1}_{k,\ell}$ and $\mathbf{3}_{[k][\ell]}$ of $\Delta(27)$ decompose to $\mathbf{1}_k$ and $\mathbf{1}_0 + \mathbf{1}_1 + \mathbf{1}_2$, respectively. The triplet components (x_1, x_2, x_3) decompose to singlets $\mathbf{1}_0 : x_1 + x_2 + x_3$, $\mathbf{1}_1 : x_1 + \omega^2 x_2 + \omega x_3$, and $\mathbf{1}_2 : x_1 + \omega x_2 + \omega^2 x_3$ of Z_3.

14.14.2 $\Delta(27) \to Z_3 \times Z_3$

The subgroup $Z_3 \times Z_3$ consists of elements of the form $\{a^m a'^n\}$ with $m, n = 0, 1, 2$. There are 9 singlet representations $\mathbf{1}_{m,n}$ and the generators a and a' are represented

14.15 General T_N

Table 14.34 Breaking pattern of T_N

T_N	Z_3	T_N	Z_N
$\mathbf{1}_k$	$\mathbf{1}_k$	$\mathbf{1}_k$	$\mathbf{1}_0$
$\mathbf{3}_{[k]}$	$\mathbf{1}_0 + \mathbf{1}_1 + \mathbf{1}_2$	$\mathbf{3}_{[k]}$	$\mathbf{1}_k + \mathbf{1}_{km} + \mathbf{1}_{km^2}$
$\bar{\mathbf{3}}_{[k]}$	$\mathbf{1}_0 + \mathbf{1}_1 + \mathbf{1}_2$	$\bar{\mathbf{3}}_{[k]}$	$\mathbf{1}_{-k} + \mathbf{1}_{-km} + \mathbf{1}_{-km^2}$

by $a = \omega^m$ and $a' = \omega^n$ on $\mathbf{1}_{m,n}$. The representations $\mathbf{1}_{k,\ell}$ and $\mathbf{3}_{[k][\ell]}$ of $\Delta(27)$ decompose to $\mathbf{1}_{\ell,\ell}$ and $\mathbf{1}_{\ell,-k-\ell} + \mathbf{1}_{k,\ell} + \mathbf{1}_{-k-\ell,k}$, respectively.

14.15 General T_N

Since the group T_N is isomorphic to $Z_N \rtimes Z_3$, Z_N and Z_3 appear as subgroups. Recall that all elements of T_N can be written in the form $a^m b^k$ with $m = 0, \ldots, N-1$, and $k = 0, 1, 2$. There are three singlets $\mathbf{1}_k$ and $2(N-1)/6$ triplets $\mathbf{3}_{[k']}$ and $\bar{\mathbf{3}}_{[k']}$ with $k' = 1, \ldots, (N-1)/6$. On the triplets $\mathbf{3}_{[k]}$, the generators a and b are represented by

$$b = \begin{pmatrix} 0 & 1 & 0 \\ 0 & 0 & 1 \\ 1 & 0 & 0 \end{pmatrix}, \quad a = \begin{pmatrix} \rho^k & 0 & 0 \\ 0 & \rho^{km} & 0 \\ 0 & 0 & \rho^{km^2} \end{pmatrix}, \quad (14.18)$$

where $\rho = e^{2\pi i/N}$. For each N, we have the value of m, as explained in Chap. 11 on T_N. The representations of $\bar{\mathbf{3}}_{[k]}$ can be obtained by changing ρ to ρ^{-1}. In the following, we obtain in detail the general breaking patterns of $T_N \to Z_3$ and $T_N \to Z_N$. A summary of the breaking patterns is shown in Table 14.34.

14.15.1 $T_N \to Z_3$

The two elements e and b generate the subgroup Z_3. Obviously, there are three singlet representations $\mathbf{1}_0, \mathbf{1}_1$, and $\mathbf{1}_2$. That is, we have $b = 1$ on $\mathbf{1}_0$, $b = \omega$ on $\mathbf{1}_1$, and $b = \omega^2$ on $\mathbf{1}_2$. The singlets $\mathbf{1}_k$ of T_N become $\mathbf{1}_k$ of Z_3. The triplets $\mathbf{3}_{[k]}$ of T_N, viz., (x_k, x_{km}, x_{km^2}), decompose to three singlets

$$\mathbf{1}_0 : x_k + x_{km} + x_{km^2}, \quad \mathbf{1}_1 : x_k + \omega^2 x_{km} + \omega x_{km^2}, \quad \mathbf{1}_2 : x_k + \omega x_{km} + \omega^2 x_{km^2}.$$

Decompositions of triplets $\bar{\mathbf{3}}_{[k]}$ are the same.

14.15.2 $T_N \to Z_N$

The subgroup Z_N comprises the elements $\{e, a, \ldots, a^{N-1}\}$. It is a normal subgroup of T_N and there are N types of irreducible singlet representation $\mathbf{1}_0, \mathbf{1}_1, \ldots, \mathbf{1}_{N-1}$.

Table 14.35 Representations of a and b in T_7

	1_0	1_1	1_2	3	$\bar{3}$
a	1	1	1	$\begin{pmatrix} \rho & 0 & 0 \\ 0 & \rho^2 & 0 \\ 0 & 0 & \rho^4 \end{pmatrix}$	$\begin{pmatrix} \rho^{-1} & 0 & 0 \\ 0 & \rho^{-2} & 0 \\ 0 & 0 & \rho^{-4} \end{pmatrix}$
b	1	ω	ω^2	$\begin{pmatrix} 0 & 1 & 0 \\ 0 & 0 & 1 \\ 1 & 0 & 0 \end{pmatrix}$	$\begin{pmatrix} 0 & 1 & 0 \\ 0 & 0 & 1 \\ 1 & 0 & 0 \end{pmatrix}$

Table 14.36 Breaking pattern of T_7

T_7	Z_3	T_7	Z_7
1_k	1_k	1_k	1_0
3	$1_0 + 1_1 + 1_2$	3	$1_1 + 1_2 + 1_4$
$\bar{3}$	$1_0 + 1_1 + 1_2$	$\bar{3}$	$1_3 + 1_5 + 1_6$

On the singlets 1_k of T_N, the generator a is represented by $a = 1$. Therefore the singlets 1_k of T_N become 1_0 of Z_N. The triplets $3_{[k]} : (x_k, x_{km}, x_{km^2})$ decompose to three singlets 1_k, 1_{km}, and 1_{km^2}. The triplets $\bar{3}_{[k]}$ decompose to 1_{-k}, 1_{-km}, and 1_{-km^2}.

14.16 T_7

All elements of the group T_7 can be expressed in the form $b^m a^n$ with $m = 0, 1, 2$, and $n = 0, \ldots, 6$, where $b^3 = e$ and $a^7 = e$. Table 14.35 shows the various representations of generators a and b for each representation of T_7. As subgroups, T_7 contains Abelian subgroups Z_3 and Z_7. The summary of the breaking pattern is shown in Table 14.36.

14.16.1 $T_7 \to Z_3$

The subgroup Z_3 consists of elements $\{e, b, b^2\}$. The three singlet representations 1_m of Z_3 with $m = 0, 1, 2$, are specified in such a way that $b = \omega^m$ on 1_m. Then the representations 1_0, 1_1, 1_2, 3, and $\bar{3}$ of T_7 decompose to 1_0, 1_1, 1_2, $1_0 + 1_1 + 1_2$, and $1_0 + 1_1 + 1_2$, respectively. Here the T_7 triplets $3 : (x_1, x_2, x_4)$ and $\bar{3} : (x_6, x_5, x_3)$ decompose to three singlets, namely, $1_0 + 1_1 + 1_2$, and their components correspond to

$$1_0 : x_1 + x_2 + x_4, \quad 1_1 : x_1 + \omega^2 x_2 + \omega x_4, \quad 1_2 : x_1 + \omega x_2 + \omega^2 x_4,$$

$$1_0 : x_6 + x_5 + x_3, \quad 1_1 : x_6 + \omega^2 x_5 + \omega x_3, \quad 1_2 : x_6 + \omega x_5 + \omega^2 x_3.$$

14.17 General $\Sigma(3N^3)$

Table 14.37 Breaking pattern of $\Sigma(3N^3)$

$\Sigma(3N^3)$	Z_3	$\Sigma(3N^3)$	$Z_N \times Z_N \times Z_N$	$\Sigma(3N^3)$	$\Delta(3N^2)$ ($N/3 \neq$ integer)
$\mathbf{1}_{k,\ell}$	$\mathbf{1}_k$	$\mathbf{1}_{k,\ell}$	$\mathbf{1}_{\ell,\ell,\ell}$	$\mathbf{1}_{k,\ell}$	$\mathbf{1}_k$
$\mathbf{3}_{[k][\ell][m]}$	$\mathbf{1}_0 + \mathbf{1}_1 + \mathbf{1}_2$	$\mathbf{3}_{[k][\ell][m]}$	$\mathbf{1}_{k,\ell,m} + \mathbf{1}_{m,k,\ell} + \mathbf{1}_{\ell,m,k}$	$\mathbf{3}_{[k][\ell][m]}$	$\mathbf{3}_{[\ell-k][k-m]}$

$\Sigma(3N^3)$		$\Delta(3N^2)$ ($N/3 =$ integer)	$\Sigma(3N^3)$		$\Sigma(3M^2)$
$\mathbf{1}_{k,\ell}$		$\mathbf{1}_{k,\ell}$	$\mathbf{1}_{k,\ell+Mn}$		$\mathbf{1}_{k,\ell}$
$\mathbf{3}_{[k][\ell][m]}$		$\mathbf{3}_{[\ell-k][k-m]}$	$\mathbf{3}_{[k+Mn][\ell+Mn'][m+Mn'']}$		$\mathbf{3}_{[k][\ell][m]}$

14.16.2 $T_7 \to Z_7$

The subgroup Z_7 consists of elements a^n with $n = 0, \ldots, 6$. The seven singlets $\mathbf{1}_m$ of Z_7 with $m = 0, \ldots, 6$ are specified in such a way that $b = \rho^m$ on $\mathbf{1}_m$, where $\rho = e^{2\pi i/7}$. Then the representations $\mathbf{1}_0$, $\mathbf{1}_1$, $\mathbf{1}_2$, $\mathbf{3}$, and $\bar{\mathbf{3}}$ of T_7 decompose to $\mathbf{1}_0$, $\mathbf{1}_0$, $\mathbf{1}_0$, $\mathbf{1}_1 + \mathbf{1}_2 + \mathbf{1}_4$, and $\mathbf{1}_3 + \mathbf{1}_5 + \mathbf{1}_6$, respectively.

14.17 General $\Sigma(3N^3)$

Recall that all elements of the group $\Sigma(3N^3)$ can be written in the form $b^k a^\ell a'^m a''^n$ with $k = 0, 1, 2$, and $\ell, m, n = 0, 1, \ldots, N-1$. As subgroups, the group $\Sigma(3N^3)$ contains Z_3, $Z_N \times Z_N \times Z_N$, $\Delta(3N^2)$, and $\Sigma(3M^3)$, where M is a divisor of N. The summary of the breaking pattern of $\Sigma(3N^3)$ is shown in Table 14.37.

14.17.1 $\Sigma(3N^2) \to Z_N \times Z_N \times Z_N$

The subgroup $Z_N \times Z_N \times Z_N$ consists of elements $a^k a'^\ell a''^m$ with $k, \ell, m = 0, \ldots, N-1$. Obviously, it is a normal subgroup of $\Sigma(3N^3)$. There are N^3 singlet representations $\mathbf{1}_{k,\ell,m}$ and the generators a, a', and a'' are represented by $a = \rho^k$, $a' = \rho^\ell$, and $a'' = \rho^m$ on $\mathbf{1}_{k,\ell,m}$. Then the representations $\mathbf{1}_{k,\ell}$ and $\mathbf{3}_{[k][\ell][m]}$ of $\Sigma(3N^3)$ decompose to $\mathbf{1}_{\ell,\ell,\ell}$ and $\mathbf{1}_{k,\ell,m} + \mathbf{1}_{m,k,\ell} + \mathbf{1}_{\ell,m,k}$, respectively.

14.17.2 $\Sigma(3N^3) \to \Delta(3N^2)$

The subgroup $\Delta(3N^2)$ consists of the elements $\tilde{a}^k \tilde{a}'^\ell b^m$ with $k, \ell = 0, \ldots, N-1$, and $m = 0, 1, 2$. There are 3 singlet representations $\mathbf{1}_k$ and $(N^2 - 1)/3$ triplet representations $\mathbf{3}_{[k][\ell]}$ for $N/3 \neq$ integer, and 9 singlet representations $\mathbf{1}_{k,\ell}$ and

Table 14.38 Representations of a, a', a'', and b in $\Sigma(81)$

	$\mathbf{1}_{k,\ell}$	$\mathbf{3}_A$	$\mathbf{3}_B$	$\mathbf{3}_C$	$\mathbf{3}_D$
a	ω^ℓ	$\begin{pmatrix} \omega & 0 & 0 \\ 0 & 1 & 0 \\ 0 & 0 & 1 \end{pmatrix}$	$\begin{pmatrix} \omega^2 & 0 & 0 \\ 0 & \omega & 0 \\ 0 & 0 & \omega \end{pmatrix}$	$\begin{pmatrix} 1 & 0 & 0 \\ 0 & \omega^2 & 0 \\ 0 & 0 & \omega^2 \end{pmatrix}$	$\begin{pmatrix} \omega^2 & 0 & 0 \\ 0 & 1 & 0 \\ 0 & 0 & \omega \end{pmatrix}$
a'	ω^ℓ	$\begin{pmatrix} 1 & 0 & 0 \\ 0 & \omega & 0 \\ 0 & 0 & 1 \end{pmatrix}$	$\begin{pmatrix} \omega & 0 & 0 \\ 0 & \omega^2 & 0 \\ 0 & 0 & \omega \end{pmatrix}$	$\begin{pmatrix} \omega^2 & 0 & 0 \\ 0 & 1 & 0 \\ 0 & 0 & \omega^2 \end{pmatrix}$	$\begin{pmatrix} \omega & 0 & 0 \\ 0 & \omega^2 & 0 \\ 0 & 0 & 1 \end{pmatrix}$
a''	ω^ℓ	$\begin{pmatrix} 1 & 0 & 0 \\ 0 & 1 & 0 \\ 0 & 0 & \omega \end{pmatrix}$	$\begin{pmatrix} \omega & 0 & 0 \\ 0 & \omega & 0 \\ 0 & 0 & \omega^2 \end{pmatrix}$	$\begin{pmatrix} \omega^2 & 0 & 0 \\ 0 & \omega^2 & 0 \\ 0 & 0 & 1 \end{pmatrix}$	$\begin{pmatrix} 1 & 0 & 0 \\ 0 & \omega & 0 \\ 0 & 0 & \omega^2 \end{pmatrix}$
b	ω^k	$\begin{pmatrix} 0 & 1 & 0 \\ 0 & 0 & 1 \\ 1 & 0 & 0 \end{pmatrix}$	$\begin{pmatrix} 0 & 1 & 0 \\ 0 & 0 & 1 \\ 1 & 0 & 0 \end{pmatrix}$	$\begin{pmatrix} 0 & 1 & 0 \\ 0 & 0 & 1 \\ 1 & 0 & 0 \end{pmatrix}$	$\begin{pmatrix} 0 & 1 & 0 \\ 0 & 0 & 1 \\ 1 & 0 & 0 \end{pmatrix}$

$(N^2 - 3)/3$ triplet representations $\mathbf{3}_{[k][\ell]}$ for $N/3$ integer. The generators $\tilde{a}$ and $\tilde{a}'$ are represented by $\tilde{a} = aa'^{-1}$ and $\tilde{a} = a'a''^{-1}$. Then the representations $\mathbf{1}_{k,\ell}$ and $\mathbf{3}_{[k][\ell][m]}$ of $\Sigma(3N^3)$ decompose to $\mathbf{1}_k$ and $\mathbf{3}_{[\ell-k][k-m]}$ for $N/3 \neq$ integer and $\mathbf{1}_{k,\ell}$ and $\mathbf{3}_{[\ell-k][k-m]}$ for $N/3$ integer.

14.17.3 $\Sigma(3N^3) \to \Sigma(3M^3)$

We consider the subgroup $\Sigma(3M^3)$, where M is a divisor of N. We denote $\tilde{a} = a^p$, $\tilde{a}' = a'^p$, and $\tilde{a}'' = a''^p$, with $p = N/M$, whence p is an integer. The subgroup $\Sigma(3M^3)$ consists of all elements of the form $b^k \tilde{a}^\ell \tilde{a}'^m \tilde{a}''^n$ with $k = 0, 1, 2$, and $\ell, m, n = 0, \ldots, M - 1$. The singlets $\mathbf{1}_{k,\ell+Mn}$ and triplets $\mathbf{3}_{[k+Mn][\ell+Mn'][m+Mn'']}$ of $\Sigma(3N^3)$ correspond to singlets $\mathbf{1}_{k,\ell}$ and triplets $\mathbf{3}_{[k][\ell][m]}$ of $\Sigma(3M^3)$, where n, n', n'' are integers.

14.18 $\Sigma(81)$

All elements of the group $\Sigma(81)$ can be written in the form $b^k a^\ell a'^m a''^n$ with $k, \ell, m, n = 0, 1, 2$, where these generators satisfy $a^3 = a'^3 = a''^3 = 1$, $aa' = a'a$, $aa'' = a''a$, $a''a' = a'a''$, $b^3 = 1$, $b^2ab = a''$, $b^2a'b = a$, and $b^2a''b = a'$. Table 14.38 shows the representations of generators b, a, a', and a'' for each representation of $\Sigma(81)$. As subgroups, the group $\Sigma(81)$ contains $Z_3 \times Z_3 \times Z_3$ and $\Delta(27)$. The breaking pattern of $\Sigma(81)$ is summarized in Table 14.39.

14.18 $\Sigma(81)$

Table 14.39 Breaking pattern of $\Sigma(81)$

$\Sigma(81)$	$Z_3 \times Z_3 \times Z_3$	$\Sigma(81)$	$\Delta(27)$
$\mathbf{1}_{k,\ell}$	$\mathbf{1}_{\ell,\ell,\ell}$	$\mathbf{1}_{k,\ell}$	$\mathbf{1}_{k,0}$
$\mathbf{3}_A$	$\mathbf{1}_{0,0,0} + \mathbf{1}_{0,1,0} + \mathbf{1}_{0,0,1}$	$\mathbf{3}_A$	$\mathbf{3}_{[0][1]}$
$\mathbf{3}_B$	$\mathbf{1}_{2,1,1} + \mathbf{1}_{1,2,1} + \mathbf{1}_{1,1,2}$	$\mathbf{3}_B$	$\mathbf{3}_{[0][1]}$
$\mathbf{3}_C$	$\mathbf{1}_{0,2,2} + \mathbf{1}_{2,0,2} + \mathbf{1}_{2,2,0}$	$\mathbf{3}_C$	$\mathbf{3}_{[0][1]}$
$\mathbf{3}_D$	$\mathbf{1}_{2,1,0} + \mathbf{1}_{0,2,1} + \mathbf{1}_{1,0,2}$	$\mathbf{3}_D$	$\mathbf{1}_{0,2} + \mathbf{1}_{1,2} + \mathbf{1}_{2,2}$
$\bar{\mathbf{3}}_A$	$\mathbf{1}_{2,0,0} + \mathbf{1}_{0,2,0} + \mathbf{1}_{0,0,2}$	$\bar{\mathbf{3}}_A$	$\mathbf{3}_{[0][2]}$
$\bar{\mathbf{3}}_B$	$\mathbf{1}_{1,2,2} + \mathbf{1}_{2,1,2} + \mathbf{1}_{2,2,1}$	$\bar{\mathbf{3}}_B$	$\mathbf{3}_{[0][2]}$
$\bar{\mathbf{3}}_C$	$\mathbf{1}_{0,1,1} + \mathbf{1}_{1,0,1} + \mathbf{1}_{1,1,0}$	$\bar{\mathbf{3}}_C$	$\mathbf{3}_{[0][2]}$
$\bar{\mathbf{3}}_D$	$\mathbf{1}_{1,2,0} + \mathbf{1}_{0,1,2} + \mathbf{1}_{2,0,1}$	$\bar{\mathbf{3}}_D$	$\mathbf{1}_{0,1} + \mathbf{1}_{1,1} + \mathbf{1}_{2,1}$

14.18.1 $\Sigma(81) \to Z_3 \times Z_3 \times Z_3$

The subgroup $Z_3 \times Z_3 \times Z_3$ consists of the elements $\{e, a, a^2, a', a'^2, a'', a''^2, \ldots\}$. There are 3^3 singlets $\mathbf{1}_{k,\ell,m}$ of $Z_3 \times Z_3 \times Z_3$ and the generators a, a', and a'' are represented on $\mathbf{1}_{k,\ell,m}$ by $a = \omega^k$, $a' = \omega^\ell$, and $a'' = \omega^m$. Then the singlet $\mathbf{1}_{k,\ell}$ of $\Sigma(81)$ decomposes to $\mathbf{1}_{\ell,\ell,\ell}$. Regarding the triplets, they become three different singlets of $Z_3 \times Z_3 \times Z_3$, as shown in detail in Table 14.39.

14.18.2 $\Sigma(81) \to \Delta(27)$

The subgroup $\Delta(27)$ consists of elements of the form $b^k \tilde{a}^m \tilde{a}'^n$, where $\tilde{a} = a^2 a''$ and $\tilde{a}' = a' a''^2$. Table 14.40 shows the representations of the generators b, $\tilde{a}$, and $\tilde{a}'$ for each representation of $\Sigma(81)$. Then the singlet $\mathbf{1}_{k,\ell}$ decomposes to $\mathbf{1}_{k,0}$ of $\Delta(27)$. Triplets $\mathbf{3}_{A,B,C}$ and $\bar{\mathbf{3}}_{A,B,C}$ become triplets $\mathbf{3}_{[0][1]}$ and $\mathbf{3}_{[0][2]}$, respectively, while $\mathbf{3}_D$

Table 14.40 Representations of b, $\tilde{a}$, and $\tilde{a}'$ of $\Delta(27)$ in $\Sigma(81)$

	$\mathbf{1}_{k,\ell}$	$\mathbf{3}_A$	$\mathbf{3}_B$	$\mathbf{3}_C$	$\mathbf{3}_D$
$\tilde{a}$	1	$\begin{pmatrix} \omega & 0 & 0 \\ 0 & 1 & 0 \\ 0 & 0 & \omega^2 \end{pmatrix}$	$\begin{pmatrix} \omega & 0 & 0 \\ 0 & 1 & 0 \\ 0 & 0 & \omega^2 \end{pmatrix}$	$\begin{pmatrix} \omega & 0 & 0 \\ 0 & 1 & 0 \\ 0 & 0 & \omega^2 \end{pmatrix}$	$\begin{pmatrix} \omega^2 & 0 & 0 \\ 0 & \omega^2 & 0 \\ 0 & 0 & \omega^2 \end{pmatrix}$
$\tilde{a}'$	1	$\begin{pmatrix} \omega^2 & 0 & 0 \\ 0 & \omega & 0 \\ 0 & 0 & 1 \end{pmatrix}$	$\begin{pmatrix} \omega^2 & 0 & 0 \\ 0 & \omega & 0 \\ 0 & 0 & 1 \end{pmatrix}$	$\begin{pmatrix} \omega^2 & 0 & 0 \\ 0 & \omega & 0 \\ 0 & 0 & 1 \end{pmatrix}$	$\begin{pmatrix} \omega^2 & 0 & 0 \\ 0 & \omega^2 & 0 \\ 0 & 0 & \omega^2 \end{pmatrix}$
b	ω^k	$\begin{pmatrix} 0 & 1 & 0 \\ 0 & 0 & 1 \\ 1 & 0 & 0 \end{pmatrix}$	$\begin{pmatrix} 0 & 1 & 0 \\ 0 & 0 & 1 \\ 1 & 0 & 0 \end{pmatrix}$	$\begin{pmatrix} 0 & 1 & 0 \\ 0 & 0 & 1 \\ 1 & 0 & 0 \end{pmatrix}$	$\begin{pmatrix} 0 & 1 & 0 \\ 0 & 0 & 1 \\ 1 & 0 & 0 \end{pmatrix}$

Table 14.41 Breaking pattern of $\Delta(6N^2)$ for $N/3 \neq$ integer

$\Delta(6N^2)$	$\Sigma(2N^2)$
1_0	1_{+0}
1_1	1_{-0}
2	$1_{+0} + 1_{-0}$
3_{1k}	$1_{+k} + 2_{0,-k}$
3_{2k}	$1_{-k} + 2_{0,-k}$
$6_{[[k],[\ell]]}$	$2_{-k-\ell,-\ell} + 2_{\ell,-k} + 2_{k,k+\ell}$

$\Delta(6N^2)$	$\Delta(3N^2)$
1_0	1_0
1_1	1_0
2	$1_1 + 1_2$
3_{1k}	$3_{[-k][k]}$
3_{2k}	$3_{[-k][k]}$
$6_{[[k],[\ell]]}$	$3_{[k][\ell]} + 3_{[-\ell][k+\ell]}$

$\Delta(6N^2)$	$\Delta(6M^2)$ ($M/3 \neq$ integer)
1_0	1_0
1_1	1_1
2	2
$3_{1(k+Mn)}$	3_{1k}
$3_{2(k+Mn)}$	3_{2k}
$6_{[[k+Mn],[\ell+Mn']]}$	$6_{[[k],[\ell]]}$

and $\bar{3}_D$ each correspond to three singlets, viz., $1_{0,2} + 1_{1,2} + 1_{2,2}$ and $1_{0,1} + 1_{2,1} + 1_{1,1}$, respectively.

14.19 General $\Delta(6N^2)$

All elements of $\Delta(6N^2)$ can be written in the form $b^k c^\ell a^m a'^n$ with $k = 0, 1, 2$, $\ell = 0, 1$, and $m, n = 0, \ldots, N - 1$, where the generators b, c, a, and a' correspond to Z_3, Z_2, Z_N, and Z'_N of $(Z_N \times Z'_N) \rtimes S_3$, respectively. We now examine the different possible breaking patterns. In particular, $\Delta(6N^2)$ can break into $\Sigma(2N^2)$ and $\Delta(3N^2)$. The breaking pattern of $\Delta(6N^2)$ is summarized in Table 14.41 for $N/3 \neq$ integer and Table 14.42 for $N/3$ integer. For further breaking of these two groups, see Sect. 14.11 for $\Sigma(2N^2)$ and Sect. 14.13 for $\Delta(3N^2)$.

14.19 General $\Delta(6N^2)$

Table 14.42 Breaking pattern of $\Delta(6N^2)$ for $N/3$ an integer

$\Delta(6N^2)$	$\Sigma(2N^2)$	$\Delta(6N^2)$	$\Delta(3N^2)$
1_0	1_{+0}	1_0	$1_{0,0}$
1_1	1_{-0}	1_1	$1_{0,0}$
2_1	$1_{+0} + 1_{-0}$	2_1	$1_{1,0} + 1_{2,0}$
2_2	$2_{2N/3,N/3}$	2_2	$1_{1,2} + 1_{2,1}$
2_3	$2_{N/3,2N/3}$	2_3	$1_{1,1} + 1_{2,2}$
2_4	$2_{N/3,2N/3}$	2_4	$1_{0,1} + 1_{0,2}$
3_{1k}	$1_{+k} + 2_{0,-k}$	3_{1k}	$3_{[-k][k]}$
3_{2k}	$1_{-k} + 2_{0,-k}$	3_{2k}	$3_{[-k][k]}$
$6_{[[k],[\ell]]}$	$2_{-k-\ell,-\ell} + 2_{\ell,-k} + 2_{k,k+\ell}$	$6_{[[k],[\ell]]}$	$3_{[k][\ell]} + 3_{[-\ell][k+\ell]}$

$\Delta(6N^2)$	$\Delta(6M^2)$ ($M/3 \neq$ integer)	$\Delta(6N^2)$	$\Delta(6M^2)$ ($M/3 =$ integer)
1_0	1_0	1_0	1_0
1_1	1_1	1_1	1_1
2_1	2	2_1	2_1
2_2	2	2_2	2_2
2_3	2	2_3	2_3
2_4	$1_0 + 1_1$	2_4	2_4
$3_{1(k+Mn)}$	3_{1k}	$3_{1(k+Mn)}$	3_{1k}
$3_{2(k+Mn)}$	3_{2k}	$3_{2(k+Mn)}$	3_{2k}
$6_{[[k+Mn],[\ell+Mn']]}$	$6_{[[k],[\ell]]}$	$6_{[[k+Mn],[\ell+Mn']]}$	$6_{[[k],[\ell]]}$

Table 14.43 Representations of $\Sigma(2N^2)$

	1_{+n}	1_{-n}	$2_{p,q}$
a^{-1}	ρ^n	ρ^n	$\begin{pmatrix} \rho^q & 0 \\ 0 & \rho^p \end{pmatrix}$
a'	ρ^n	ρ^n	$\begin{pmatrix} \rho^p & 0 \\ 0 & \rho^q \end{pmatrix}$
b	1	-1	$\begin{pmatrix} 0 & 1 \\ 1 & 0 \end{pmatrix}$

14.19.1 $\Delta(6N^2) \to \Sigma(2N^2)$

The subgroup $\Sigma(2N^2)$ consists of elements of the form $\{c^\ell a^{-m} a'^n\}$. There are $2N$ singlets and $N(N-1)/2$ doublets for $\Sigma(2N^2)$. The representations of the generators are summarized in Table 14.43. When $N/3 \neq$ integer, the singlets 1_0 and 1_1 correspond to 1_{+0} and 1_{-0}, the doublet 2 becomes $1_{+0} + 1_{-0}$, and the triplets

Table 14.44 Representations of $\Delta(3N^2)$ for $3N \neq$ integer (*left*) and $3N$ integer (*right*)

	$\mathbf{1}_k$	$\mathbf{3}_{[k][\ell]}$		$\mathbf{1}_{k,\ell}$	$\mathbf{3}_{[k][\ell]}$
a	1	$\begin{pmatrix} \rho^\ell & 0 & 0 \\ 0 & \rho^k & 0 \\ 0 & 0 & \rho^{-k-\ell} \end{pmatrix}$	a	ω^ℓ	$\begin{pmatrix} \rho^\ell & 0 & 0 \\ 0 & \rho^k & 0 \\ 0 & 0 & \rho^{-k-\ell} \end{pmatrix}$
a'	1	$\begin{pmatrix} \rho^{-k-\ell} & 0 & 0 \\ 0 & \rho^\ell & 0 \\ 0 & 0 & \rho^k \end{pmatrix}$	a'	ω^ℓ	$\begin{pmatrix} \rho^{-k-\ell} & 0 & 0 \\ 0 & \rho^\ell & 0 \\ 0 & 0 & \rho^k \end{pmatrix}$
b	ω^k	$\begin{pmatrix} 0 & 1 & 0 \\ 0 & 0 & 1 \\ 1 & 0 & 0 \end{pmatrix}$	b	ω^k	$\begin{pmatrix} 0 & 1 & 0 \\ 0 & 0 & 1 \\ 1 & 0 & 0 \end{pmatrix}$

$\mathbf{3}_{1k}$ and $\mathbf{3}_{2k}$ decompose to $\mathbf{1}_{+k} + \mathbf{2}_{0,-k}$ and $\mathbf{1}_{-k} + \mathbf{2}_{0,-k}$, respectively. The sextet $\mathbf{6}_{[[k],[\ell]]}$ decomposes to $\mathbf{2}_{-k-\ell,-\ell} + \mathbf{2}_{\ell,-k} + \mathbf{2}_{k,k+\ell}$. The decomposition of $\mathbf{3}_{2k}$, written (x_1, x_2, x_3), is non-trivial. It decomposes to $(x_2)_{\mathbf{1}_{-k}} + (x_1, -x_3)_{\mathbf{2}_{0,-k}}$. Similarly, when $N/3$ is an integer, the singlets $\mathbf{1}_0$ and $\mathbf{1}_1$ correspond to $\mathbf{1}_{+0}$ and $\mathbf{1}_{-0}$, and doublets $\mathbf{2}_1$, $\mathbf{2}_2$, $\mathbf{2}_3$, and $\mathbf{2}_4$ become $\mathbf{1}_{+0} + \mathbf{1}_{-0}$, $\mathbf{2}_{2N/3,N/3}$, $\mathbf{2}_{N/3,2N/3}$, and $\mathbf{2}_{N/3,2N/3}$, respectively. The triplets $\mathbf{3}_{1k}$ and $\mathbf{3}_{2k}$ decompose to $\mathbf{1}_{+k} + \mathbf{2}_{0,-k}$ and $\mathbf{1}_{-k} + \mathbf{2}_{0,-k}$, while the sextet $\mathbf{6}_{[[k],[\ell]]}$ decomposes to $\mathbf{2}_{-k-\ell,-\ell} + \mathbf{2}_{\ell,-k} + \mathbf{2}_{k,k+\ell}$.

14.19.2 $\Delta(6N^2) \to \Delta(3N^2)$

The subgroup $\Delta(3N^2)$ consists of elements of the form $\{b^k a^m a'^n\}$. The representations of the generators are summarized in Table 14.44. When $N/3 \neq$ integer, the singlets $\mathbf{1}_0$ and $\mathbf{1}_1$ correspond to $\mathbf{1}_0$ and $\mathbf{1}_0$, the doublet $\mathbf{2}$ becomes $\mathbf{1}_1 + \mathbf{1}_2$, and the triplets $\mathbf{3}_{1k}$ and $\mathbf{3}_{2k}$ correspond to $\mathbf{3}_{[-k][k]}$ and $\mathbf{3}_{[-k][k]}$, respectively. The sextet $\mathbf{6}_{[[k],[\ell]]}$ decomposes to $\mathbf{3}_{[k][\ell]} + \mathbf{3}_{[\ell][k+\ell]}$. When $N/3$ is an integer, the singlets $\mathbf{1}_0$ and $\mathbf{1}_1$ correspond to $\mathbf{1}_{0,0}$ and $\mathbf{1}_{0,0}$, and the doublets $\mathbf{2}_1$, $\mathbf{2}_2$, $\mathbf{2}_3$, and $\mathbf{2}_4$ become $\mathbf{1}_{1,0} + \mathbf{1}_{2,0}$, $\mathbf{1}_{1,2} + \mathbf{1}_{2,1}$, $\mathbf{1}_{1,1} + \mathbf{1}_{2,2}$, and $\mathbf{1}_{0,1} + \mathbf{1}_{0,2}$, respectively. The triplets $\mathbf{3}_{1k}$ and $\mathbf{3}_{2k}$ correspond to $\mathbf{3}_{[-k][k]}$ and $\mathbf{3}_{[-k][k]}$, while the sextet $\mathbf{6}_{[[k],[\ell]]}$ decomposes to $\mathbf{3}_{[k][\ell]} + \mathbf{3}_{[\ell][k+\ell]}$.

14.19.3 $\Delta(6N^2) \to \Delta(6M^2)$

We consider the subgroup $\Delta(6M^2)$, where M is a divisor of N. We define $\tilde{a} = a^p$ and $\tilde{a}' = a'^p$ with $p = N/M$, whence p is an integer. The subgroup $\Delta(6M^2)$ consists of elements of the form $b^k \tilde{c}^\ell \tilde{a}^m \tilde{a}'^n$ with $k = 0, 1, 2$, $\ell = 0, 1$, and $m, n = 0, \ldots, M - 1$. There are three types of combination (N, M), i.e., (1) both $N/3$ and $M/3$ are integers, (2) $N/3$ is an integer, but $M/3$ is not an integer, and (3) neither $N/3$ nor $M/3$ is an integer.

14.20 $\Delta(54)$

Table 14.45 Representations of a, a', b, and c in $\Delta(54)$

	1_0 1_1 2_1		2_2	2_3	2_4	3_{1k}	3_{2k}
a	1	1	$\begin{pmatrix} 1 & 0 \\ 0 & 1 \end{pmatrix}$	$\begin{pmatrix} \omega^2 & 0 \\ 0 & \omega \end{pmatrix}$	$\begin{pmatrix} \omega & 0 \\ 0 & \omega^2 \end{pmatrix}$	$\begin{pmatrix} \omega & 0 \\ 0 & \omega^2 \end{pmatrix}$	$\begin{pmatrix} \omega^k & 0 & 0 \\ 0 & \omega^{2k} & 0 \\ 0 & 0 & 1 \end{pmatrix}$	$\begin{pmatrix} \omega^k & 0 & 0 \\ 0 & \omega^{2k} & 0 \\ 0 & 0 & 1 \end{pmatrix}$
a'	1	1	$\begin{pmatrix} 1 & 0 \\ 0 & 1 \end{pmatrix}$	$\begin{pmatrix} \omega^2 & 0 \\ 0 & \omega \end{pmatrix}$	$\begin{pmatrix} \omega & 0 \\ 0 & \omega^2 \end{pmatrix}$	$\begin{pmatrix} \omega & 0 \\ 0 & \omega^2 \end{pmatrix}$	$\begin{pmatrix} 1 & 0 & 0 \\ 0 & \omega^k & 0 \\ 0 & 0 & \omega^{2k} \end{pmatrix}$	$\begin{pmatrix} 1 & 0 & 0 \\ 0 & \omega^k & 0 \\ 0 & 0 & \omega^{2k} \end{pmatrix}$
b	1	1	$\begin{pmatrix} \omega & 0 \\ 0 & \omega^2 \end{pmatrix}$	$\begin{pmatrix} \omega & 0 \\ 0 & \omega^2 \end{pmatrix}$	$\begin{pmatrix} \omega & 0 \\ 0 & \omega^2 \end{pmatrix}$	$\begin{pmatrix} 1 & 0 \\ 0 & 1 \end{pmatrix}$	$\begin{pmatrix} 0 & 1 & 0 \\ 0 & 0 & 1 \\ 1 & 0 & 0 \end{pmatrix}$	$\begin{pmatrix} 0 & 1 & 0 \\ 0 & 0 & 1 \\ 1 & 0 & 0 \end{pmatrix}$
c	1	-1	$\begin{pmatrix} 0 & 1 \\ 1 & 0 \end{pmatrix}$	$\begin{pmatrix} 0 & 1 \\ 1 & 0 \end{pmatrix}$	$\begin{pmatrix} 0 & 1 \\ 1 & 0 \end{pmatrix}$	$\begin{pmatrix} 0 & 1 \\ 1 & 0 \end{pmatrix}$	$\begin{pmatrix} 0 & 0 & 1 \\ 0 & 1 & 0 \\ 1 & 0 & 0 \end{pmatrix}$	$\begin{pmatrix} 0 & 0 & -1 \\ 0 & -1 & 0 \\ -1 & 0 & 0 \end{pmatrix}$

When both $N/3$ and $M/3$ are integers, each representation of $\Delta(6N^2)$ decomposes to representations of $\Delta(6M^2)$ as follows: the singlets 1_0 and 1_1 and the doublet 2_k remain the same, while the triplets $3_{1(k+Mn)}$ and $3_{2(k+Mn)}$ decompose to 3_{1k} and 3_{2k}, and the sextet $6_{[[k+Mn],[\ell+Mn']]}$ corresponds to $6_{[[k],[\ell]]}$, where n and n' are integers.

Next we consider the case when $N/3$ integer and $M/3 \neq$ integer, where $p = N/M$ must be $3n$. In this case, the singlets 1_0 and 1_1 remain the same representations, while the doublets 2_1, 2_2, and 2_3 correspond to 2, and 2_4 decomposes to $1_0 + 1_1$ because p has a factor of three. The triplets $3_{1(k+Mn)}$ and $3_{2(k+Mn)}$ correspond to 3_{1k} and 3_{2k}, and the sextet $6_{[[k+Mn],[\ell+Mn']]}$ corresponds to $6_{[[k],[\ell]]}$.

The last case is when neither $N/3$ nor $M/3$ is an integer. Then the singlets 1_0 and 1_1 and the doublet 2 remain the same. The triplets $3_{1(k+Mn)}$ and $3_{2(k+Mn)}$ decompose to 3_{1k} and 3_{2k}, while the sextet $6_{[[k+Mn],[\ell+Mn']]}$ corresponds to $6_{[[k],[\ell]]}$.

14.20 $\Delta(54)$

All elements of the group $\Delta(54)$ can be expressed in the form $b^k c^\ell a^m a'^n$ with $k, m, n = 0, 1, 2$, and $\ell = 0, 1$. Here, the generators a and a' correspond to Z_3 and Z'_3 of $(Z_3 \times Z'_3) \rtimes S_3$, respectively, while b and c correspond to Z_3 and Z_2 in S_3 of $(Z_3 \times Z'_3) \rtimes S_3$, respectively. Table 14.45 shows the representations of the generators b, c, a, and a' for each representation of $\Delta(54)$. Regarding subgroups, the group $\Delta(54)$ contains $S_3 \times Z_3$, $\Sigma(18)$, and $\Delta(27)$. The breaking pattern of $\Delta(54)$ is summarized in Table 14.46.

Table 14.46 Breaking pattern of $\Delta(54)$

$\Delta(54)$	$S_3 \times Z_3$	$\Delta(54)$	$\Sigma(18)$	$\Delta(54)$	$\Delta(27)$
1_0	1_0	1_0	1_{+0}	1_0	$1_{0,0}$
1_1	$1'_0$	1_1	1_{-0}	1_1	$1_{0,0}$
2_1	2_0	2_1	$2_{0,0}$	2_1	$1_{1,0} + 1_{2,0}$
2_2	2_0	2_2	$2_{2,1}$	2_2	$1_{1,1} + 1_{2,2}$
2_3	2_0	2_3	$2_{1,2}$	2_3	$1_{1,2} + 1_{2,1}$
2_4	$1_0 + 1'_0$	2_4	$2_{1,2}$	2_4	$1_{0,1} + 1_{0,2}$
3_{1k}	$1_k + 2_k$	3_{1k}	$1_{+k} + 2_{0,2k}$	3_{1k}	$3_{[0][k]}$
3_{2k}	$1'_k + 2_k$	3_{2k}	$1_{-k} + 2_{2k,0}$	3_{2k}	$3_{[0][k]}$

Table 14.47 Representations of b and c of S_3 in $\Delta(54)$

	1	**1'**	**2**
b	1	1	$\begin{pmatrix} \omega & 0 \\ 0 & \omega^2 \end{pmatrix}$
c	1	-1	$\begin{pmatrix} 0 & 1 \\ 1 & 0 \end{pmatrix}$

14.20.1 $\Delta(54) \to S_3 \times Z_3$

The group $\Delta(54)$ contains $S_3 \times Z_3$ as a subgroup. The subgroup S_3 consists of elements $\{e, b, c, b^2, bc, b^2c\}$. The Z_3 part of $S_3 \times Z_3$ consists of elements $\{e, aa'^2, a^2a'\}$, where $(aa'^2)^3 = e$ and the element aa'^2 commutes with all the S_3 elements. Representations r_k for $S_3 \times Z_3$ are specified by representations r of S_3 and the Z_3 charge k, where $r = \mathbf{1}, \mathbf{1'}, \mathbf{2}$ and $k = 0, 1, 2$. That is, the element aa'^2 is represented by $aa'^2 = \omega^k$ on r_k for $k = 0, 1, 2$. For the decomposition of $\Delta(54)$ to $S_3 \times Z_3$, it is convenient to use the basis for S_3 representations $\mathbf{1}, \mathbf{1'}$, and $\mathbf{2}$, which is shown in Table 14.47. Then the representations $\mathbf{1}_0, \mathbf{1}_1, \mathbf{2}_1, \mathbf{2}_2, \mathbf{2}_3, \mathbf{2}_4, \mathbf{3}_{1(k)}$, and $\mathbf{3}_{2(k)}$ of $\Delta(54)$ decompose to representations $\mathbf{1}_0, \mathbf{1}'_0, \mathbf{2}_0, \mathbf{2}_0, \mathbf{2}_0, \mathbf{1}_0 + \mathbf{1}'_0, \mathbf{1}_k + \mathbf{2}_k$, and $\mathbf{1}'_k + \mathbf{2}_k$ of $S_3 \times Z_3$, for $k = 1, 2$. Components of S_3 doublets and singlets obtained from $\Delta(54)$ triplets are the same as those considered in the decomposition for $S_4 \to S_3$.

14.20.2 $\Delta(54) \to \Sigma(18)$

We consider the subgroup $\Sigma(18)$, which consists of elements $\tilde{b}^\ell \tilde{a}^m a'^n$ with $\ell = 0, 1$, and $m, n = 0, 1, 2$, where $\tilde{b} = c$ and $\tilde{a} = a^2$. Table 14.48 shows the representations of the generators $\tilde{a}, a'$, and $\tilde{b}$ for each representation of $\Delta(54)$. Then the representations $\mathbf{1}_0, \mathbf{1}_1, \mathbf{2}_1, \mathbf{2}_2, \mathbf{2}_3, \mathbf{2}_4, \mathbf{3}_{1k}$, and $\mathbf{3}_{2k}$ of $\Delta(54)$ decompose to representations

14.20 Δ(54)

Table 14.48 Representations of $\tilde{b}$, $\tilde{a}$, and a' of $\Sigma(18)$ in $\Delta(54)$

	1_0	1_1	2_1	2_2	2_3	2_4	3_{1k}	3_{2k}
$\tilde{a}$	1	1	$\begin{pmatrix}1 & 0\\ 0 & 1\end{pmatrix}$	$\begin{pmatrix}\omega & 0\\ 0 & \omega^2\end{pmatrix}$	$\begin{pmatrix}\omega^2 & 0\\ 0 & \omega\end{pmatrix}$	$\begin{pmatrix}\omega^2 & 0\\ 0 & \omega\end{pmatrix}$	$\begin{pmatrix}\omega^{2k} & 0 & 0\\ 0 & \omega^k & 0\\ 0 & 0 & 1\end{pmatrix}$	$\begin{pmatrix}\omega^{2k} & 0 & 0\\ 0 & \omega^k & 0\\ 0 & 0 & 1\end{pmatrix}$
a'	1	1	$\begin{pmatrix}1 & 0\\ 0 & 1\end{pmatrix}$	$\begin{pmatrix}\omega^2 & 0\\ 0 & \omega\end{pmatrix}$	$\begin{pmatrix}\omega & 0\\ 0 & \omega^2\end{pmatrix}$	$\begin{pmatrix}\omega & 0\\ 0 & \omega^2\end{pmatrix}$	$\begin{pmatrix}1 & 0 & 0\\ 0 & \omega^k & 0\\ 0 & 0 & \omega^{2k}\end{pmatrix}$	$\begin{pmatrix}1 & 0 & 0\\ 0 & \omega^k & 0\\ 0 & 0 & \omega^{2k}\end{pmatrix}$
$\tilde{b}$	1	-1	$\begin{pmatrix}0 & 1\\ 1 & 0\end{pmatrix}$	$\begin{pmatrix}0 & 1\\ 1 & 0\end{pmatrix}$	$\begin{pmatrix}0 & 1\\ 1 & 0\end{pmatrix}$	$\begin{pmatrix}0 & 1\\ 1 & 0\end{pmatrix}$	$\begin{pmatrix}0 & 0 & 1\\ 0 & 1 & 0\\ 1 & 0 & 0\end{pmatrix}$	$\begin{pmatrix}0 & 0 & -1\\ 0 & -1 & 0\\ -1 & 0 & 0\end{pmatrix}$

1_{+0}, 1_{-0}, $2_{0,0}$, $2_{2,1}$, $2_{1,2}$, $2_{1,2}$, $1_{+k} + 2_{0,2k}$, and $1_{-k} + 2_{2k,0}$ of $\Sigma(18)$, respectively. The decomposition of triplet components is obtained as follows:

$$\begin{pmatrix}x_1\\ x_2\\ x_3\end{pmatrix}_{3_{1k}} \to (x_2)_{1_{+k}} \oplus \begin{pmatrix}x_1\\ x_3\end{pmatrix}_{2_{0,2k}}, \quad \begin{pmatrix}x_1\\ x_2\\ x_3\end{pmatrix}_{3_{2k}} \to (x_2)_{1_{-k}} \oplus \begin{pmatrix}x_3\\ -x_1\end{pmatrix}_{2_{2k,0}}.$$

(14.19)

14.20.3 Δ(54) → Δ(27)

We consider the subgroup $\Delta(27)$, which consists of elements of the form $b^k a^m a'^n$, with $k, m, n = 0, 1, 2$. Using Table 14.45, it is found that the representations 1_0, 1_1, 2_1, 2_2, 2_3, 2_4, 3_{1k}, and 3_{2k} of $\Delta(54)$ decompose to representations $1_{0,0}$, $1_{0,0}$, $1_{1,0} + 1_{2,0}$, $1_{1,1} + 1_{2,2}$, $1_{1,2} + 1_{2,1}$, $1_{0,2} + 1_{0,1}$, $3_{[0][k]}$, and $3_{[0][k]}$ of $\Delta(27)$, respectively.

Chapter 15
Anomalies

15.1 Generic Aspects

Several interesting applications of Abelian and non-Abelian discrete symmetries have been studied in areas of particle physics such as flavor physics (see Chap. 16). In general, symmetries at the tree level can be broken by quantum effects, that is, anomalies. When symmetries are anomalous, symmetry-breaking terms are induced. On the other hand, symmetries should be anomaly-free to be exact, even including any quantum effects.

Anomalies of continuous symmetries, in particular gauge symmetries, have been well studied. However, anomalies of non-Abelian discrete symmetries, or indeed Abelian ones, may not be so well-known. Here we review anomalies of Abelian and non-Abelian discrete symmetries.

We study the gauge theory with a (non-Abelian) gauge group G_g and a set of fermions $\Psi = [\psi^{(1)}, \ldots, \psi^{(M)}]$. We assume that their Lagrangian is invariant under the following chiral transformation:

$$\Psi(x) \to U\Psi(x), \tag{15.1}$$

where $U = \exp(i\alpha P_L)$ and $\alpha = \alpha^A T_A$, with T_A the generators of the transformation, α^A the transformation parameters, and P_L the left-chiral projector. The above transformation is not necessarily a gauge transformation. The fermions $\Psi(x)$ carry the (irreducible) M-plet representation $\mathbf{R}^M$. For the moment, we consider an Abelian flavor symmetry and we suppose that $\Psi(x)$ corresponds to (non-trivial) singlets under the flavor symmetry, while they correspond to the $\mathbf{R}^M$ representation under the gauge group G_g. Since the generators T_A and also α are represented on $\mathbf{R}^M$ as an $(M \times M)$ matrix, we use the notation $T_A(\mathbf{R}^M)$ and $\alpha(\mathbf{R}^M) = \alpha^A T_A(\mathbf{R}^M)$.

For our purposes, the path integral approach is convenient. Thus, we use Fujikawa's method [1, 2] to derive anomalies of continuous and discrete symmetries (see, e.g., [3]). We calculate the transformation of the path integral measure:

$$\mathcal{D}\Psi \mathcal{D}\bar{\Psi} \to \mathcal{D}\Psi \mathcal{D}\bar{\Psi} J(\alpha), \tag{15.2}$$

where the Jacobian $J(\alpha)$ is written as

$$J(\alpha) = \exp\left[i \int d^4 x \mathcal{A}(x;\alpha)\right]. \tag{15.3}$$

The anomaly function $\mathcal{A}$ consists of a gauge part and a gravitational part [4–6]:

$$\mathcal{A} = \mathcal{A}_{\text{gauge}} + \mathcal{A}_{\text{grav}}. \tag{15.4}$$

The gauge part is given by

$$\mathcal{A}_{\text{gauge}}(x;\alpha) = \frac{1}{32\pi^2} \text{Tr}\left[\alpha(\boldsymbol{R}^M) F^{\mu\nu}(x) \widetilde{F}_{\mu\nu}(x)\right], \tag{15.5}$$

where $F^{\mu\nu}$ denotes the field strength of the gauge fields given by $F_{\mu\nu} = [D_\mu, D_\nu]$, and $\widetilde{F}_{\mu\nu}$ denotes its dual given by $\widetilde{F}^{\mu\nu} = \varepsilon^{\mu\nu\rho\sigma} F_{\rho\sigma}$. The trace Tr runs over all internal indices. When the transformation corresponds to a continuous symmetry, this anomaly can be calculated by the triangle diagram with external lines of two gauge bosons and one current corresponding to the symmetry for (15.1).

Similarly, the gravitational part is obtained by [4–6]

$$\mathcal{A}_{\text{grav}} = -\mathcal{A}_{\text{grav}}^{\text{Weyl fermion}} \, \text{tr}\left[\alpha(\boldsymbol{R}^{(M)})\right], \tag{15.6}$$

where tr denotes the trace for the $(M \times M)$ matrix $T_A(\boldsymbol{R}^M)$. The contribution of a single Weyl fermion to the gravitational anomaly is given by [4–6]

$$\mathcal{A}_{\text{grav}}^{\text{Weyl fermion}} = \frac{1}{384\pi^2} \frac{1}{2} \varepsilon^{\mu\nu\rho\sigma} R_{\mu\nu}{}^{\lambda\gamma} R_{\rho\sigma\lambda\gamma}. \tag{15.7}$$

When other sets of M_i-plet fermions Ψ_{M_i} are included in a theory, the total gauge and gravity anomalies are obtained by summing to give $\sum_{\Psi_{M_i}} \mathcal{A}_{\text{gauge}}$ and $\sum_{\Psi_{M_i}} \mathcal{A}_{\text{grav}}$.

We evaluate these anomalies by using the following index theorems [4, 5]:

$$\int d^4 x \frac{1}{32\pi^2} \varepsilon^{\mu\nu\rho\sigma} F^a_{\mu\nu} F^b_{\rho\sigma} \, \text{tr}[t_a t_b] \in \mathbb{Z}, \tag{15.8a}$$

$$\frac{1}{2}\int d^4 x \frac{1}{384\pi^2} \frac{1}{2} \varepsilon^{\mu\nu\rho\sigma} R_{\mu\nu}{}^{\lambda\gamma} R_{\rho\sigma\lambda\gamma} \in \mathbb{Z}, \tag{15.8b}$$

where t_a are generators of G_g in the fundamental representation. We use the convention that $\text{tr}[t_a t_b] = \delta_{ab}/2$. The factor $1/2$ in (15.8b) follows from Rohlin's theorem [7], as discussed in [8]. Of course, these indices are independent of each other. The path integral includes all possible configurations corresponding to different index numbers.

First of all, we study anomalies of the continuous $U(1)$ symmetry. We consider a theory with a (non-Abelian) gauge symmetry G_g as well as the continuous $U(1)$ symmetry, which may be gauged. This theory includes fermions with $U(1)$ charges

15.1 Generic Aspects

$q^{(f)}$ and representations $\boldsymbol{R}^{(f)}$. Those anomalies vanish if and only if the Jacobian is trivial, i.e., $J(\alpha) = 1$ for an arbitrary value of α. Using the index theorems, one finds that the anomaly-free conditions require

$$A_{U(1)-G_g-G_g} \equiv \sum_{\boldsymbol{R}^{(f)}} q^{(f)} T_2(\boldsymbol{R}^{(f)}) = 0, \tag{15.9}$$

for the mixed $U(1)-G_g-G_g$ anomaly, and

$$A_{U(1)-\text{grav}-\text{grav}} \equiv \sum_f q^{(f)} = 0, \tag{15.10}$$

for the $U(1)$–gravity–gravity anomaly. Here, $T_2(\boldsymbol{R}^{(f)})$ is the Dynkin index of the $\boldsymbol{R}^f$ representation, i.e.,

$$\text{tr}[t_a(\boldsymbol{R}^{(f)}) t_b(\boldsymbol{R}^{(f)})] = \delta_{ab} T_2(\boldsymbol{R}^{(f)}). \tag{15.11}$$

Next, we study anomalies of the Abelian discrete symmetry, i.e., the Z_N symmetry. For the Z_N symmetry, we write $\alpha = 2\pi Q_N/N$, where Q_N is the Z_N charge operator and its eigenvalues are integers. Here we denote the Z_N charges of the fermions by $q_N^{(f)}$. Then we can evaluate the $Z_N-G_g-G_g$ and Z_N–gravity–gravity anomalies in a similar way to the above $U(1)$ anomalies. However, the important difference is that α is a discrete parameter, whereas α is a continuous parameter in the $U(1)$ symmetry. Hence, the anomaly-free conditions, i.e., $J(\alpha) = 1$, for a discrete transformation require

$$A_{Z_N-G_g-G_g} = \frac{1}{N} \sum_{\boldsymbol{R}^{(f)}} q_N^{(f)} (2T_2(\boldsymbol{R}^{(f)})) \in \mathbb{Z}, \tag{15.12}$$

for the $Z_N-G_g-G_g$ anomaly, and

$$A_{Z_N-\text{grav}-\text{grav}} = \frac{2}{N} \sum_f q_N^{(f)} \dim \boldsymbol{R}^{(f)} \in \mathbb{Z}, \tag{15.13}$$

for the Z_N–gravity–gravity anomaly. These anomaly-free conditions reduce to

$$\sum_{\boldsymbol{R}^{(f)}} q_N^{(f)} T_2(\boldsymbol{R}^{(f)}) = 0 \mod N/2, \tag{15.14a}$$

$$\sum_f q_N^{(f)} \dim \boldsymbol{R}^{(f)} = 0 \mod N/2. \tag{15.14b}$$

Note that the Z_2 symmetry is always free from the Z_2–gravity–gravity anomaly.

Finally, we study anomalies of non-Abelian discrete symmetries G [3, 9]. A discrete group G comprises a finite number of elements g_i. Hence, the non-Abelian discrete symmetry is anomaly-free if and only if the Jacobian vanishes for the transformation corresponding to each element g_i. Furthermore, recall that $(g_i)^{N_i} = 1$.

That is, each element g_i in the non-Abelian discrete group generates a Z_{N_i} symmetry. Thus, the analysis of non-Abelian discrete anomalies reduces to that of Abelian discrete anomalies. The field basis can be chosen so that g_i is represented in a diagonal form. In such a basis, each field has a definite Z_{N_i} charge $q_{N_i}^{(f)}$. The anomaly-free conditions for the g_i transformation are written as

$$\sum_{R^{(f)}} q_{N_i}^{(f)} T_2(R^{(f)}) = 0 \mod N_i/2, \tag{15.15a}$$

$$\sum_f q_{N_i}^{(f)} \dim R^{(f)} = 0 \mod N_i/2. \tag{15.15b}$$

If these conditions are satisfied for all of $g_i \in G$, there are no anomalies of the full non-Abelian symmetry G. Otherwise, the non-Abelian symmetry is broken to its subgroup by quantum effects, where the subgroup does not include anomalous g_i elements. Furthermore, the non-Abelian symmetry is completely broken if all elements $g_i \in G$ except the identity are anomalous.

In principle, we can investigate anomalies of non-Abelian discrete symmetries G following the above procedure. However, we give a practically simpler way to analyze these anomalies [3, 9]. Here, we reconsider a transformation similar to (15.1) for a set of fermions $\Psi = [\psi^{(1)}, \ldots, \psi^{(Md_\alpha)}]$, which correspond to the R^M irreducible representation of the gauge group G_g and the r^α irreducible representation of the non-Abelian discrete symmetry G with dimension d_α. Let U correspond to one of the group elements $g_i \in G$, which is represented by the matrix $D_\alpha(g_i)$ on r^α. Then the Jacobian is proportional to its determinant $\det D(g_i)$. Thus, the representations with $\det D_\alpha(g_i) = 1$ do not contribute to anomalies. Therefore, non-trivial Jacobians, i.e., anomalies, originate from representations with $\det D_\alpha(g_i) \neq 1$.

Note that $\det D_\alpha(g_i) = \det D_\alpha(gg_ig^{-1})$ for $g \in G$, that is, the determinant is constant on a conjugacy class. It is thus useful to calculate the determinants of elements on each irreducible representation. Such a determinant for the conjugacy class C_i can be written

$$\det(C_i)_\alpha = e^{2\pi i q_{\hat{N}_i}^\alpha / \hat{N}_i}, \tag{15.16}$$

on the irreducible representation r^α. Note that $\hat{N}_i$ is a divisor of N_i, where N_i is the order of g_i in the conjugacy class C_i, i.e., $g^{N_i} = e$, so the $q_{\hat{N}_i}^\alpha$ are normalized to be integers for all the irreducible representations r^α.

We consider the $Z_{\hat{N}_i}$ symmetries and their anomalies. Then we obtain anomaly-free conditions similar to (15.15a), (15.15b). That is, the anomaly-free conditions for the conjugacy classes C_i can be written

$$\sum_{r^{(\alpha)}, R^{(f)}} q_{\hat{N}_i}^{\alpha(f)} T_2(R^{(f)}) = 0 \mod \hat{N}_i/2, \tag{15.17a}$$

$$\sum_{\alpha, f} q_{\hat{N}_i}^{\alpha(f)} \dim R^{(f)} = 0 \mod \hat{N}_i/2, \tag{15.17b}$$

15.2 Explicit Calculations

Table 15.1 Determinants on S_3 representations

	1	**1'**	**2**
$\det(C_1)$	1	1	1
$\det(C_2)$	1	1	1
$\det(C_3)$	1	−1	−1

for the theory including fermions with the $R^{(f)}$ representations of the gauge group G_g and the $r^{\alpha(f)}$ representations of the flavor group G, which correspond to the $Z_{\hat{N}_i}$ charges $q_{\hat{N}_i}^{\alpha(f)}$. Note that the fermion fields with the d_α-dimensional representation r^α contribute $q_{\hat{N}_i}^{\alpha(f)} T_2(R^{(f)})$ and $q_{\hat{N}_i}^{\alpha(f)} \dim R^{(f)}$ to these anomalies, but not $d_\alpha q_{\hat{N}_i}^{\alpha(f)} T_2(R^{(f)})$ and $d_\alpha q_{\hat{N}_i}^{\alpha(f)} \dim R^{(f)}$. If these conditions are satisfied for all conjugacy classes of G, the full non-Abelian symmetry G is free from anomalies. Otherwise, the non-Abelian symmetry is broken by quantum effects.

As we will see below, in concrete examples, the above anomaly-free conditions often lead to the same conditions between different conjugacy classes. Note that, when $\hat{N}_i = 2$, the symmetry is always free from the mixed gravitational anomalies. In what follows, we shall investigate some concrete examples of groups explicitly.

15.2 Explicit Calculations

Here, we apply the above considerations of anomalies to concrete groups.

15.2.1 S_3

As shown in Sect. 3.1, the group S_3 has three conjugacy classes $C_1 = \{e\}$, $C_2 = \{ab, ba\}$, and $C_3 = \{a, b, bab\}$, and three irreducible representations **1**, **1'**, and **2**. Note that the determinants of elements are constant in a conjugacy class. The determinants of elements in singlet representations are equal to characters, and the determinants of elements in a trivial singlet representation **1** are obviously always equal to 1. On the doublet representation **2**, the determinants of representation matrices in C_1, C_2, and C_3 are found to be 1, 1, and −1, respectively. These determinants are shown in Table 15.1.

From these results, it is found that only the conjugacy class C_3 is relevant to anomalies and only the Z_2 symmetry can be anomalous. Under such a Z_2 symmetry, the trivial singlet has vanishing Z_2 charge, while the other representations **1'** and **2** have Z_2 charges $q_2 = 1$, that is,

$$Z_2 \text{ even}: \quad \mathbf{1}, \qquad\qquad\qquad (15.18)$$
$$Z_2 \text{ odd}: \quad \mathbf{1'}, \mathbf{2}$$

Table 15.2 Determinants on S_4 representations

	1	**1'**	**2**	**3**	**3'**
$\det(C_1)$	1	1	1	1	1
$\det(C_3)$	1	1	1	1	1
$\det(C_6)$	1	−1	−1	−1	1
$\det(C_6')$	1	−1	−1	−1	1
$\det(C_8)$	1	1	1	1	1

The anomaly-free conditions for the Z_2–G_g–G_g mixed anomaly (15.17a), (15.17b) thus take the form

$$\sum_{\mathbf{1'}}\sum_{\boldsymbol{R}^{(f)}} T_2(\boldsymbol{R}^{(f)}) + \sum_{\mathbf{2}}\sum_{\boldsymbol{R}^{(f)}} T_2(\boldsymbol{R}^{(f)}) = 0 \quad \text{mod } 1. \tag{15.19}$$

Note that a doublet **2** contributes to the anomaly coefficient, not by $2T_2(\boldsymbol{R}^{(f)})$ but by $T_2(\boldsymbol{R}^{(f)})$, which is the same as **1'**. To show this explicitly, we have written the summations on **1'** and **2** separately.

15.2.2 S_4

Similarly, we can study anomalies of S_4. As shown in Sect. 3.2, the group S_4 has five conjugacy classes, C_1, C_3, C_6, C_6', and C_8, and five irreducible representations **1**, **1'**, **2**, **3**, and **3'**. The determinants of group elements in each representation are shown in Table 15.2. These results imply that only the Z_2 symmetry can be anomalous. Under such a Z_2 symmetry, each representation has the following behavior:

$$\begin{aligned} Z_2 \text{ even:} & \quad \mathbf{1}, \mathbf{3'}, \\ Z_2 \text{ odd:} & \quad \mathbf{1'}, \mathbf{2}, \mathbf{3}. \end{aligned} \tag{15.20}$$

The anomaly-free conditions for the Z_2–G_g–G_g mixed anomaly (15.15a), (15.15b) are thus

$$\sum_{\mathbf{1'}}\sum_{\boldsymbol{R}^{(f)}} T_2(\boldsymbol{R}^{(f)}) + \sum_{\mathbf{2}}\sum_{\boldsymbol{R}^{(f)}} T_2(\boldsymbol{R}^{(f)}) + \sum_{\mathbf{3}}\sum_{\boldsymbol{R}^{(f)}} T_2(\boldsymbol{R}^{(f)}) = 0 \quad \text{mod } 1.$$

$$\tag{15.21}$$

15.2.3 A_4

As shown in Sect. 4.1, there are four conjugacy classes, C_1, C_3, C_4, and C_4', and four irreducible representations **1**, **1'**, **1''**, and **3**. The determinants of group elements in

15.2 Explicit Calculations

Table 15.3 Determinants on A_4 representations

	1	**1'**	**1''**	**3**
$\det(C_1)$	1	1	1	1
$\det(C_3)$	1	1	1	1
$\det(C_4)$	1	ω	ω^2	1
$\det(C'_4)$	1	ω^2	ω	1

Table 15.4 Determinants on A_5 representations

	1	**3**	**3'**	**4**	**5**
$\det(C_1)$	1	1	1	1	1
$\det(C_{15})$	1	1	1	1	1
$\det(C_{20})$	1	1	1	1	1
$\det(C_{12})$	1	1	1	1	1
$\det(C'_{12})$	1	1	1	1	1

each representation are shown in Table 15.3, where $\omega = e^{2\pi i/3}$. These results imply that only the Z_3 symmetry can be anomalous. Under such a Z_3 symmetry, each representation has the following Z_3 charge q_3:

$$\begin{aligned} q_3 = 0: & \quad \mathbf{1}, \mathbf{3}, \\ q_3 = 1: & \quad \mathbf{1'}, \\ q_3 = 2: & \quad \mathbf{1''}. \end{aligned} \tag{15.22}$$

This corresponds to the Z_3 symmetry for the conjugacy class C_4. There is another Z_3 symmetry for the conjugacy class C'_4, but it is not independent of the former Z_3. The anomaly-free conditions are thus

$$\sum_{\mathbf{1'}}\sum_{R^{(f)}} T_2(R^{(f)}) + 2\sum_{\mathbf{1''}}\sum_{R^{(f)}} T_2(R^{(f)}) = 0 \mod 3/2, \tag{15.23}$$

for the Z_3–G_g–G_g anomaly and

$$\sum_{\mathbf{1'}}\sum_{R^{(f)}} \dim R^{(f)} + 2\sum_{\mathbf{1''}}\sum_{R^{(f)}} \dim R^{(f)} = 0 \mod 3/2, \tag{15.24}$$

for the Z_3–gravity–gravity anomaly.

15.2.4 A_5

We study anomalies of A_5. As shown in Sect. 4.2, there are five conjugacy classes C_1, C_{15}, C_{20}, C_{12}, and C'_{12}, and five irreducible representations **1**, **3**, **3'**, **4**, and

5. The determinants of group elements in each representation are shown in Table 15.4. That is, the determinants of all the A_5 elements are equal to unity on any representation. This result can be understood as follows. All the elements of A_5 can be expressed as products of $s = a$ and $t = bab$. The generators s and t are written as real matrices in all the representations **1**, **3**, **3′**, **4**, and **5**. We thus find that $\det(t) = 1$, since $t^5 = e$. Similarly, since $s^2 = b^3 = e$, the possible values are $\det(s) = \pm 1$ and $\det(b) = \omega^k$, with $k = 0, 1, 2$. By imposing $\det(bab) = \det(t) = 1$, we find $\det(s) = \det(b) = 1$. Thus, it turns out that $\det(g) = 1$ for all the A_5 elements on any representation and the A_5 symmetry is therefore always anomaly-free.

15.2.5 T'

As shown in Chap. 5, the group T' has seven conjugacy classes C_1, C_1', C_4, C_4', C_4'', C_4''', and C_6, and seven irreducible representations **1**, **1′**, **1″**, **2**, **2′**, **2″**, and **3**. The determinants of group elements on each representation are shown in Table 15.5. These results imply that only the Z_3 symmetry can be anomalous. Under such a Z_3 symmetry, each representation has the following Z_3 charge q_3:

$$\begin{aligned} q_3 &= 0: \quad \mathbf{1}, \mathbf{2}, \mathbf{3}, \\ q_3 &= 1: \quad \mathbf{1'}, \mathbf{2''}, \\ q_3 &= 2: \quad \mathbf{1''}, \mathbf{2'}. \end{aligned} \tag{15.25}$$

This corresponds to the Z_3 symmetry for the conjugacy class C_4. There are other Z_3 symmetries for the conjugacy classes C_4', C_4'', and C_4''', but these are not independent of the former Z_3. The anomaly-free conditions are therefore

$$\sum_{\mathbf{1'}}\sum_{R^{(f)}} T_2(R^{(f)}) + 2\sum_{\mathbf{1''}}\sum_{R^{(f)}} T_2(R^{(f)}) + \sum_{\mathbf{2''}}\sum_{R^{(f)}} T_2(R^{(f)})$$
$$+ 2\sum_{\mathbf{2'}}\sum_{R^{(f)}} T_2(R^{(f)}) = 0 \mod 3/2, \tag{15.26}$$

for the Z_3–G_g–G_g anomaly and

$$\sum_{\mathbf{1'}}\sum_{R^{(f)}} \dim R^{(f)} + 2\sum_{\mathbf{1''}}\sum_{R^{(f)}} \dim R^{(f)} + \sum_{\mathbf{2''}}\sum_{R^{(f)}} \dim R^{(f)}$$
$$+ 2\sum_{\mathbf{2'}}\sum_{R^{(f)}} \dim R^{(f)} = 0 \mod 3/2, \tag{15.27}$$

for the Z_3–gravity–gravity anomaly.

15.2 Explicit Calculations

Table 15.5 Determinants on T' representations

	1	1'	1''	2	2'	2''	3
$\det(C_1)$	1	1	1	1	1	1	1
$\det(C'_1)$	1	1	1	1	1	1	1
$\det(C_4)$	1	ω	ω^2	1	ω^2	ω	1
$\det(C'_4)$	1	ω^2	ω	1	ω	ω^2	1
$\det(C''_4)$	1	ω	ω^2	1	ω^2	ω	1
$\det(C'''_4)$	1	ω^2	ω	1	ω	ω^2	1
$\det(C_6)$	1	1	1	1	1	1	1

15.2.6 D_N (N Even)

We now study anomalies of D_N with N even. As shown in Chap. 6, the group D_N with N even has four singlets $\mathbf{1}_{\pm\pm}$ and $(N/2 - 1)$ doublets $\mathbf{2}_k$. All elements of D_N can be written as products of two elements a and b. Their determinants on $\mathbf{2}_k$ are $\det(a) = 1$ and $\det(b) = -1$. Similarly, we can obtain the determinants of a and b on the four singlets $\mathbf{1}_{\pm\pm}$. Indeed, the four singlets are classified by the values of $\det(b)$ and $\det(ab)$, that is, $\det(b) = 1$ for $\mathbf{1}_{+\pm}$, $\det(b) = -1$ for $\mathbf{1}_{-\pm}$, $\det(ab) = 1$ for $\mathbf{1}_{\pm+}$, and $\det(ab) = -1$ for $\mathbf{1}_{\pm-}$. Thus, the determinants of b and ab are essential for anomalies. These determinants are summarized in Table 15.6. This implies that two Z_2 symmetries can be anomalous. One Z_2 corresponds to b and the other Z'_2 corresponds to ab. Under this $Z_2 \times Z'_2$ symmetry, each representation has the following behavior:

$$\begin{aligned} Z_2 \text{ even}: &\quad \mathbf{1}_{+\pm}, \\ Z_2 \text{ odd}: &\quad \mathbf{1}_{-\pm}, \mathbf{2}_k, \end{aligned} \tag{15.28}$$

$$\begin{aligned} Z'_2 \text{ even}: &\quad \mathbf{1}_{\pm+}, \\ Z'_2 \text{ odd}: &\quad \mathbf{1}_{\pm-}, \mathbf{2}_k. \end{aligned} \tag{15.29}$$

The anomaly-free conditions are then

$$\sum_{\mathbf{1}_{-\pm}} \sum_{R^{(f)}} T_2(R^{(f)}) + \sum_{\mathbf{2}_k} \sum_{R^{(f)}} T_2(R^{(f)}) = 0 \mod 1, \tag{15.30}$$

for the Z_2-G_g-G_g anomaly and

$$\sum_{\mathbf{1}_{\pm-}} \sum_{R^{(f)}} T_2(R^{(f)}) + \sum_{\mathbf{2}_k} \sum_{R^{(f)}} T_2(R^{(f)}) = 0 \mod 1, \tag{15.31}$$

for the Z'_2-G_g-G_g anomaly.

Table 15.6 Determinants on D_N representations for N even

	1_{++}	1_{+-}	1_{-+}	1_{--}	2_k
det(b)	1	1	−1	−1	−1
det(ab)	1	−1	1	−1	−1

Table 15.7 Determinants on D_N representations for N odd

	1_+	1_-	2_k
det(b)	1	−1	−1
det(a)	1	1	1

15.2.7 D_N (N Odd)

Similarly, we study anomalies of D_N with N odd. As shown in Chap. 6, the group D_N with N odd has two singlets $1_\pm$ and $(N-1)/2$ doublets 2_k. Similarly to D_N with N even, all elements of D_N with N odd can be written as products of two elements a and b. The determinants of a are $\det(a) = 1$ on all representations $1_\pm$ and 2_k. The determinants of b are $\det b = 1$ on 1_+ and $\det(b) = -1$ on 1_- and 2_k. These are shown in Table 15.7. Thus, only the Z_2 symmetry corresponding to b can be anomalous. Under such a Z_2 symmetry, each representation has the following behavior:

$$Z_2 \text{ even}: \quad 1_+, \qquad Z_2 \text{ odd}: \quad 1_-, 2_k. \tag{15.32}$$

The anomaly-free condition is then

$$\sum_{1_-}\sum_{R^{(f)}} T_2\left(R^{(f)}\right) + \sum_{2_k}\sum_{R^{(f)}} T_2\left(R^{(f)}\right) = 0 \mod 1, \tag{15.33}$$

for the Z_2-G_g-G_g anomaly.

15.2.8 Q_N (N = 4n)

We study anomalies of Q_N with $N = 4n$. As shown in Chap. 7, the group Q_N with $N = 4n$ has four singlets $1_{\pm\pm}$ and $(N/2 - 1)$ doublets 2_k. All elements of Q_N can be written as products of a and b. The determinant of a is $\det(a) = 1$ on all the doublets 2_k. On the other hand, the determinant of b is $\det(b) = 1$ on the doublets 2_k with k odd and $\det(b) = -1$ on the doublets 2_k with k even. Similarly to D_N with N even, the four singlets $1_{\pm\pm}$ are classified by the values of $\det(b)$ and $\det(ab)$, that is, $\det(b) = 1$ for 1_{++}, $\det(b) = -1$ for $1_{-\pm}$, $\det(b) = 1$ for $1_{\pm+}$, and $\det(b) = -1$ for $1_{\pm-}$. Thus, the determinants of b and ab are essential for anomalies. These determinants are summarized in Table 15.8. Similarly to D_N with

15.2 Explicit Calculations

Table 15.8 Determinants on Q_N representations for $N/2$ even

	$\mathbf{1}_{++}$	$\mathbf{1}_{+-}$	$\mathbf{1}_{-+}$	$\mathbf{1}_{--}$	$\mathbf{2}_{k\,odd}$	$\mathbf{2}_{k\,even}$
det(b)	1	1	−1	−1	1	−1
det(ab)	1	−1	1	−1	1	−1

N even, two Z_2 symmetries can be anomalous. One Z_2 corresponds to b and the other Z_2' corresponds to ab. Under this $Z_2 \times Z_2'$ symmetry, each representation has the following behavior:

$$Z_2 \text{ even}: \quad \mathbf{1}_{+\pm}, \mathbf{2}_{k\,odd}, \qquad (15.34)$$
$$Z_2 \text{ odd}: \quad \mathbf{1}_{-\pm}, \mathbf{2}_{k\,even},$$

$$Z_2' \text{ even}: \quad \mathbf{1}_{\pm+}, \mathbf{2}_{k\,odd}, \qquad (15.35)$$
$$Z_2' \text{ odd}: \quad \mathbf{1}_{\pm-}, \mathbf{2}_{k\,even}.$$

The anomaly-free conditions are then

$$\sum_{\mathbf{1}_{-\pm}} \sum_{R^{(f)}} T_2(R^{(f)}) + \sum_{\mathbf{2}_{k\,even}} \sum_{R^{(f)}} T_2(R^{(f)}) = 0 \mod 1, \qquad (15.36)$$

for the Z_2-G_g-G_g anomaly and

$$\sum_{\mathbf{1}_{\pm-}} \sum_{R^{(f)}} T_2(R^{(f)}) + \sum_{\mathbf{2}_{k\,even}} \sum_{R^{(f)}} T_2(R^{(f)}) = 0 \mod 1, \qquad (15.37)$$

for the Z_2'-g-G_g anomaly.

15.2.9 Q_N ($N = 4n + 2$)

Similarly, we study anomalies of Q_N with $N = 4n + 2$. As shown in Chap. 7, the group Q_N with $N = 4n + 2$ has four singlets $\mathbf{1}_{\pm\pm}$ and $(N/2 - 1)$ doublets $\mathbf{2}_k$. All elements of Q_N can be expressed as products of a and b. The determinant of a is found to be $\det(a) = 1$ on all doublets $\mathbf{2}_k$. On the other hand, the determinants of b are $\det(b) = 1$ on the doublets $\mathbf{2}_k$ with k odd and $\det(b) = -1$ on the doublets $\mathbf{2}_k$ with k even. For all singlets, it is found that $\chi_\alpha(a) = \chi_\alpha(b^2)$, i.e., $\det(a) = \det(b^2)$. This implies that the determinants of b are more important for anomalies than the determinants of a. Indeed, the determinants of b are $\det(b) = 1$ on $\mathbf{1}_{++}$, $\det(b) = i$ on $\mathbf{1}_{+-}$, $\det(b) = -i$ on $\mathbf{1}_{-+}$, and $\det(b) = -1$ on $\mathbf{1}_{--}$. These determinants are summarized in Table 15.9. This implies that only the Z_4 symmetry corresponding to b can be anomalous. Under such a Z_4 symmetry, each representation has the

Table 15.9 Determinants on Q_N representations for $N/2$ odd

	1_{++}	1_{+-}	1_{-+}	1_{--}	$2_{k\,\text{odd}}$	$2_{k\,\text{even}}$
det(b)	1	i	$-i$	-1	1	-1
det(a)	1	-1	-1	1	1	1

following Z_4 charge q_4:

$$\begin{aligned} q_4 &= 0: \quad \mathbf{1}_{++}, \mathbf{2}_{k\,\text{odd}}, \\ q_4 &= 1: \quad \mathbf{1}_{+-}, \\ q_4 &= 2: \quad \mathbf{1}_{--}, \mathbf{2}_{k\,\text{even}}, \\ q_4 &= 3: \quad \mathbf{1}_{-+}. \end{aligned} \quad (15.38)$$

This includes the Z_2 symmetry corresponding to a and the Z_2 charge q_2 for each representation is defined as $q_2 = q_4$ mod 2. The anomaly-free conditions are

$$\sum_{\mathbf{1}_{+-}}\sum_{\mathbf{R}^{(f)}} T_2(\mathbf{R}^{(f)}) + 2\sum_{\mathbf{1}_{--}}\sum_{\mathbf{R}^{(f)}} T_2(\mathbf{R}^{(f)}) + 3\sum_{\mathbf{1}_{-+}}\sum_{\mathbf{R}^{(f)}} T_2(\mathbf{R}^{(f)})$$

$$+ 2\sum_{\mathbf{2}_{k\,\text{even}}}\sum_{\mathbf{R}^{(f)}} T_2(\mathbf{R}^{(f)}) = 0 \quad \text{mod } 2, \quad (15.39)$$

for the Z_4–G_g–G_g anomaly and

$$\sum_{\mathbf{1}_{+-}}\sum_{\mathbf{R}^{(f)}} \dim \mathbf{R}^{(f)} + 2\sum_{\mathbf{1}_{--}}\sum_{\mathbf{R}^{(f)}} \dim \mathbf{R}^{(f)} + 3\sum_{\mathbf{1}_{-+}}\sum_{\mathbf{R}^{(f)}} \dim \mathbf{R}^{(f)}$$

$$+ 2\sum_{\mathbf{2}_{k\,\text{even}}}\sum_{\mathbf{R}^{(f)}} \dim \mathbf{R}^{(f)} = 0 \quad \text{mod } 2, \quad (15.40)$$

for the Z_4–gravity–gravity anomaly. Similarly, we obtain the anomaly-free condition on the Z_2 symmetry corresponding to a as

$$\sum_{\mathbf{1}_{+-}}\sum_{\mathbf{R}^{(f)}} T_2(\mathbf{R}^{(f)}) + \sum_{\mathbf{1}_{-+}}\sum_{\mathbf{R}^{(f)}} T_2(\mathbf{R}^{(f)}) = 0 \quad \text{mod } 1, \quad (15.41)$$

for the Z_2–G_g–G_g anomaly.

15.2.10 QD_{2N}

As shown in Chap. 8, the group QD_{2N} has four singlets $\mathbf{1}_{\pm\pm}$ and $(N/2 - 1)$ doublets $\mathbf{2}_k$. All elements of QD_{2N} can be written as products of a and b. The determinants of b are $\det(b) = -1$ on all of doublets $\mathbf{2}_k$. On the other hand, the determinants of a are found to be $\det(a) = -1$ on the doublets $\mathbf{2}_k$ with k odd and $\det(a) = 1$ on the doublets $\mathbf{2}_k$ with k even. For singlets, it is found that $\det(a) = 1$ for $\mathbf{1}_{\pm+}$, $\det(a) =$

15.2 Explicit Calculations

Table 15.10 Determinants on QD_{2N} representations

	$\mathbf{1}_{++}$	$\mathbf{1}_{-+}$	$\mathbf{1}_{+-}$	$\mathbf{1}_{--}$	$\mathbf{2}_{k\,\text{odd}}$	$\mathbf{2}_{k\,\text{even}}$
$\det(b)$	1	1	-1	-1	-1	-1
$\det(a)$	1	-1	1	-1	-1	1

-1 for $\mathbf{1}_{\pm-}$, $\det(b) = 1$ for $\mathbf{1}_{++}$, and $\det(b) = -1$ for $\mathbf{1}_{-\pm}$. These determinants are summarized in Table 15.10. This implies that two Z_2 symmetries can be anomalous. One Z_2 corresponds to a and the other Z'_2 corresponds to b. Under this $Z_2 \times Z'_2$ symmetry, each representation has the following behavior:

$$Z_2 \text{ even}: \quad \mathbf{1}_{+\pm}, \mathbf{2}_{k\,\text{even}}, \qquad (15.42)$$
$$Z_2 \text{ odd}: \quad \mathbf{1}_{-\pm}, \mathbf{2}_{k\,\text{odd}},$$

$$Z'_2 \text{ even}: \quad \mathbf{1}_{\pm+}, \qquad (15.43)$$
$$Z'_2 \text{ odd}: \quad \mathbf{1}_{\pm-}, \mathbf{2}_k.$$

The anomaly-free conditions are then

$$\sum_{\mathbf{1}_{-\pm}} \sum_{\boldsymbol{R}^{(f)}} T_2(\boldsymbol{R}^{(f)}) + \sum_{\mathbf{2}_{k\,\text{odd}}} \sum_{\boldsymbol{R}^{(f)}} T_2(\boldsymbol{R}^{(f)}) = 0 \mod 1, \qquad (15.44)$$

for the Z_2-G_g-G_g anomaly and

$$\sum_{\mathbf{1}_{\pm-}} \sum_{\boldsymbol{R}^{(f)}} T_2(\boldsymbol{R}^{(f)}) + \sum_{\mathbf{2}_k} \sum_{\boldsymbol{R}^{(f)}} T_2(\boldsymbol{R}^{(f)}) = 0 \mod 1, \qquad (15.45)$$

for the Z'_2-G_g-G_g anomaly.

15.2.11 $\Sigma(2N^2)$

We study anomalies of $\Sigma(2N^2)$. As shown in Chap. 9, the group $\Sigma(2N^2)$ has $2N$ singlets $\mathbf{1}_{\pm n}$ and $N(N-1)/2$ doublets $\mathbf{2}_{p,q}$. All elements of $\Sigma(2N^2)$ can be written as products of a, a', and b. Their determinants for each representation are shown in Table 15.11, where $\rho = e^{2\pi i/N}$. We then find that only the Z_2 symmetry corresponding to b and the Z_N symmetry corresponding to a can be anomalous. The other Z_N symmetry corresponding to a' is not independent of the Z_N symmetry for a. Under the Z_2 symmetry, each representation has the following behavior:

$$Z_2 \text{ even}: \quad \mathbf{1}_{+n}, \qquad (15.46)$$
$$Z_2 \text{ odd}: \quad \mathbf{1}_{-n}, \mathbf{2}_{p,q},$$

and under the Z_N symmetry corresponding to a each representation has the following Z_N charge q_N:

$$q_N = n: \quad \mathbf{1}_{\pm n}, \qquad (15.47)$$
$$q_N = p + q: \quad \mathbf{2}_{p,q}.$$

Table 15.11 Determinants on $\Sigma(2N^2)$ representations

	$\mathbf{1}_{+n}$	$\mathbf{1}_{-n}$	$\mathbf{2}_k$
det(b)	1	-1	-1
det(a)	ρ^n	ρ^n	ρ^{p+q}
det(a')	ρ^n	ρ^n	ρ^{p+q}

The anomaly-free condition is then

$$\sum_{\mathbf{1}_{-n}}\sum_{\mathbf{R}^{(f)}} T_2(\mathbf{R}^{(f)}) + \sum_{\mathbf{2}_{p,q}}\sum_{\mathbf{R}^{(f)}} T_2(\mathbf{R}^{(f)}) = 0 \quad \text{mod } 1, \tag{15.48}$$

for the Z_2-G_g-G_g anomaly. Similarly, the anomaly-free conditions for the Z_N symmetry are

$$\sum_{\mathbf{1}_{\pm n}}\sum_{\mathbf{R}^{(f)}} nT_2(\mathbf{R}^{(f)}) + \sum_{\mathbf{2}_{p,q}}\sum_{\mathbf{R}^{(f)}} (p+q)T_2(\mathbf{R}^{(f)}) = 0 \quad \text{mod } N/2, \tag{15.49}$$

for the Z_N-G_g-G_g anomaly and

$$\sum_{\mathbf{1}_{\pm n}}\sum_{\mathbf{R}^{(f)}} n \dim \mathbf{R}^{(f)} + \sum_{\mathbf{2}_{p,q}}\sum_{\mathbf{R}^{(f)}} (p+q)\dim \mathbf{R}^{(f)} = 0 \quad \text{mod } N/2, \tag{15.50}$$

for the Z_N-gravity-gravity anomaly.

15.2.12 $\Delta(3N^2)$ ($N/3 \neq$ Integer)

We study anomalies of $\Delta(3N^2)$ when $N/3 \neq$ integer. As shown in Chap. 10, the group $\Delta(3N^2)$ with $N/3 \neq$ integer has three singlets $\mathbf{1}_0$, $\mathbf{1}_1$, and $\mathbf{1}_2$, and $(N^2-1)/3$ triplets $\mathbf{3}_{[k][\ell]}$. All elements of $\Delta(3N^2)$ can be written as products of a, a', and b. It is found that $\det(a) = \det(a') = 1$ on all representations. These elements are thus irrelevant to anomalies. On the other hand, the determinant of b is found to be $\det(b) = 1$ for all $\mathbf{3}_{[k][\ell]}$ and $\mathbf{1}_0$, $\det(b) = \omega$ for $\mathbf{1}_1$, and $\det(b) = \omega^2$ for $\mathbf{1}_2$, with $\omega = e^{2\pi i/3}$, as shown in Table 15.12. This implies that only the Z_3 symmetry corresponding to b can be anomalous. Under such a Z_3 symmetry, each representation has the following Z_3 charge q_3:

$$\begin{aligned} q_3 &= 0: \quad \mathbf{1}_0, \mathbf{3}_{[k][\ell]}, \\ q_3 &= 1: \quad \mathbf{1}_1, \\ q_3 &= 2: \quad \mathbf{1}_2. \end{aligned} \tag{15.51}$$

The anomaly-free conditions are then

$$\sum_{\mathbf{1}_1}\sum_{\mathbf{R}^{(f)}} T_2(\mathbf{R}^{(f)}) + 2\sum_{\mathbf{1}_2}\sum_{\mathbf{R}^{(f)}} T_2(\mathbf{R}^{(f)}) = 0 \quad \text{mod } 3/2, \tag{15.52}$$

15.2 Explicit Calculations

Table 15.12 Determinants on $\Delta(3N^2)$ representations when $N/3 \neq$ integer

	$\mathbf{1}_k$	$\mathbf{3}_{[k][\ell]}$
$\det(b)$	ω^k	1
$\det(a)$	1	1
$\det(a')$	1	1

for the Z_3-G_g-G_g anomaly and

$$\sum_{\mathbf{1}_1}\sum_{\mathbf{R}^{(f)}} \dim \mathbf{R}^{(f)} + 2\sum_{\mathbf{1}_2}\sum_{\mathbf{R}^{(f)}} \dim \mathbf{R}^{(f)} = 0 \quad \text{mod } 3/2, \tag{15.53}$$

for the Z_3-gravity-gravity anomaly.

15.2.13 $\Delta(3N^2)$ ($N/3$ Integer)

Similarly, we study anomalies of $\Delta(3N^2)$ when $N/3$ is an integer. As shown in Chap. 10, when $N/3$ is an integer, the group $\Delta(3N^2)$ has a total of nine singlets $\mathbf{1}_{k,\ell}$ and $(N^2 - 3)/3$ triplets $\mathbf{3}_{[k][\ell]}$. All elements of $\Delta(3N^2)$ can be expressed as products of a, a', and b. On all triplet representations $\mathbf{3}_{[k][\ell]}$, their determinants are found to be $\det(a) = \det(a') = \det(b) = 1$. On the other hand, it is found that $\det(a) = \det(a')$ on all nine singlets. Furthermore, the nine singlets are classified by the values of $\det(a) = \det(a')$ and $\det(b)$. That is, the determinants of $\det(a) = \det(a')$ and $\det(b)$ are $\det(a) = \det(a') = \omega^\ell$ and $\det(b) = \omega^k$ on $\mathbf{1}_{k,\ell}$. These results are shown in Table 15.13. This implies that two independent Z_3 symmetries can be anomalous. One corresponds to b and the other corresponds to a. For the Z_3 symmetry corresponding to b, each representation has the following Z_3 charge q_3:

$$\begin{aligned} q_3 &= 0: \quad \mathbf{1}_{0,\ell}, \mathbf{3}_{[k][\ell]}, \\ q_3 &= 1: \quad \mathbf{1}_{1,\ell}, \\ q_3 &= 2: \quad \mathbf{1}_{2,\ell}, \end{aligned} \tag{15.54}$$

while for the Z_3' symmetry corresponding to a, each representation has the following Z_3 charge q_3':

$$\begin{aligned} q_3' &= 0: \quad \mathbf{1}_{k,0}, \mathbf{3}_{[k][\ell]}, \\ q_3' &= 1: \quad \mathbf{1}_{k,1}, \\ q_3' &= 2: \quad \mathbf{1}_{k,2}. \end{aligned} \tag{15.55}$$

The anomaly-free conditions are then

$$\sum_{\mathbf{1}_{1,\ell}}\sum_{\mathbf{R}^{(f)}} T_2(\mathbf{R}^{(f)}) + 2\sum_{\mathbf{1}_{2,\ell}}\sum_{\mathbf{R}^{(f)}} T_2(\mathbf{R}^{(f)}) = 0 \quad \text{mod } 3/2, \tag{15.56}$$

Table 15.13 Determinants on $\Delta(3N^2)$ representations when $N/3$ is an integer

	$\mathbf{1}_{k,\ell}$	$\mathbf{3}_{[k][\ell]}$
$\det(b)$	ω^k	1
$\det(a)$	ω^ℓ	1
$\det(a')$	ω^ℓ	1

for the Z_3–G_g–G_g anomaly and

$$\sum_{\mathbf{1}_{1,\ell}}\sum_{\boldsymbol{R}^{(f)}} \dim \boldsymbol{R}^{(f)} + 2\sum_{\mathbf{1}_{2,\ell}}\sum_{\boldsymbol{R}^{(f)}} \dim \boldsymbol{R}^{(f)} = 0 \mod 3/2, \qquad (15.57)$$

for the Z_3–gravity–gravity anomaly. Similarly, for the Z'_3 symmetry, the anomaly-free conditions are

$$\sum_{\mathbf{1}_{k,1}}\sum_{\boldsymbol{R}^{(f)}} T_2(\boldsymbol{R}^{(f)}) + 2\sum_{\mathbf{1}_{k,2}}\sum_{\boldsymbol{R}^{(f)}} T_2(\boldsymbol{R}^{(f)}) = 0 \mod 3/2, \qquad (15.58)$$

for the Z'_3–G_g–G_g anomaly and

$$\sum_{\mathbf{1}_{k,1}}\sum_{\boldsymbol{R}^{(f)}} \dim \boldsymbol{R}^{(f)} + 2\sum_{\mathbf{1}_{k,2}}\sum_{\boldsymbol{R}^{(f)}} \dim \boldsymbol{R}^{(f)} = 0 \mod 3/2, \qquad (15.59)$$

for the Z'_3–gravity–gravity anomaly.

15.2.14 T_N

As shown in Chap. 11, the group T_N has three singlets $\mathbf{1}_{0,1,2}$ and $(N-1)/3$ triplets $\mathbf{3}(\bar{\mathbf{3}})_m$. All elements of T_N can be written as products of a and b, where a and b correspond to the generators of Z_N and Z_3, respectively. It is found that $\det(a) = 1$ on all representations, so these elements are irrelevant to anomalies. On the other hand, the determinant of b is found to be $\det(b) = 1$ for any triplet $\mathbf{3}(\bar{\mathbf{3}})_m$ and $\det(b) = \omega^k$ for $\mathbf{1}_k$ ($k = 0, 1, 2$), as shown in Table 15.14. These results imply that only the Z_3 symmetry corresponding to b can be anomalous. Under such a Z_3 symmetry, each representation has the following Z_3 charge q_3:

$$\begin{aligned} q_3 &= 0: \quad \mathbf{1}_0, \mathbf{3}, \bar{\mathbf{3}}, \\ q_3 &= 1: \quad \mathbf{1}_1, \\ q_3 &= 2: \quad \mathbf{1}_2. \end{aligned} \qquad (15.60)$$

The anomaly-free conditions are

$$\sum_{\mathbf{1}_1}\sum_{\boldsymbol{R}^{(f)}} T_2(\boldsymbol{R}^{(f)}) + 2\sum_{\mathbf{1}_2}\sum_{\boldsymbol{R}^{(f)}} T_2(\boldsymbol{R}^{(f)}) = 0 \mod 3/2, \qquad (15.61)$$

15.2 Explicit Calculations

Table 15.14 Determinants on T_N representations

	$\mathbf{1}_0$	$\mathbf{1}_1$	$\mathbf{1}_2$	$\mathbf{3}_m$	$\bar{\mathbf{3}}_m$
$\det(a)$	1	1	1	1	1
$\det(b)$	1	ω	ω^2	1	1

for the Z_3–G_g–G_g anomaly and

$$\sum_{\mathbf{1}_1}\sum_{\mathbf{R}^{(f)}} \dim \mathbf{R}^{(f)} + 2\sum_{\mathbf{1}_2}\sum_{\mathbf{R}^{(f)}} \dim \mathbf{R}^{(f)} = 0 \mod 3/2, \qquad (15.62)$$

for the Z_3–gravity–gravity anomaly.

15.2.15 $\Sigma(3N^3)$

We now study anomalies of $\Sigma(3N^3)$, which has $3N$ singlets $\mathbf{1}_{k,\ell}$ and $N(N^2-1)/3$ triplets $\mathbf{3}_{[\ell][m][n]}$. All elements of $\Sigma(3N^3)$ can be written as products of a, a', a'', and b. Their determinants are shown for each representation in Table 15.15, where $\rho = e^{2\pi i/N}$. It turns out that only the Z_3 symmetry corresponding to b and the Z_N symmetry corresponding to a can be anomalous. Other Z_N symmetries corresponding to a' and a'' are not independent of the Z_N symmetry for a. For the Z_3 symmetry corresponding to b, each representation has the following Z_3 charge q_3:

$$\begin{aligned} q_3 &= 0: \quad \mathbf{1}_{0,\ell}, \mathbf{3}_{[\ell][m][n]}, \\ q_3 &= 1: \quad \mathbf{1}_{1,\ell}, \\ q_3 &= 2: \quad \mathbf{1}_{2,\ell}, \end{aligned} \qquad (15.63)$$

and under the Z_N symmetry corresponding to a, each representation has the following Z_N charge q_N:

$$\begin{aligned} q_N &= \ell: & \mathbf{1}_{k,\ell}, \\ q_N &= \ell + m + n: & \mathbf{3}_{[\ell][m][n]}. \end{aligned} \qquad (15.64)$$

The anomaly-free condition is then

$$\sum_{\mathbf{1}_{1,\ell}}\sum_{\mathbf{R}^{(f)}} T_2(\mathbf{R}^{(f)}) + \sum_{\mathbf{1}_{2,\ell}}\sum_{\mathbf{R}^{(f)}} T_2(\mathbf{R}^{(f)}) = 0 \mod 3/2, \qquad (15.65)$$

for the Z_3–G_g–G_g anomaly and

$$\sum_{\mathbf{1}_{1,\ell}}\sum_{\mathbf{R}^{(f)}} \dim \mathbf{R}^{(f)} + 2\sum_{\mathbf{1}_{2,\ell}}\sum_{\mathbf{R}^{(f)}} \dim \mathbf{R}^{(f)} = 0 \mod 3/2, \qquad (15.66)$$

Table 15.15 Determinants on $\Sigma(3N^3)$ representations

	$\mathbf{1}_{k,\ell}$	$\mathbf{3}_{[\ell][m][n]}$
$\det(b)$	ω^k	1
$\det(a)$	ρ^ℓ	$\rho^{\ell+m+n}$
$\det(a')$	ρ^ℓ	$\rho^{\ell+m+n}$
$\det(a'')$	ρ^ℓ	$\rho^{\ell+m+n}$

for the Z_3–gravity–gravity anomaly. Similarly, the anomaly-free conditions for the Z_N symmetry are

$$\sum_{\mathbf{1}_{k,\ell}}\sum_{\mathbf{R}^{(f)}} \ell T_2(\mathbf{R}^{(f)}) + \sum_{\mathbf{3}_{[\ell][m][n]}}\sum_{\mathbf{R}^{(f)}} (\ell+m+n) T_2(\mathbf{R}^{(f)}) = 0 \mod N/2,$$

(15.67)

for the Z_N–G_g–G_g anomaly and

$$\sum_{\mathbf{1}_{k,\ell}}\sum_{\mathbf{R}^{(f)}} \ell \dim \mathbf{R}^{(f)} + \sum_{\mathbf{3}_{[\ell][m][n]}}\sum_{\mathbf{R}^{(f)}} (\ell+m+n) \dim \mathbf{R}^{(f)} = 0 \mod N/2,$$

(15.68)

for the Z_N–gravity–gravity anomaly.

15.2.16 $\Delta(6N^2)$ ($N/3 \neq$ Integer)

We study anomalies of $\Delta(6N^2)$ when $N/3 \neq$ integer. As shown in Chap. 13, when $N/3 \neq$ integer, the group has two singlets $\mathbf{1}_{0,1}$, one doublet $\mathbf{2}$, $2(N-1)$ triplets $\mathbf{3}_{1k}$ and $\mathbf{3}_{2k}$, and $N(N-3)/6$ sextets $\mathbf{6}_{[[k],[\ell]]}$. All elements can be written as products of a, a', b, and c. The determinants of a, a', and b on any representation are found to be $\det(a) = \det(a') = \det(b) = 1$. The determinants of c for $\mathbf{1}_0$ and $\mathbf{3}_{2k}$ are $\det(c) = 1$, while the other representations lead to $\det(c) = -1$. These results are shown in Table 15.16. This implies that only the Z_2 symmetry corresponding to the generator c can be anomalous. Under such a Z_2 symmetry, each representation has the following Z_2 charge q_2:

$$\begin{aligned} q_2 &= 0: \quad \mathbf{1}_0, \mathbf{3}_{2k}, \\ q_2 &= 1: \quad \mathbf{1}_1, \mathbf{2}, \mathbf{3}_{1k}, \mathbf{6}_{[[k],[\ell]]}. \end{aligned}$$

(15.69)

The anomaly-free conditions are then

$$\sum_{\mathbf{1}_1}\sum_{\mathbf{R}^{(f)}} T_2(\mathbf{R}^{(f)}) + \sum_{\mathbf{2}}\sum_{\mathbf{R}^{(f)}} T_2(\mathbf{R}^{(f)}) + \sum_{\mathbf{3}_{1k}}\sum_{\mathbf{R}^{(f)}} T_2(\mathbf{R}^{(f)}) + \sum_{\mathbf{6}_{[[k],[\ell]]}}\sum_{\mathbf{R}^{(f)}} T_2(\mathbf{R}^{(f)})$$
$$= 0 \mod 1,$$

(15.70)

for the Z_2–G_g–G_g anomaly.

15.3 Comments on Anomalies

Table 15.16 Determinants on representations of $\Delta(6N^2)$ with $3N \neq$ integer

	1_0	1_1	2	3_{1k}	3_{2k}	$6_{[[k],[\ell]]}$
det(a)	1	1	1	1	1	1
det(a')	1	1	1	1	1	1
det(b)	1	1	1	1	1	1
det(c)	1	-1	-1	-1	1	-1

Table 15.17 Determinants on representations of $\Delta(6N^2)$ with $3N$ integer

	1_0	1_1	2_1	2_2	2_3	2_4	3_{1k}	3_{2k}	$6_{[[k],[\ell]]}$
det(a)	1	1	1	1	1	1	1	1	1
det(a')	1	1	1	1	1	1	1	1	1
det(b)	1	1	1	1	1	1	1	1	1
det(c)	1	-1	-1	-1	-1	-1	-1	1	-1

15.2.17 $\Delta(6N^2)$ ($N/3$ Integer)

In the same way, we study anomalies of $\Delta(6N^2)$ when $N/3$ is an integer. As shown in Chap. 13, when $N/3$ is an integer, the group has two singlets $1_{0,1}$, four doublets $2_{1,2,3,4}$, $2(N-1)$ triplets 3_{1k} and 3_{2k}, and $(N^2 - 3N + 2)/6$ sextets $6_{[[k],[\ell]]}$. All elements of $\Delta(54)$ can be written as products of a, a', b, and c. The determinants of a, a', and b on any representation are $\det(a) = \det(a') = \det(b) = 1$. The determinants of c for 1_0 and 3_{2k} are found to be $\det(c) = 1$, while the other representations lead to $\det(c) = -1$. These results are shown in Table 15.17. This implies that only the Z_2 symmetry corresponding to the generator c can be anomalous. Under such a Z_2 symmetry, each representation has the following Z_2 charge q_2:

$$q_2 = 0: \quad 1_0, 3_{2k}, \qquad\qquad (15.71)$$
$$q_2 = 1: \quad 1_1, 2_{1,2,3,4}, 3_{1k}, 6_{[[k],[\ell]]}.$$

The anomaly-free conditions are then

$$\sum_{1_1} \sum_{R^{(f)}} T_2(R^{(f)}) + \sum_{2_k} \sum_{R^{(f)}} T_2(R^{(f)}) + \sum_{3_{1k}} \sum_{R^{(f)}} T_2(R^{(f)}) + \sum_{6_{[[k],[\ell]]}} \sum_{R^{(f)}} T_2(R^{(f)})$$
$$= 0 \mod 1, \qquad\qquad (15.72)$$

for the Z_2-G_g-G_g anomaly.

Similarly, we can analyze anomalies for other non-Abelian discrete symmetries.

15.3 Comments on Anomalies

Finally, we comment on symmetry breaking by quantum effects. When a discrete (flavor) symmetry is anomalous, breaking terms can appear in the Lagrangian, e.g., by instanton effects such as

$$\frac{1}{M^n}\Lambda^m\Phi_1\cdots\Phi_k,$$

where Λ is a dynamical scale and M is a typical (cutoff) scale. Within the framework of string theory discrete anomalies and also anomalies of continuous gauge symmetries can be canceled by the Green–Schwarz (GS) mechanism [10], unless discrete symmetries are accidental. In the GS mechanism, dilaton and moduli fields, i.e., the so-called GS fields Φ_{GS}, transform non-linearly under anomalous transformation. The anomaly cancellation due to the GS mechanism imposes certain relations among anomalies (see, e.g., [3]).[1] Stringy non-perturbative effects, but also field-theoretical effects, induce terms in the Lagrangian such as

$$\frac{1}{M^n}e^{-a\Phi_{GS}}\Phi_1\cdots\Phi_k.$$

The GS fields Φ_{GS}, i.e., dilaton/moduli fields, are expected to develop non-vanishing vacuum expectation values, and the above terms correspond to breaking terms of discrete symmetries.

The above breaking terms may be small. Such approximate discrete symmetries with small breaking terms may be useful in particle physics,[2] if breaking terms are controllable. Alternatively, if exact symmetries are necessary, one has to arrange matter fields and their quantum numbers in such a way that models are free from anomalies.

References

1. Fujikawa, K.: Phys. Rev. Lett. **42**, 1195 (1979)
2. Fujikawa, K.: Phys. Rev. D **21**, 2848 (1980)
3. Araki, T., Kobayashi, T., Kubo, J., Ramos-Sanchez, S., Ratz, M., Vaudrevange, P.K.S.: Nucl. Phys. B **805**, 124 (2008). arXiv:0805.0207 [hep-th]
4. Alvarez-Gaume, L., Witten, E.: Nucl. Phys. B **234**, 269 (1984)
5. Alvarez-Gaume, L., Ginsparg, P.H.: Ann. Phys. **161**, 423 (1985)
6. Fujikawa, K., Ojima, S., Yajima, S.: Phys. Rev. D **34**, 3223 (1986)
7. Rohlin, V.: Dokl. Akad. Nauk **128**, 980–983 (1959)
8. Csaki, C., Murayama, H.: Nucl. Phys. B **515**, 114–162 (1998). arXiv:hep-th/9710105
9. Araki, T.: Prog. Theor. Phys. **117**, 1119–1138 (2007). arXiv:hep-ph/0612306
10. Green, M.B., Schwarz, J.H.: Phys. Lett. B **149**, 117–122 (1984)
11. Kobayashi, T., Nakano, H.: Nucl. Phys. B **496**, 103–131 (1997). arXiv:hep-th/9612066
12. Fukuoka, H., Kubo, J., Suematsu, D.: Phys. Lett. B **678**, 401 (2009). arXiv:0905.2847 [hep-ph]

---

[1] See also [11].

[2] For some applications, see, e.g., [12].

Chapter 16
Non-Abelian Discrete Symmetry in Quark/Lepton Flavor Models

Non-Abelian discrete groups have been adopted for the flavor models of the quarks and leptons. In this chapter, we present some typical flavor models based on these discrete symmetries. The examples will illustrate how such models are built. However, before discussing the models themselves, we briefly review the main features of recent experimental data regarding neutrino flavor mixing.

16.1 Neutrino Flavor Mixing and Neutrino Mass Matrix

The recent experimental data on neutrino oscillations has stimulated work on the non-Abelian discrete symmetry of flavors. Both the atmospheric neutrino mixing angle θ_{23} and the solar neutrino mixing angle θ_{12} are quite large. In particular, θ_{23} is almost maximal. These neutrino mixing angles are defined in the neutrino mixing matrix U by

$$U = \begin{pmatrix} c_{13}c_{12} & c_{13}s_{12} & s_{13}e^{-i\delta} \\ -c_{23}s_{12} - s_{23}s_{13}c_{12}e^{i\delta} & c_{23}c_{12} - s_{23}s_{13}s_{12}e^{i\delta} & s_{23}c_{13} \\ s_{23}s_{12} - c_{23}s_{13}c_{12}e^{i\delta} & -s_{23}c_{12} - c_{23}s_{13}s_{12}e^{i\delta} & c_{23}c_{13} \end{pmatrix}, \quad (16.1)$$

where c_{ij} and s_{ij} denote $\cos\theta_{ij}$ and $\sin\theta_{ij}$, respectively.

The global fit of the neutrino experimental data in Table 16.1 [1–3], indicates the tri-bimaximal mixing matrix U_{tribi} for three lepton flavors [4–7] as follows:

$$U_{\text{tribi}} = \begin{pmatrix} \frac{2}{\sqrt{6}} & \frac{1}{\sqrt{3}} & 0 \\ -\frac{1}{\sqrt{6}} & \frac{1}{\sqrt{3}} & -\frac{1}{\sqrt{2}} \\ -\frac{1}{\sqrt{6}} & \frac{1}{\sqrt{3}} & \frac{1}{\sqrt{2}} \end{pmatrix}, \quad (16.2)$$

which favors the non-Abelian discrete symmetry for the lepton flavor. Indeed, various types of models leading to tri-bimaximal mixing have been proposed on the basis of non-Abelian discrete flavor symmetries, as can be seen, e.g., in [8, 9].

Table 16.1 Summary of neutrino oscillation parameters. For Δm^2_{31}, $\sin^2\theta_{23}$, and $\sin^2\theta_{13}$, the upper (lower) row corresponds to the normal (inverted) neutrino mass hierarchy. In [10], they assume the new reactor anti-neutrino fluxes [11] and include short-baseline reactor neutrino experiments in the fit

Parameter	Best fit $\pm 1\sigma$	2σ	3σ
Δm^2_{sol} [10^{-5} eV^2]	$7.59^{+0.20}_{-0.18}$	7.24–7.99	7.09–8.19
Δm^2_{atm} [10^{-3} eV^2]	$2.50^{+0.09}_{-0.16}$	2.25–2.68	2.14–2.76
	$-(2.40^{+0.08}_{-0.09})$	$-(2.23$–$2.58)$	$-(2.13$–$2.67)$
$\sin^2\theta_{12}$	$0.312^{+0.017}_{-0.015}$	0.28–0.35	0.27–0.36
$\sin^2\theta_{23}$	$0.52^{+0.06}_{-0.07}$	0.41–0.61	0.39–0.64
	0.52 ± 0.06	0.42–0.61	
$\sin^2\theta_{13}$	$0.013^{+0.007}_{-0.005}$	0.004–0.028	0.001–0.035
	$0.016^{+0.008}_{-0.006}$	0.005–0.031	0.001–0.039

In tri-bimaximal mixing, θ_{13} vanishes. However, the T2K collaboration presented evidence at 2.5σ for a non-zero value of the reactor angle θ_{13} [12]. Quantitatively, it was found that

$$0.03(0.04) < \sin^2 2\theta_{13} < 0.28(0.34), \quad 90\,\%\ \text{C.L.}, \qquad (16.3)$$

for $|\Delta m^2_{32}| = 2.4\times 10^{-3}$ eV^2, $\sin^2 2\theta_{23} = 1$, and $\delta = 0$, in the normal (inverted) hierarchy of neutrino masses. Finally, Daya Bay reported the following result [13]:

$$\sin^2 2\theta_{13} = 0.092 \pm 0.016 \pm 0.005, \quad 68\,\%\ \text{C.L.} \qquad (16.4)$$

Thus the theoretical estimate of the neutrino mixing angles is an important subject. To begin with, we introduce typical models to reproduce the tri-bimaximal mixing of neutrino flavors in the flavor model with non-Abelian discrete symmetry.

The neutrino mass matrix with tri-bimaximal mixing of flavors is expressed as the sum of three simple mass matrices in the flavor diagonal basis of the charged lepton. In terms of neutrino mass eigenvalues m_1, m_2, and m_3, the neutrino mass matrix is given by

$$\begin{aligned} M_\nu &= U^*_{\text{tribi}} \begin{pmatrix} m_1 & 0 & 0 \\ 0 & m_2 & 0 \\ 0 & 0 & m_3 \end{pmatrix} U^\dagger_{\text{tribi}} \\ &= \frac{m_1+m_3}{2}\begin{pmatrix}1&0&0\\0&1&0\\0&0&1\end{pmatrix} + \frac{m_2-m_1}{3}\begin{pmatrix}1&1&1\\1&1&1\\1&1&1\end{pmatrix} + \frac{m_1-m_3}{2}\begin{pmatrix}1&0&0\\0&0&1\\0&1&0\end{pmatrix}. \end{aligned}$$

$$(16.5)$$

This neutrino mass matrix is easily realized in some non-Abelian discrete symmetries. In the following sections, we shall thus present a simple realization which

Table 16.2 Assignments of $SU(2)$, A_4, and Z_3 representations, where $\omega = e^{2\pi i/3}$

	(l_e, l_μ, l_τ)	e^c	μ^c	τ^c	$h_{u,d}$	ϕ_l	ϕ_v	ξ	ξ'	ξ''
$SU(2)$	2	1	1	1	2	1	1	1	1	1
A_4	3	1	$1''$	$1'$	1	3	3	1	$1'$	$1''$
Z_3	ω	ω^2	ω^2	ω^2	1	1	ω	ω	ω	ω

arises from the 5D non-renormalizable operators [14], or the seesaw mechanism [15–19].

16.2 A_4 Flavor Symmetry

Simple models realizing tri-bimaximal mixing have been proposed using the non-Abelian finite group A_4 [20–67]. The A_4 flavor model considered by Alterelli et al. [27, 33] realizes tri-bimaximal flavor mixing. The deviation from tri-bimaximal mixing can also be predicted. Actually, we have investigated the deviation from tri-bimaximal mixing including higher dimensional operators in the effective model [44, 63].

16.2.1 Realizing Tri-Bimaximal Mixing of Flavors

We begin by presenting the $A_4 \times Z_3$ flavor model with supersymmetry, including right-handed neutrinos [27, 33]. In the non-Abelian finite group A_4, there are twelve group elements and four irreducible representations: **1**, **1'**, **1''**, and **3**. The A_4 and Z_3 charge assignments of leptons, Higgs fields, and SM-singlets are listed in Table 16.2. Under the A_4 symmetry, the chiral superfields for three families of the left-handed lepton doublet $l = (l_e, l_\mu, l_\tau)$ are assumed to transform according to **3**, while the right-handed ones of the charged leptons e^c, μ^c, and τ^c are assigned with **1**, **1''**, **1'**, respectively. The third row of Table 16.2 shows how each chiral multiplet transforms under Z_3, where $\omega = e^{2\pi i/3}$. We assume flavons ϕ_l and ϕ_v which are A_4 triplets. In addition to these triplet flavons, we can consider singlet flavons ξ, ξ', ξ'', which are **1**, **1'**, **1''**, respectively. The flavor symmetry is spontaneously broken by the vacuum expectation values (VEVs) of two **3**'s, $\phi_l = (\phi_{l1}, \phi_{l2}, \phi_{l3})$, $\phi_v = (\phi_{v1}, \phi_{v2}, \phi_{v3})$, and by singlets, ξ, ξ', ξ'', which are $SU(2)_L \times U(1)_Y$ singlets.

In order to realize tri-bimaximal mixing, we consider the case where $\langle \xi' \rangle = \langle \xi'' \rangle = 0$. The superpotential of the lepton sector which respects the gauge and the flavor symmetry is described by

$$w_\ell = y^e e^c l \phi_l h_d / \Lambda + y^\mu \mu^c l \phi_l h_d / \Lambda + y^\tau \tau^c l \phi_l h_d / \Lambda \\ + \left(y^\nu_{\phi_v} \phi_v + y^\nu_\xi \xi + y^\nu_{\xi'} \xi' + y^\nu_{\xi''} \xi'' \right) l l h_u h_u / \Lambda^2, \quad (16.6)$$

where y^e, y^μ, y^τ, $y^\nu_{\phi_v}$, y^ν_ξ, $y^\nu_{\xi'}$, and $y^\nu_{\xi''}$ are the dimensionless coupling constants, and Λ is the cutoff scale. Hereafter, we follow the convention that the chiral superfield and its lowest component are denoted by the same letter. Decompositions into the A_4 singlet are given using the basis in Appendix C.2:

$$e^c l \phi_l \to e^c (l_e \phi_{l1} + l_\mu \phi_{l3} + l_\tau \phi_{l2}),$$

$$ll\phi_v \to \begin{pmatrix} 2l_e l_e - l_\mu l_\tau - l_\tau l_\mu \\ 2l_\tau l_\tau - l_e l_\mu - l_\mu l_e \\ 2l_\mu l_\mu - l_e l_\tau - l_\tau l_e \end{pmatrix} \phi_v$$

$$\to (2l_e l_e - l_\mu l_\tau - l_\tau l_\mu)\phi_{v1} + (2l_\tau l_\tau - l_e l_\mu - l_\mu l_e)\phi_{v3} \quad (16.7)$$
$$+ (2l_\mu l_\mu - l_e l_\tau - l_\tau l_e)\phi_{v2},$$

$$ll\xi \to (l_e l_e + l_\mu l_\tau + l_\tau l_\mu)\xi,$$

$$ll\xi' \to (l_\mu l_\mu + l_e l_\tau + l_\tau l_e)\xi',$$

$$ll\xi'' \to (l_\tau l_\tau + l_e l_\mu + l_\mu l_e)\xi''.$$

We now suppose the following vacuum alignments:

$$\langle \phi_l \rangle = \alpha_l \Lambda (1, 0, 0), \qquad \langle \phi_v \rangle = \alpha_v \Lambda (1, 1, 1), \tag{16.8}$$

with $\langle \xi \rangle = \alpha_\xi \Lambda$. We omit the discussion of the origin of these vacuum alignments, since the purpose of this section is not to present details of the model, but rather to apply the A_4 group to neutrino mixing.

Using these vacuum alignments and (16.7), we can obtain mass matrices for the charged leptons and neutrinos. Then, the diagonal charged lepton mass matrix is given by

$$M_l = \alpha_l v_d \begin{pmatrix} y^e & 0 & 0 \\ 0 & y^\mu & 0 \\ 0 & 0 & y^\tau \end{pmatrix}, \tag{16.9}$$

where $\langle h_{u,d} \rangle = v_{u,d}$. Furthermore, the effective neutrino mass matrix is given by

$$M_v = \frac{y^v_{\phi_v} \alpha_v v_u^2}{3\Lambda} \begin{pmatrix} 2 & -1 & -1 \\ -1 & 2 & -1 \\ -1 & -1 & 2 \end{pmatrix} + \frac{y^v_{\phi_\xi} \alpha_\xi v_u^2}{\Lambda} \begin{pmatrix} 1 & 0 & 0 \\ 0 & 0 & 1 \\ 0 & 1 & 0 \end{pmatrix}$$

$$= a \begin{pmatrix} 1 & 0 & 0 \\ 0 & 1 & 0 \\ 0 & 0 & 1 \end{pmatrix} + b \begin{pmatrix} 1 & 1 & 1 \\ 1 & 1 & 1 \\ 1 & 1 & 1 \end{pmatrix} + c \begin{pmatrix} 1 & 0 & 0 \\ 0 & 0 & 1 \\ 0 & 1 & 0 \end{pmatrix}, \tag{16.10}$$

where

$$a = \frac{y^v_{\phi_v} \alpha_v v_u^2}{\Lambda}, \qquad b = -\frac{y^v_{\phi_v} \alpha_v v_u^2}{3\Lambda}, \qquad c = \frac{y^v_\xi \alpha_\xi v_u^2}{\Lambda}. \tag{16.11}$$

Thus, tri-bimaximal mixing is easily derived in the $A_4 \times Z_3$ flavor model.

16.2.2 Breaking Tri-Bimaximal Mixing

It should be emphasized that the A_4 flavor symmetry does not necessarily give tri-bimaximal mixing at the leading order, even if the relevant alignments of the VEVs are realized. Certainly, for the neutrino mass matrix with three flavors, the A_4 symmetry can give the mass matrix with the $(2,3)$ off-diagonal matrix due to the A_4 singlet flavon, 1, in addition to the unit matrix and the democratic matrix, which leads to the tri-bimaximal mixing of flavors. However, the $(1,3)$ off-diagonal matrix and the $(1,2)$ off-diagonal matrix also appear at the leading order as long as the VEV of the ξ' or ξ'' flavons does not vanish [45]:

$$\begin{pmatrix} 0 & 0 & 1 \\ 0 & 1 & 0 \\ 1 & 0 & 0 \end{pmatrix} \text{ for } \xi', \qquad \begin{pmatrix} 0 & 1 & 0 \\ 1 & 0 & 0 \\ 0 & 0 & 1 \end{pmatrix} \text{ for } \xi''. \tag{16.12}$$

Tri-bimaximal mixing is broken at the leading order in such a case.

Let us consider the case of non-vanishing $\langle \xi' \rangle$, but still vanishing $\langle \xi'' \rangle$. The charged lepton mass matrix is still diagonal, as in (16.9). The effective neutrino mass matrix is modified to

$$M_\nu = \frac{y^\nu_{\phi_\nu} \alpha_\nu v_u^2}{3\Lambda} \begin{pmatrix} 2 & -1 & -1 \\ -1 & 2 & -1 \\ -1 & -1 & 2 \end{pmatrix} + \frac{y^\nu_{\phi_\xi} \alpha_\xi v_u^2}{\Lambda} \begin{pmatrix} 1 & 0 & 0 \\ 0 & 0 & 1 \\ 0 & 1 & 0 \end{pmatrix}$$

$$+ \frac{y^\nu_{\phi_{\xi'}} \alpha_{\xi'} v_u^2}{\Lambda} \begin{pmatrix} 0 & 0 & 1 \\ 0 & 1 & 0 \\ 1 & 0 & 0 \end{pmatrix}$$

$$= a \begin{pmatrix} 1 & 0 & 0 \\ 0 & 1 & 0 \\ 0 & 0 & 1 \end{pmatrix} + b \begin{pmatrix} 1 & 1 & 1 \\ 1 & 1 & 1 \\ 1 & 1 & 1 \end{pmatrix} + c \begin{pmatrix} 1 & 0 & 0 \\ 0 & 0 & 1 \\ 0 & 1 & 0 \end{pmatrix} + d \begin{pmatrix} 0 & 0 & 1 \\ 0 & 1 & 0 \\ 1 & 0 & 0 \end{pmatrix}, \tag{16.13}$$

where

$$a = \frac{y^\nu_{\phi_\nu} \alpha_\nu v_u^2}{\Lambda}, \qquad b = -\frac{y^\nu_{\phi_\nu} \alpha_\nu v_u^2}{3\Lambda}, \qquad c = \frac{y^\nu_\xi \alpha_\xi v_u^2}{\Lambda}, \qquad d = \frac{y^\nu_{\xi'} \alpha_{\xi'} v_u^2}{\Lambda}. \tag{16.14}$$

Therefore, the tri-bimaximal mixing is broken in the A_4 flavor model.

As can be seen from (16.13) and (16.14), the non-vanishing d is generated through the coupling $ll\xi' h_u h_u$. Since the relation $a = -3b$ is given in this model, the predicted regions of the lepton mixing angles are reduced compared with the one in the previous section. In the case where the parameters a, c, d are real, they are fixed by the three neutrino masses m_1, m_2, and m_3. That is, $\sin\theta_{13}$ can be plotted as a function of the total mass $\sum m_i$.

Figure 16.1 shows the predicted $\sin\theta_{13}$ versus $\sum m_i$, where the normal hierarchy of the neutrino masses is taken. The leptonic mixing is almost tri-bimaximal, that

Fig. 16.1 The $\sum m_i$ dependence of $\sin\theta_{13}$ for normal mass hierarchy, where *horizontal lines* denote Daya Bay data with 3 σ

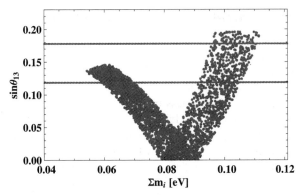

Fig. 16.2 The $\sum m_i$ dependence of $\sin\theta_{13}$ for inverted mass hierarchy, where *horizontal lines* denote Daya Bay data with 3 σ

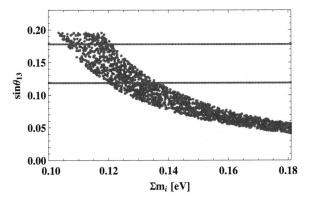

is, $\sin\theta_{13} = 0$, in the regime where $\sum m_i \simeq 0.08$–0.09 eV. In the case where $m_3 \gg m_2, m_1$, that is, $\sum m_i \simeq 0.05$ eV, $\sin\theta_{13}$ is expected to be around 0.15.

We can also predict $\sin\theta_{13}$ versus $\sum m_i$ in the case of the inverted hierarchy of neutrino masses. We get a different prediction for $\sin\theta_{13}$, as shown in Fig. 16.2. The predicted maximal value of $\sin\theta_{13}$ is 0.2 at $\sum m_i \simeq 0.1$ eV, which corresponds to $m_3 \ll m_2, m_1$.

In conclusion, the $A_4 \times Z_3$ model predicts $\sin\theta_{13} = 0.15 - 0.2$ for the cases with $m_3 \gg m_2, m_1$ and $m_3 \ll m_2, m_1$.

Finally, we comment on flavor models with other non-Abelian discrete symmetries which give the non-vanishing d effectively. One is the flavor model based on the group $\Delta(27)$, as described by Grimus and Lavoura [68]. Trimaximal mixing is enforced by the softly broken discrete symmetry. In this model, we find the relation $d = e^{i\pi/3} c$, where a, b, c, d are complex. As shown in [68], a large value of $\sin\theta_{13}$ is expected. Another example is the flavor twisting model in the 5D framework [69, 70]. In this model, flavor symmetry breaking is triggered by the boundary conditions of the bulk right-handed neutrino in the fifth spatial dimension. The parameters a, b, c, d involve the bulk neutrino masses and the volume of the extra dimension. In the case of the S_4 flavor symmetry [70], there is one relation among these four parameters, so that the general allowed region is further restricted as in

16.3 S_4 Flavor Model

In this section, we present a $S_4 \times Z_4$ flavor model to unify the quarks and leptons in the framework of the $SU(5)$ GUT [71]. The S_4 group has 24 distinct elements and five irreducible representations **1**, **1'**, **2**, **3**, and **3'**. In $SU(5)$, matter fields are unified into $10(q_1, u^c, e^c)_L$ and $\bar{5}(d^c, l_e)_L$ dimensional representations. The 5-dimensional, $\bar{5}$-dimensional, and 45-dimensional Higgs of $SU(5)$, H_5, $H_{\bar{5}}$, and H_{45} are assigned **1** of S_4. Three generations of $\bar{5}$, which are denoted by F_i, are assigned **3** of S_4. On the other hand, the third generation of the 10-dimensional representation is assigned **1** of S_4, so that the top quark Yukawa coupling is allowed at the tree level, while the first and the second generations are assigned **2** of S_4. These 10-dimensional representations are denoted by T_3 and (T_1, T_2), respectively. These assignments of S_4 for $\bar{5}$ and **10** lead to the completely different structure of the quark and lepton mass matrices.

Right-handed neutrinos, which are $SU(5)$ gauge singlets, are also assigned **1'** and **2** for N_τ^c and (N_e^c, N_μ^c), respectively. These assignments are essential to realize the tri-bimaximal mixing of neutrino flavors. Assignments of $SU(5)$, S_4, Z_4, and $U(1)_{FN}$ representations are summarized in Table 16.3. Taking the vacuum alignments of the relevant gauge singlet scalars, we predict the quark mixing as well as the tri-bimaximal mixing of leptons. In particular, the Cabibbo angle is predicted to be around $15°$ under the relevant vacuum alignments.

We introduce new scalars χ_i ($i = 1$–14), which are assumed to be $SU(5)$ gauge singlets. Those flavons are summarized in Table 16.3. In order to obtain the natural hierarchy among quark and lepton masses, the Froggatt–Nielsen mechanism [72] is introduced as an additional $U(1)_{FN}$ flavor symmetry. Θ denotes the Froggatt–Nielsen flavon. The particle assignments of $SU(5)$, S_4 and Z_4, and $U(1)_{FN}$ are presented in Table 16.3.

We can now write down the superpotential respecting the S_4, Z_4, and $U(1)_{FN}$ symmetries in terms of the S_4 cutoff scale Λ and the $U(1)_{FN}$ cutoff scale $\bar{\Lambda}$. The $SU(5)$ invariant superpotential of the Yukawa sector up to the linear terms of χ_i is given by

$$\begin{aligned} w_{SU(5)} = \big[& y_1^u (T_1, T_2) \otimes T_3 \otimes (\chi_1, \chi_2) \otimes H_5/\Lambda + y_2^u T_3 \otimes T_3 \otimes H_5 \\ & + y_1^N (N_e^c, N_\mu^c) \otimes (N_e^c, N_\mu^c) \otimes \Theta^2/\bar{\Lambda} \\ & + y_2^N (N_e^c, N_\mu^c) \otimes (N_e^c, N_\mu^c) \otimes (\chi_3, \chi_4) + M N_\tau^c \otimes N_\tau^c \\ & + y_1^D (N_e^c, N_\mu^c) \otimes (F_1, F_2, F_3) \otimes (\chi_5, \chi_6, \chi_7) \otimes H_5 \otimes \Theta/(\Lambda\bar{\Lambda}) \\ & + y_2^D N_\tau^c \otimes (F_1, F_2, F_3) \otimes (\chi_5, \chi_6, \chi_7) \otimes H_5/\Lambda \end{aligned}$$

Table 16.3 Assignments of $SU(5)$, S_4, Z_4, and $U(1)_{FN}$ representations

	(T_1, T_2)	T_3	(F_1, F_2, F_3)	(N^c_e, N^c_μ)	N^c_τ	H_5	$H_{\bar{5}}$	H_{45}	Θ
$SU(5)$	10	10	$\bar{5}$	1	1	5	$\bar{5}$	45	1
S_4	2	1	3	2	$1'$	1	1	1	1
Z_4	$-i$	-1	i	1	1	1	1	-1	1
$U(1)_{FN}$	1	0	0	1	0	0	0	0	-1

	(χ_1, χ_2)	(χ_3, χ_4)	(χ_5, χ_6, χ_7)	$(\chi_8, \chi_9, \chi_{10})$	$(\chi_{11}, \chi_{12}, \chi_{13})$	χ_{14}
$SU(5)$	1	1	1	1	1	1
S_4	2	2	$3'$	3	3	1
Z_4	$-i$	1	$-i$	-1	i	i
$U(1)_{FN}$	-1	-2	0	0	0	-1

$$+ y_1 (T_1, T_2) \otimes (F_1, F_2, F_3) \otimes (\chi_8, \chi_9, \chi_{10}) \otimes H_{45} \otimes \Theta / (\Lambda \bar{\Lambda})$$
$$+ y_2 T_3 \otimes (F_1, F_2, F_3) \otimes (\chi_{11}, \chi_{12}, \chi_{13}) \otimes H_{\bar{5}}/\Lambda]_1, \quad (16.15)$$

where $[\ ]_1$ denotes the only trivial singlet components of S_4 extracted from tensor products. Parameters y_1^u, y_2^u, y_1^N, y_2^N, y_1^D, y_2^D, y_1, and y_2 are Yukawa couplings. We take the basis II in Appendix B.2 for the multiplication rules to get the S_4 singlet:

$$\begin{pmatrix} T_1 \\ T_2 \end{pmatrix}_2 \otimes \begin{pmatrix} \chi_1 \\ \chi_2 \end{pmatrix}_2 \to (T_1 \chi_1 + T_2 \chi_2)_1,$$

$$\begin{pmatrix} N^c_e \\ N^c_\mu \end{pmatrix}_2 \otimes \begin{pmatrix} N^c_e \\ N^c_\mu \end{pmatrix}_2 \to (N^c_e N^c_e + N^c_\mu N^c_\mu)_1,$$

$$\begin{pmatrix} N^c_e \\ N^c_\mu \end{pmatrix}_2 \otimes \begin{pmatrix} N^c_e \\ N^c_\mu \end{pmatrix}_2 \otimes \begin{pmatrix} \chi_3 \\ \chi_4 \end{pmatrix}_2$$

$$\to \begin{pmatrix} N^c_e N^c_\mu + N^c_\mu N^c_e \\ N^c_e N^c_e - N^c_\mu N^c_\mu \end{pmatrix} \otimes \begin{pmatrix} \chi_3 \\ \chi_4 \end{pmatrix}_2$$

$$\to [(N^c_e N^c_\mu + N^c_\mu N^c_e)\chi_3 + (N^c_e N^c_e - N^c_\mu N^c_\mu)\chi_4]_1,$$

$$\begin{pmatrix} N^c_e \\ N^c_\mu \end{pmatrix}_2 \otimes \begin{pmatrix} F_1 \\ F_2 \\ F_3 \end{pmatrix}_3 \otimes \begin{pmatrix} \chi_5 \\ \chi_6 \\ \chi_7 \end{pmatrix}_{3'}$$

$$\to \begin{pmatrix} N^c_e \\ N^c_\mu \end{pmatrix}_2 \otimes \begin{pmatrix} \frac{1}{\sqrt{6}}(2F_1 \chi_5 - F_2 \chi_6 - F_3 \chi_7) \\ \frac{1}{\sqrt{2}}(F_2 \chi_6 - F_3 \chi_7) \end{pmatrix}_2$$

$$\to \left[\frac{N^c_e}{\sqrt{6}}(2F_1 \chi_5 - F_2 \chi_6 - F_3 \chi_7) + \frac{N^c_\mu}{\sqrt{2}}(F_2 \chi_6 - F_3 \chi_7)\right]_1,$$

16.3 S_4 Flavor Model

$$\begin{pmatrix} F_1 \\ F_2 \\ F_3 \end{pmatrix}_3 \otimes \begin{pmatrix} \chi_5 \\ \chi_6 \\ \chi_7 \end{pmatrix}_{3'} \to (F_1\chi_5 + F_2\chi_6 + F_3\chi_7)_{1'},$$

$$\begin{pmatrix} T_1 \\ T_2 \end{pmatrix}_2 \otimes \begin{pmatrix} F_1 \\ F_2 \\ F_3 \end{pmatrix}_3 \otimes \begin{pmatrix} \chi_8 \\ \chi_9 \\ \chi_{10} \end{pmatrix}_3$$

$$\to \begin{pmatrix} T_1 \\ T_2 \end{pmatrix}_2 \otimes \begin{pmatrix} \tfrac{1}{\sqrt{2}}(F_2\chi_9 - F_3\chi_{10}) \\ \tfrac{1}{\sqrt{6}}(2F_1\chi_8 - F_2\chi_9 - F_3\chi_{10}) \end{pmatrix}_2$$

$$\to \left[\frac{T_1}{\sqrt{2}}(F_2\chi_9 - F_3\chi_{10}) + \frac{T_2}{\sqrt{6}}(-2F_1\chi_8 + F_2\chi_9 + F_3\chi_{10}) \right]_1,$$

$$\begin{pmatrix} F_1 \\ F_2 \\ F_3 \end{pmatrix}_3 \otimes \begin{pmatrix} \chi_8 \\ \chi_9 \\ \chi_{10} \end{pmatrix}_3 \to (F_1\chi_8 + F_2\chi_9 + F_3\chi_{10})_1.$$

(16.16)

We discuss the quark and lepton mass matrices and flavor mixing based on this superpotential. Furthermore, we take into account the next order superpotential in the numerical study of flavor mixing and CP violation.

Let us start by extracting the lepton sector from the superpotential $w_{SU(5)}^{(0)}$. Denoting Higgs doublets by h_u and h_d, the superpotential of the Yukawa sector respecting the $S_4 \times Z_4 \times U(1)_{FN}$ symmetry is given for charged leptons by

$$w_l = -3y_1 \left[\frac{e^c}{\sqrt{2}}(l_\mu \chi_9 - l_\tau \chi_{10}) + \frac{\mu^c}{\sqrt{6}}(-2l_e\chi_8 + l_\mu\chi_9 + l_\tau\chi_{10}) \right] h_{45}\Theta/(\Lambda\bar{\Lambda})$$
$$+ y_2 \tau^c (l_e \chi_{11} + l_\mu \chi_{12} + l_\tau \chi_{13}) h_d/\Lambda. \tag{16.17}$$

For right-handed Majorana neutrinos, the superpotential is given by

$$w_N = y_1^N (N_e^c N_e^c + N_\mu^c N_\mu^c) \Theta^2/\bar{\Lambda}$$
$$+ y_2^N \left[(N_e^c N_\mu^c + N_\mu^c N_e^c)\chi_3 + (N_e^c N_e^c - N_\mu^c N_\mu^c)\chi_4 \right] + M N_\tau^c N_\tau^c, \tag{16.18}$$

and for Dirac neutrino Yukawa couplings, the superpotential is

$$w_D = y_1^D \left[\frac{N_e^c}{\sqrt{6}}(2l_e\chi_5 - l_\mu\chi_6 - l_\tau\chi_7) + \frac{N_\mu^c}{\sqrt{2}}(l_\mu\chi_6 - l_\tau\chi_7) \right] h_u\Theta/(\Lambda\bar{\Lambda})$$
$$+ y_2^D N_\tau^c (l_e\chi_5 + l_\mu\chi_6 + l_\tau\chi_7) h_u/\Lambda. \tag{16.19}$$

Higgs doublets h_u, h_d and gauge singlet scalars Θ and χ_i are assumed to develop their VEVs as follows:

$$\langle h_u \rangle = v_u, \qquad \langle h_d \rangle = v_d, \qquad \langle h_{45} \rangle = v_{45}, \qquad \langle \Theta \rangle = \theta,$$
$$\langle (\chi_3, \chi_4) \rangle = (u_3, u_4), \qquad \langle (\chi_5, \chi_6, \chi_7) \rangle = (u_5, u_6, u_7), \qquad (16.20)$$
$$\langle (\chi_8, \chi_9, \chi_{10}) \rangle = (u_8, u_9, u_{10}), \qquad \langle (\chi_{11}, \chi_{12}, \chi_{13}) \rangle = (u_{11}, u_{12}, u_{13}),$$

which are assumed to be real. We then obtain the mass matrix for charged leptons as

$$M_l = -3y_1 \lambda v_{45} \begin{pmatrix} 0 & \alpha_9/\sqrt{2} & -\alpha_{10}/\sqrt{2} \\ -2\alpha_8/\sqrt{6} & \alpha_9/\sqrt{6} & \alpha_{10}/\sqrt{6} \\ 0 & 0 & 0 \end{pmatrix} + y_2 v_d \begin{pmatrix} 0 & 0 & 0 \\ 0 & 0 & 0 \\ \alpha_{11} & \alpha_{12} & \alpha_{13} \end{pmatrix}, \qquad (16.21)$$

while the right-handed Majorana neutrino mass matrix is given by

$$M_N = \begin{pmatrix} y_1^N \lambda^2 \bar{\Lambda} + y_2^N \alpha_4 \Lambda & y_2^N \alpha_3 \Lambda & 0 \\ y_2^N \alpha_3 \Lambda & y_1^N \lambda^2 \bar{\Lambda} - y_2^N \alpha_4 \Lambda & 0 \\ 0 & 0 & M \end{pmatrix}. \qquad (16.22)$$

Note that the (1, 3), (2, 3), (3, 1), and (3, 3) elements of the right-handed Majorana neutrino mass matrix vanish. These are the so-called SUSY zeros. The Dirac mass matrix for the neutrinos is

$$M_D = y_1^D \lambda v_u \begin{pmatrix} 2\alpha_5/\sqrt{6} & -\alpha_6/\sqrt{6} & -\alpha_7/\sqrt{6} \\ 0 & \alpha_6/\sqrt{2} & -\alpha_7/\sqrt{2} \\ 0 & 0 & 0 \end{pmatrix} + y_2^D v_u \begin{pmatrix} 0 & 0 & 0 \\ 0 & 0 & 0 \\ \alpha_5 & \alpha_6 & \alpha_7 \end{pmatrix}, \qquad (16.23)$$

where we denote $\alpha_i \equiv u_i/\Lambda$ and $\lambda \equiv \theta/\bar{\Lambda}$.

If we can take the vacuum alignment to be

$$(u_8, u_9, u_{10}) = (0, u_9, 0), \qquad (u_{11}, u_{12}, u_{13}) = (0, 0, u_{13}),$$

that is, $\alpha_8 = \alpha_{10} = \alpha_{11} = \alpha_{12} = 0$, we obtain

$$M_l = \begin{pmatrix} 0 & -3y_1 \lambda \alpha_9 v_{45}/\sqrt{2} & 0 \\ 0 & -3y_1 \lambda \alpha_9 v_{45}/\sqrt{6} & 0 \\ 0 & 0 & y_2 \alpha_{13} v_d \end{pmatrix}, \qquad (16.24)$$

and $M_l^\dagger M_l$ is then obtained as follows:

$$M_l^\dagger M_l = v_d^2 \begin{pmatrix} 0 & 0 & 0 \\ 0 & 6|\bar{y}_1 \lambda \alpha_9|^2 & 0 \\ 0 & 0 & |y_2|^2 \alpha_{13}^2 \end{pmatrix}, \qquad (16.25)$$

16.3 S_4 Flavor Model

where we replace $y_1 v_{45}$ by $\bar{y}_1 v_d$. That is, the left-handed mixing angles of the charged lepton mass matrix vanish. The charged lepton masses are given by

$$m_e^2 = 0, \qquad m_\mu^2 = 6|\bar{y}_1 \lambda \alpha_9|^2 v_d^2, \qquad m_\tau^2 = |y_2|^2 \alpha_{13}^2 v_d^2. \tag{16.26}$$

It is remarkable that the electron mass vanishes. The electron mass is obtained in the next order.

Taking the vacuum alignment $(u_3, u_4) = (0, u_4)$ and $(u_5, u_6, u_7) = (u_5, u_5, u_5)$ in (16.22), the right-handed Majorana mass matrix for the neutrinos turns out to be

$$M_N = \begin{pmatrix} y_1^N \lambda^2 \bar{\Lambda} + y_2^N \alpha_4 \Lambda & 0 & 0 \\ 0 & y_1^N \lambda^2 \bar{\Lambda} - y_2^N \alpha_4 \Lambda & 0 \\ 0 & 0 & M \end{pmatrix}, \tag{16.27}$$

and the Dirac mass matrix for the neutrinos is

$$M_D = y_1^D \lambda v_u \begin{pmatrix} 2\alpha_5/\sqrt{6} & -\alpha_5/\sqrt{6} & -\alpha_5/\sqrt{6} \\ 0 & \alpha_5/\sqrt{2} & -\alpha_5/\sqrt{2} \\ 0 & 0 & 0 \end{pmatrix} + y_2^D v_u \begin{pmatrix} 0 & 0 & 0 \\ 0 & 0 & 0 \\ \alpha_5 & \alpha_5 & \alpha_5 \end{pmatrix}. \tag{16.28}$$

Using the seesaw mechanism $M_\nu = M_D^T M_N^{-1} M_D$, the left-handed Majorana neutrino mass matrix can be written as

$$M_\nu = \begin{pmatrix} a + \tfrac{2}{3}b & a - \tfrac{1}{3}b & a - \tfrac{1}{3}b \\ a - \tfrac{1}{3}b & a + \tfrac{1}{6}b + \tfrac{1}{2}c & a + \tfrac{1}{6}b - \tfrac{1}{2}c \\ a - \tfrac{1}{3}b & a + \tfrac{1}{6}b - \tfrac{1}{2}c & a + \tfrac{1}{6}b + \tfrac{1}{2}c \end{pmatrix}, \tag{16.29}$$

where

$$a = \frac{(y_2^D \alpha_5 v_u)^2}{M}, \qquad b = \frac{(y_1^D \alpha_5 v_u \lambda)^2}{y_1^N \lambda^2 \bar{\Lambda} + y_2^N \alpha_4 \Lambda}, \qquad c = \frac{(y_1^D \alpha_5 v_u \lambda)^2}{y_1^N \lambda^2 \bar{\Lambda} - y_2^N \alpha_4 \Lambda}. \tag{16.30}$$

The neutrino mass matrix is decomposed as

$$M_\nu = \frac{b+c}{2}\begin{pmatrix} 1 & 0 & 0 \\ 0 & 1 & 0 \\ 0 & 0 & 1 \end{pmatrix} + \frac{3a-b}{3}\begin{pmatrix} 1 & 1 & 1 \\ 1 & 1 & 1 \\ 1 & 1 & 1 \end{pmatrix} + \frac{b-c}{2}\begin{pmatrix} 1 & 0 & 0 \\ 0 & 0 & 1 \\ 0 & 1 & 0 \end{pmatrix}, \tag{16.31}$$

which gives the tri-bimaximal mixing matrix U_{tribi} and mass eigenvalues as follows:

$$m_1 = b, \qquad m_2 = 3a, \qquad m_3 = c. \tag{16.32}$$

We now discuss the quark sector. For down-type quarks, we can write the superpotential as follows:

$$w_d = y_1 \left[\frac{1}{\sqrt{2}} (s^c \chi_9 - b^c \chi_{10}) q_1 + \frac{1}{\sqrt{6}} (-2d^c \chi_8 + s^c \chi_9 + b^c \chi_{10}) q_2 \right] h_{45} \Theta/(\Lambda \bar{\Lambda})$$

$$+ y_2 (d^c \chi_{11} + s^c \chi_{12} + b^c \chi_{13}) q_3 h_d / \Lambda. \tag{16.33}$$

Since the vacuum alignment is fixed in the lepton sector, as can be seen from (16.20), the down-type quark mass matrix is given to leading order by

$$M_d = v_d \begin{pmatrix} 0 & 0 & 0 \\ \bar{y}_1 \lambda \alpha_9/\sqrt{2} & \bar{y}_1 \lambda \alpha_9/\sqrt{6} & 0 \\ 0 & 0 & y_2 \alpha_{13} \end{pmatrix}, \quad (16.34)$$

where we denote $\bar{y}_1 v_d = y'_1 v_{45}$. Then, we have

$$M_d^\dagger M_d = v_d^2 \begin{pmatrix} \frac{1}{2}|\bar{y}_1 \lambda \alpha_9|^2 & \frac{1}{2\sqrt{3}}|\bar{y}_1 \lambda \alpha_9|^2 & 0 \\ \frac{1}{2\sqrt{3}}|\bar{y}_1 \lambda \alpha_9|^2 & \frac{1}{6}|\bar{y}_1 \lambda \alpha_9|^2 & 0 \\ 0 & 0 & |y_2|^2 \alpha_{13}^2 \end{pmatrix}. \quad (16.35)$$

This matrix can be diagonalized by the orthogonal matrix

$$U_d = \begin{pmatrix} \cos 60° & \sin 60° & 0 \\ -\sin 60° & \cos 60° & 0 \\ 0 & 0 & 1 \end{pmatrix}. \quad (16.36)$$

The down-type quark masses are given by

$$m_d^2 = 0, \quad m_s^2 = \frac{2}{3}|\bar{y}_1 \lambda \alpha_9|^2 v_d^2, \quad m_b^2 \approx |y_2|^2 \alpha_{13}^2 v_d^2, \quad (16.37)$$

which correspond to the charged lepton masses in (16.26). The down quark mass vanishes, like the electron mass, but tiny masses appear in the next order.

A realistic CKM mixing matrix is obtained in the quark sector by including the next order terms of the superpotential, in which terms quadratic in the χ_i, such as $d^c q_1 \chi_1 \chi_5$, are dominant. We obtain the next order down-type quark mass matrix elements $\bar{\epsilon}_{ij}$, which are given in terms of Yukawa couplings and VEVs of flavons. The magnitudes of the $\bar{\epsilon}_{ij}$'s are $\mathcal{O}(\tilde{\alpha}^2)$, where $\tilde{\alpha}$ is a linear combination of α_i's. The down-type quark mass matrix is written in terms of $\bar{\epsilon}_{ij}$ as

$$M_d \simeq \begin{pmatrix} \bar{\epsilon}_{11} & \bar{\epsilon}_{21} & \bar{\epsilon}_{31} \\ \frac{\sqrt{3}m_s}{2} + \bar{\epsilon}_{12} & \frac{m_s}{2} + \bar{\epsilon}_{22} & \bar{\epsilon}_{32} \\ \bar{\epsilon}_{13} & \bar{\epsilon}_{23} & m_b + \bar{\epsilon}_{33} \end{pmatrix}, \quad (16.38)$$

where ϵ_{ij} should be the order of m_d.

By rotating $M_d^\dagger M_d$ with the mixing matrix U_d in (16.36), we have

$$U_d^\dagger M_d^\dagger M_d U_d \simeq \begin{pmatrix} \mathcal{O}(m_d^2) & \mathcal{O}(m_d m_s) & \mathcal{O}(m_d m_b) \\ \mathcal{O}(m_d m_s) & \mathcal{O}(m_s^2) & \mathcal{O}(m_d m_b) \\ \mathcal{O}(m_d m_b) & \mathcal{O}(m_d m_b) & m_b^2 \end{pmatrix}. \quad (16.39)$$

Then we get the mixing angles θ_{12}^d, θ_{13}^d, θ_{23}^d in the mass matrix of (16.39) as

$$\theta_{12}^d = \mathcal{O}\left(\frac{m_d}{m_s}\right), \quad \theta_{13}^d = \mathcal{O}\left(\frac{m_d}{m_b}\right), \quad \theta_{23}^d = \mathcal{O}\left(\frac{m_d}{m_b}\right), \quad (16.40)$$

16.3 S_4 Flavor Model

where CP violating phases are neglected.

We now consider the up-type quark sector. Here the superpotential respecting $S_4 \times Z_4 \times U(1)_{FN}$ is given by

$$w_u = y_1^u\left[(u^c\chi_1 + c^c\chi_2)q_3 + t^c(q_1\chi_1 + q_2\chi_2)\right]h_u/\Lambda + y_2^u t^c q_3 h_u. \quad (16.41)$$

We denote their VEVs by

$$\langle(\chi_1, \chi_2)\rangle = (u_1, u_2). \quad (16.42)$$

We then obtain the mass matrix for up-type quarks as

$$M_u = v_u \begin{pmatrix} 0 & 0 & y_1^u\alpha_1 \\ 0 & 0 & y_1^u\alpha_2 \\ y_1^u\alpha_1 & y_1^u\alpha_2 & y_2^u \end{pmatrix}. \quad (16.43)$$

The next order terms of the superpotential are also important to predict the CP violation in the quark sector. The relevant superpotential is given to the next highest order by

$$\Delta w_u = \left[y_{\Delta_a}^u (T_1, T_2) \otimes (T_1, T_2) \otimes (\chi_1, \chi_2) \otimes (\chi_1, \chi_2) \otimes H_5/\Lambda^2 \right.$$
$$+ y_{\Delta_b}^u (T_1, T_2) \otimes (T_1, T_2) \otimes \chi_{14} \otimes \chi_{14} \otimes H_5/\Lambda^2$$
$$\left.+ y_{\Delta_c}^u T_3 \otimes T_3 \otimes (\chi_8, \chi_9, \chi_{10}) \otimes (\chi_8, \chi_9, \chi_{10}) \otimes H_5/\Lambda^2\right]_1. \quad (16.44)$$

The multiplication rule to get the S_4 singlet is

$$\begin{pmatrix} T_1 \\ T_2 \end{pmatrix}_2 \otimes \begin{pmatrix} T_1 \\ T_2 \end{pmatrix}_2 \otimes \begin{pmatrix} \chi_1 \\ \chi_2 \end{pmatrix}_2 \otimes \begin{pmatrix} \chi_1 \\ \chi_2 \end{pmatrix}_2$$
$$\to (T_1 T_1 + T_2 T_2)_1 \otimes (\chi_1\chi_1 + \chi_2\chi_2)_1$$
$$\oplus \begin{pmatrix} T_1 T_2 + T_2 T_1 \\ T_1 T_1 - T_2 T_2 \end{pmatrix}_2 \otimes \begin{pmatrix} \chi_1\chi_2 + \chi_2\chi_1 \\ \chi_1\chi_1 - \chi_2\chi_2 \end{pmatrix}_2$$
$$\to \left[(T_1 T_1 + T_2 T_2)(\chi_1\chi_1 + \chi_2\chi_2)\right]_1$$
$$\oplus \left[(T_1 T_2 + T_2 T_1)(\chi_1\chi_2 + \chi_2\chi_1)\right.$$
$$\left.+ (T_1 T_1 - T_2 T_2)(\chi_1\chi_1 - \chi_2\chi_2)\right]_1. \quad (16.45)$$

We obtain the following mass matrix including the next order terms:

$$M_u = v_u \begin{pmatrix} 2y_{\Delta_{a1}}^u \alpha_1^2 + y_{\Delta_b}^u \alpha_{14}^2 & y_{\Delta_{a2}}^u \alpha_1^2 & y_1^u\alpha_1 \\ y_{\Delta_{a2}}^u \alpha_1^2 & 2y_{\Delta_{a1}}^u \alpha_1^2 + y_{\Delta_b}^u \alpha_{14}^2 & y_1^u\alpha_1 \\ y_1^u\alpha_1 & y_1^u\alpha_1 & y_2^u + y_{\Delta_c}^u \alpha_9^2 \end{pmatrix}, \quad (16.46)$$

where we take the alignment $\alpha_1 = \alpha_2$. After rotating the mass matrix M_u through $\theta_{12} = 45°$, we get

$$\hat{M}_u \approx v_u \begin{pmatrix} (2y^u_{\Delta_{a1}} - y^u_{\Delta_{a2}})\alpha_1^2 + y^u_{\Delta_b}\alpha_{14}^2 & 0 & 0 \\ 0 & (2y^u_{\Delta_{a1}} + y^u_{\Delta_{a2}})\alpha_1^2 + y^u_{\Delta_b}\alpha_{14}^2 & \sqrt{2}y^u_1 \alpha_1 \\ 0 & \sqrt{2}y^u_1 \alpha_1 & y^u_2 \end{pmatrix}. \tag{16.47}$$

This mass matrix is taken to be real by removing phases. The matrix is diagonalized by the orthogonal transformation $V_u^T \hat{M}_u V_F$, where

$$V_u \simeq \begin{pmatrix} 1 & 0 & 0 \\ 0 & r_t & r_c \\ 0 & -r_c & r_t \end{pmatrix}, \quad r_c = \sqrt{\frac{m_c}{m_c + m_t}}, \quad r_t = \sqrt{\frac{m_t}{m_c + m_t}}. \tag{16.48}$$

Now we can calculate the CKM matrix. Mixing matrices of up- and down-type quarks are summarized by

$$U_u \simeq \begin{pmatrix} \cos 45° & \sin 45° & 0 \\ -\sin 45° & \cos 45° & 0 \\ 0 & 0 & 1 \end{pmatrix} \begin{pmatrix} 1 & 0 & 0 \\ 0 & e^{-i\rho} & 0 \\ 0 & 0 & 1 \end{pmatrix} \begin{pmatrix} 1 & 0 & 0 \\ 0 & r_t & r_c \\ 0 & -r_c & r_t \end{pmatrix},$$

$$U_d \simeq \begin{pmatrix} \cos 60° & \sin 60° & 0 \\ -\sin 60° & \cos 60° & 0 \\ 0 & 0 & 1 \end{pmatrix} \begin{pmatrix} 1 & \theta^d_{12} & \theta^d_{13} \\ -\theta^d_{12} - \theta^d_{13}\theta^d_{23} & 1 & \theta^d_{23} \\ -\theta^d_{13} + \theta^d_{12}\theta^d_{23} & -\theta^d_{23} - \theta^d_{12}\theta^d_{13} & 1 \end{pmatrix}. \tag{16.49}$$

Therefore, the CKM matrix is given by $U_u^\dagger U_d$. The relevant mixing elements are

$$V_{us} \approx \theta^d_{12}\cos 15° + \sin 15°,$$
$$V_{ub} \approx \theta^d_{13}\cos 15° + \theta^d_{23}\sin 15°,$$
$$V_{cb} \approx -r_t \theta^d_{13} e^{i\rho}\sin 15° + r_t \theta^d_{23} e^{i\rho}\cos 15° - r_c,$$
$$V_{td} \approx -r_c \sin 15° e^{i\rho} - r_c(\theta^d_{12} + \theta^d_{13}\theta^d_{23})e^{i\rho}\cos 15° + r_t(-\theta^d_{13} + \theta^d_{12}\theta^d_{23}). \tag{16.50}$$

We can reproduce the experimental values with a parameter set

$$\rho = 123°, \quad \theta^d_{12} = -0.0340, \quad \theta^d_{13} = 0.00626, \quad \theta^d_{23} = -0.00880, \tag{16.51}$$

putting typical GUT scale masses $m_u = 1.04 \times 10^{-3}$ GeV, $m_c = 302 \times 10^{-3}$ GeV, $m_t = 129$ GeV [73].

In terms of the phase ρ, we can also estimate the magnitude of CP violation through the Jarlskog invariant J_{CP} [74], which is given by

$$|J_{CP}| = |\text{Im}\{V_{us}V_{cs}^* V_{ub} V_{cb}^*\}| \approx 3.06 \times 10^{-5}. \tag{16.52}$$

Our prediction is consistent with experimental values $J_{CP} = 3.05^{+0.19}_{-0.20}$.

16.4 Alternative Flavor Mixing

Table 16.4 Assignments of $SU(2)$ and A_5 representations

	L	$\bar{e}$	H	ξ	ψ	χ
$SU(2)$	2	1	2	1	1	1
A_5	3	3'	1	5	5	4

16.4 Alternative Flavor Mixing

In the previous section, we presented flavor models reproducing tri-bimaximal mixing of lepton flavors. However, there are other flavor mixing patterns like the golden ratio, trimaximal, and bimaximal for lepton flavor mixing.

Let us begin with the golden ratio which appears in the solar mixing angle θ_{12}. The golden ratio can be derived from the A_5 [75] and D_{10} [76] models. One example is proposed as $\tan\theta_{12} = 1/\phi$, where $\phi = (1+\sqrt{5})/2 \simeq 1.62$ [77]. The rotational icosahedral group, which is isomorphic to A_5, the alternating group of five elements, provides a natural context for the golden ratio $\cos\theta_{12} = \phi/2$ [75]. In this model, the solar angle is related to the golden ratio, the atmospheric angle is maximal, and the reactor angle vanishes. The particle contents are summarized in Table 16.4.

In this model, ξ couples to the $LHLH$ operator and ψ and χ couple to the $L\bar{e}H$ operator. This setup is derived from additional symmetries such as Z_n, which forbid ξ and ψ from coupling to $L\bar{e}H$ and $LHLH$, respectively. For simplicity, it is assumed that the tree level LL term is forbidden by such additional symmetries.

With these assumptions, the mass terms can be written as follows:

$$\mathcal{L}_{\text{mass}} = \frac{\alpha_{ijk}}{MM'} L_i H L_j H \xi_k + \frac{\beta_{ijk}}{M'} L_i \bar{e}_j H \psi_k + \frac{\gamma_{ijl}}{M'} L_i \bar{e}_j H \chi_l + \text{h.c.}, \quad (16.53)$$

in which M' represents the scale of flavor symmetry breaking, and α_{ijk}, β_{ijk}, and γ_{ijl} are dimensionless couplings that encode the tensor product decomposition of the icosahedral symmetry. In principle, M' bears no relation to M. Taking the VEV alignment to be

$$\langle \xi \rangle = \frac{\sqrt{3}}{2\alpha}\left(\frac{1}{\sqrt{15}}(m_2 - m_1), \frac{2}{\sqrt{15}}(m_2 - m_1), 0, 0, -(m_1 + m_2)\right), \quad (16.54)$$

the neutrino mass matrix becomes

$$M_\nu = \frac{1}{\sqrt{5}} \begin{pmatrix} \phi m_1 + \frac{1}{\phi} m_2 & m_2 - m_1 & 0 \\ m_2 - m_1 & \frac{1}{\phi} m_1 + \phi m_2 & 0 \\ 0 & 0 & -\sqrt{5}(m_1 + m_2) \end{pmatrix}, \quad (16.55)$$

where $\phi = (1+\sqrt{5})/2$. The neutrino mass eigenvalues are then m_1, m_2, and $m_3 = -(m_1 + m_2)$. The neutrino mixing matrix U_ν, defined by $U_\nu M_\nu U_\nu^T$, satisfies

$\theta_{12} = \tan^{-1}(1/\phi) = 31.72°$:

$$U_\nu = \begin{pmatrix} \sqrt{\frac{\phi}{\sqrt{5}}} & -\sqrt{\frac{1}{\sqrt{5}\phi}} & 0 \\ \sqrt{\frac{1}{\sqrt{5}\phi}} & \sqrt{\frac{\phi}{\sqrt{5}}} & 0 \\ 0 & 0 & -i \end{pmatrix}. \tag{16.56}$$

The maximal mixing between the second and third families is derived from the charged lepton sector. To leading order in the flavon fields, only m_τ is nonvanishing. Taking the VEV alignment to be

$$\langle \psi \rangle = \frac{m_\tau}{2\sqrt{6}\beta}\left(-\sqrt{\frac{5}{3}}, 0, -\frac{2}{\sqrt{3}}\phi, 0, 1\right), \quad \langle \chi \rangle = \frac{m_\tau}{3\sqrt{2}\gamma}\left(0, \frac{1}{\phi}, 0, 1\right), \tag{16.57}$$

the charged lepton mass matrix takes the form

$$M_e = \frac{1}{\sqrt{2}}\begin{pmatrix} 0 & 0 & 0 \\ 0 & 0 & m_\tau \\ 0 & 0 & m_\tau \end{pmatrix}. \tag{16.58}$$

The left-handed states are diagonalized by the mixing matrix U_e given by

$$U_e = \begin{pmatrix} 1 & 0 & 0 \\ 0 & \frac{1}{\sqrt{2}} & -\frac{1}{\sqrt{2}} \\ 0 & \frac{1}{\sqrt{2}} & \frac{1}{\sqrt{2}} \end{pmatrix}. \tag{16.59}$$

In (16.56) and (16.59), the lepton mixing matrix U is obtained as

$$U = U_e U_\nu^\dagger = \begin{pmatrix} \sqrt{\frac{\phi}{\sqrt{5}}} & \sqrt{\frac{1}{\sqrt{5}\phi}} & 0 \\ -\frac{1}{\sqrt{2}}\sqrt{\frac{1}{\sqrt{5}\phi}} & \frac{1}{\sqrt{2}}\sqrt{\frac{\phi}{\sqrt{5}}} & -\frac{1}{\sqrt{2}} \\ -\frac{1}{\sqrt{2}}\sqrt{\frac{1}{\sqrt{5}\phi}} & \frac{1}{\sqrt{2}}\sqrt{\frac{\phi}{\sqrt{5}}} & \frac{1}{\sqrt{2}} \end{pmatrix}\mathcal{P}, \tag{16.60}$$

where $\mathcal{P} = \text{Diag}(1, 1, i)$ is the Majorana phase matrix. In conclusion, the lepton mixing matrix has a vanishing reactor mixing angle, a maximal atmospheric mixing angle, and a solar angle given by $\theta_{12} = \tan^{-1}(1/\phi)$.

We now discuss another mixing pattern, namely, trimaximal lepton mixing [68], defined by $|U_{\alpha 2}|^2 = 1/3$ for $\alpha = e, \mu$, whence the mixing matrix U_{tri} is given using an arbitrary angle θ and a phase η by

$$U_{\text{tri}} = \begin{pmatrix} \sqrt{\frac{2}{6}} & \frac{1}{\sqrt{3}} & 0 \\ -\frac{1}{\sqrt{6}} & \frac{1}{\sqrt{3}} & -\frac{1}{\sqrt{2}} \\ -\frac{1}{\sqrt{6}} & \frac{1}{\sqrt{3}} & \frac{1}{\sqrt{2}} \end{pmatrix}\begin{pmatrix} \cos\theta & 0 & \sin\theta e^{-i\eta} \\ 0 & 1 & 0 \\ -\sin\theta e^{i\eta} & 0 & \cos\theta \end{pmatrix}. \tag{16.61}$$

This corresponds to a two-parameter lepton flavor mixing matrix. In Sect. 16.2.2, we presented a model for the lepton sector in which trimaximal mixing is based on

16.4 Alternative Flavor Mixing

Table 16.5 Assignments of $SU(2)$, S_4, Z_4, and $U(1)_{FN}$ representations

	l	e^c	μ^c	τ^c	$h_{u,d}$	θ	φ_l	χ_l	ξ_ν	φ_ν
$SU(2)$	2	1	1	1	2	1	1	1	1	1
S_4	3	1	$1'$	1	1	1	3	$3'$	1	3
Z_4	1	-1	$-i$	$-i$	1	1	i	i	1	1
$U(1)_{FN}$	0	2	1	0	0	-1	0	0	0	0

the group A_4, or the group $\Delta(27)$. An interesting feature of the $\Delta(27)$ model is that no vacuum alignment is required. The other model is based on the S_3 [69] or S_4 [70] discrete symmetry, where the symmetry breaking is triggered by the boundary conditions of the bulk right-handed neutrino in the fifth spatial dimension.

Finally, we consider bimaximal mixing [78], with solar angle $\theta_{12} = \pi/4$, atmospheric angle $\theta_{23} = \pi/4$, and reactor angle $\theta_{13} = 0$. The mixing matrix U_{bi} is

$$U_{bi} = \begin{pmatrix} \frac{1}{\sqrt{2}} & \frac{1}{\sqrt{2}} & 0 \\ -\frac{1}{2} & \frac{1}{2} & -\frac{1}{\sqrt{2}} \\ -\frac{1}{2} & \frac{1}{2} & \frac{1}{\sqrt{2}} \end{pmatrix}. \tag{16.62}$$

The $S_4 \times Z_4$ flavor model is presented in [79]. The particle contents are summarized in Table 16.5.

In the charged lepton sector, the superpotential is given by

$$w_l = \frac{y_e^{(1)}}{\Lambda^2} \frac{\theta^2}{\Lambda^2} e^c (l\varphi_l \varphi_l) h_d + \frac{y_e^{(2)}}{\Lambda^2} \frac{\theta^2}{\Lambda^2} e^c (l\chi_l \chi_l) h_d + \frac{y_e^{(3)}}{\Lambda^2} \frac{\theta^2}{\Lambda^2} e^c (l\varphi_l \chi_l) h_d$$
$$+ \frac{y_\mu}{\Lambda} \frac{\theta}{\Lambda} \mu^c (l\chi_l)' h_d + \frac{y_\tau}{\Lambda} \tau^c (l\varphi_l) h_d, \tag{16.63}$$

where $y_e^{(i)}$, y_μ, and y_τ are Yukawa couplings. In the neutrino sector, the effective superpotential is

$$w_\nu^{\text{eff}} = \frac{M}{\Lambda}(lh_u lh_u) + \frac{a}{\Lambda^2}(lh_u lh_u)\xi_\nu + \frac{b}{\Lambda^2}(lh_u lh_u \varphi_l), \tag{16.64}$$

where M, a, and b are given in mass units. Taking the VEV alignment and VEVs to be

$$\frac{\langle \varphi_l \rangle}{\Lambda} = A(0,1,0), \quad \frac{\langle \chi_l \rangle}{\Lambda} = B(0,0,1), \quad \frac{\langle \varphi_\nu \rangle}{\Lambda} = C(0,1,-1),$$
$$\frac{\langle \xi_\nu \rangle}{\Lambda} = D, \quad \frac{\langle \theta \rangle}{\Lambda} = t, \quad \langle h_{u,d} \rangle = v_{u,d}, \tag{16.65}$$

the charged lepton mass matrix is diagonal:

$$m_l = \begin{pmatrix} (y_e^{(1)} B^2 - y_e^{(2)} A^2 + y_e^{(3)} AB) t^2 & 0 & 0 \\ 0 & y_\mu Bt & 0 \\ 0 & 0 & y_\tau A \end{pmatrix} v_d, \tag{16.66}$$

and the effective neutrino mass matrix is

$$m_\nu^{\text{eff}} = \begin{pmatrix} 2M+2aD & -2bC & -2bC \\ -2bC & 0 & 2M+2aD \\ -2bC & 2M+2aD & 0 \end{pmatrix} \frac{v_u^2}{\Lambda}. \qquad (16.67)$$

The neutrino mixing is then bimaximal, as can be seen from (16.62), and the neutrino mass eigenvalues are

$$m_1 = 2|M+aD-\sqrt{2}bC|\frac{v_u^2}{\Lambda}, \qquad m_2 = 2|M+aD+\sqrt{2}bC|\frac{v_u^2}{\Lambda},$$

$$m_3 = 2|M+aD|\frac{v_u^2}{\Lambda}. \qquad (16.68)$$

Note that bimaximal neutrino mixing is also studied in the context of the quark–lepton complementarity of mixing angles [80] in the S_4 model [81].

16.5 Comments on Other Applications

Supersymmetric extension is an interesting candidate for physics beyond the standard model. Even if the theory is supersymmetric at high energy, supersymmetry must break above the weak scale. Supersymmetry breaking induces soft supersymmetry breaking terms such as gaugino masses, sfermion masses, and scalar trilinear couplings, i.e., the so-called A-terms. Flavor symmetries control not only quark/lepton mass matrices, but also squark/slepton masses and their A-terms. Suppose the flavor symmetries are exact. When the three families have different quantum numbers under flavor symmetries, the squark/slepton mass-squared matrices are diagonal. Furthermore, when three (two) of the three families correspond to triplets (doublets) of flavor symmetries, their diagonal squark/slepton masses are degenerate. That is, the sfermion mass-squared matrix $(m^2)_{ij}$ is

$$(m^2)_{ij} = \begin{pmatrix} m^2 & 0 & 0 \\ 0 & m^2 & 0 \\ 0 & 0 & m^2 \end{pmatrix}, \qquad (16.69)$$

when the three families correspond to a triplet. On the other hand, the sfermion mass-squared matrix $(m^2)_{ij}$ becomes

$$(m^2)_{ij} = \begin{pmatrix} m^2 & 0 & 0 \\ 0 & m^2 & 0 \\ 0 & 0 & m'^2 \end{pmatrix}, \qquad (16.70)$$

when the first two families correspond to a doublet and the third family corresponds to a singlet. These patterns would become an interesting prediction of a certain class of flavor models, and it could be tested if the supersymmetry breaking scale is reachable by collider experiments.

Flavor symmetries have similar effects on A-terms. These results are very important to suppress flavor changing neutral currents, which are strongly constrained by experiments. However, the flavor symmetry must break to lead to realistic quark/lepton mass matrices. Such breaking effects deform the above predictions. How much results are changed depends on breaking patterns. If masses of superpartners are $\mathcal{O}(100)$ GeV, some models may be ruled out, e.g., by experiments on flavor changing neutral currents (see, e.g., [24, 52, 82–85]).

The application to dark matter (DM) models is another interesting topic. It is known that DM should have a global symmetry to be stabilized or long-lived, even after electroweak breaking. This kind of symmetry could arise naturally from the non-Abelian flavor symmetries [86–92].

Indirect detections of DM were recently reported by PAMELA [93] and Fermi-LAT [94], where the positron excess and the total flux of electrons and positrons are observed in cosmic rays. These observations can be explained by scattering/decay of TeV-scale DM particles. Since PAMELA measured negative results for the anti-proton excess [95], leptophilic DM is preferable.

Even so, if the main final state of scattering or decay of DM is $\tau^{\pm}$, this annihilation/decay mode is disfavored because it will overproduce gamma-rays in the final radiation state [96]. This may indicate that, if the cosmic-ray anomalies are induced by DM scattering or decay, these processes also reflect the flavor structure of the model.

In decaying DM models, no anti-proton excess in cosmic rays implies that the lifetime of the DM particle should be $\mathcal{O}(10^{26})$ s. This long lifetime is achieved if the TeV-scale DM (e.g., a gauge singlet fermion X) decays into leptons by dimension six operators $\bar{L}E\bar{L}X/\Lambda^2$ suppressed by GUT scale $\Lambda \sim 10^{16}$ GeV. In this case, the lifetime of the DM is estimated as $\Gamma^{-1} \sim [(\text{TeV})^5/\Lambda^4]^{-1} \sim 10^{26}$ s. However, it could be difficult in general to induce only such an operator and forbid the other undesirable operators, e.g., $\bar{L}EX$ or other four-dimensional interacting terms of X. On the other hand, the non-Abelian discrete symmetries can play an important role in well selecting the operator [97, 98]. [Moreover, it has been shown in [97] that it is impossible to adapt $U(1)$ symmetries.] The bosonic DM case can be considered in a similar way [99–101].

16.6 Comment on Origins of Flavor Symmetries

What is the origin of non-Abelian flavor symmetries? Some of them are symmetries of geometrical solids, so their origin may be geometrical symmetries in extra dimensions. For example, it is found that the two-dimensional orbifold T^2/Z_2 with proper values of moduli has discrete symmetries such as A_4 and S_4 [102, 103] (see also [104]).

Superstring theory is a promising candidate for a unified theory including gravity, and predicts an extra six dimensions. Superstring theory on a certain type of six-dimensional compact space realizes a discrete flavor symmetry. Such a string theory leads to stringy selection rules for allowed couplings among matter fields in four-dimensional effective field theory. Such stringy selection rules and geometrical symmetries along with broken (continuous) gauge symmetries result in discrete flavor symmetries in superstring theory. For example, discrete flavor symmetries in heterotic orbifold models are studied in [83, 105, 106], and D_4 and $\Delta(54)$ are realized. Magnetized/intersecting D-brane models also realize the same flavor symmetries, together with other types such as $\Delta(27)$ [107–109]. Different types of non-Abelian flavor symmetries may be derived in other string models. These studies are thus quite important.

Alternatively, discrete flavor symmetries may originate from continuous (gauge) symmetries [110–112].

At any rate, the experimental data of quark/lepton masses and mixing angles have no symmetry. Thus, non-Abelian flavor symmetries must be broken to reproduce the experimentally observed masses and mixing angles. The breaking direction is important, because the forms of mass matrices are determined by the direction along which the flavor symmetries break. Hence, we need a proper breaking direction to derive realistic values of quark/lepton masses and mixing angles.

One way to fix the breaking direction is to analyze the potential minima of scalar fields with non-trivial representations of flavor symmetries. The number of potential minima may be finite, and realistic breaking would occur in one of them. At least, this is the conventional approach.

Another scenario to fix the breaking direction could be realized in theories with extra dimensions. One can impose the boundary conditions of matter fermions [69] and/or flavon scalars [113–115] in bulk in such a way that zero modes for some components of the irreducible multiplets are projected out, that is, breaking the symmetry. If a proper component of a flavon multiplet remains, that can realize a realistic breaking direction.

References

1. Schwetz, T., Tortola, M.A., Valle, J.W.F.: New J. Phys. **10**, 113011 (2008). arXiv:0808.2016 [hep-ph]
2. Fogli, G.L., Lisi, E., Marrone, A., Palazzo, A., Rotunno, A.M.: Phys. Rev. Lett. **101**, 141801 (2008). arXiv:0806.2649 [hep-ph]
3. Fogli, G.L., Lisi, E., Marrone, A., Palazzo, A., Rotunno, A.M.: Nucl. Phys. Proc. Suppl. **188**, 27 (2009)
4. Harrison, P.F., Perkins, D.H., Scott, W.G.: Phys. Lett. B **530**, 167 (2002). arXiv:hep-ph/0202074
5. Harrison, P.F., Scott, W.G.: Phys. Lett. B **535**, 163 (2002). arXiv:hep-ph/0203209
6. Harrison, P.F., Scott, W.G.: Phys. Lett. B **557**, 76 (2003). arXiv:hep-ph/0302025
7. Harrison, P.F., Scott, W.G.: arXiv:hep-ph/0402006
8. Altarelli, G., Feruglio, F.: arXiv:1002.0211 [hep-ph]
9. Ishimori, H., Kobayashi, T., Ohki, H., Shimizu, Y., Okada, H., Tanimoto, M.: Prog. Theor. Phys. Suppl. **183**, 1–163 (2010). arXiv:1003.3552 [hep-th]

References

10. Schwetz, T., Tortola, M., Valle, J.W.F.: New J. Phys. **13**, 063004 (2011). arXiv:1103.0734 [hep-ph]
11. Mueller, T.A., Lhuillier, D., Fallot, M., Letourneau, A., Cormon, S., Fechner, M., Giot, L., Lasserre, T., et al.: Phys. Rev. C **83**, 054615 (2011). arXiv:1101.2663 [hep-ex]
12. Abe, K., et al. (T2K Collaboration): Phys. Rev. Lett. **107**, 041801 (2011). arXiv:1106.2822 [hep-ex]
13. An, F.P., et al. (Daya Bay Collaboration): Phys. Rev. Lett. **108**, 171803 (2012). arXiv:1203.1669 [hep-ex]
14. Weinberg, S.: Phys. Rev. Lett. **43**, 1566 (1979)
15. Minkowski, P.: Phys. Lett. B **67**, 421 (1977)
16. Yanagida, T.: In: Proc. of the Workshop on Unified Theory and Baryon Number in the Universe, KEK, March (1979)
17. Gell-Mann, M., Ramond, P., Slansky, R. In: Supergravity, Stony Brook, September (1979)
18. Glashow, S.L.: In: Lévy, M., et al. (eds.) Quarks and Leptons, Cargèse, p. 707. Plenum, New York (1980)
19. Mohapatra, R.N., Senjanovic, G.: Phys. Rev. Lett. **44**, 912 (1980)
20. Ma, E., Rajasekaran, G.: Phys. Rev. D **64**, 113012 (2001). arXiv:hep-ph/0106291
21. Babu, K.S., Enkhbat, T., Gogoladze, I.: Phys. Lett. B **555**, 238 (2003). arXiv:hep-ph/0204246
22. Babu, K.S., Ma, E., Valle, J.W.F.: Phys. Lett. B **552**, 207 (2003). arXiv:hep-ph/0206292
23. Ma, E.: Mod. Phys. Lett. A **17**, 2361 (2002). arXiv:hep-ph/0211393
24. Babu, K.S., Kobayashi, T., Kubo, J.: Phys. Rev. D **67**, 075018 (2003). arXiv:hep-ph/0212350
25. Hirsch, M., Romao, J.C., Skadhauge, S., Valle, J.W.F., Villanova del Moral, A.: Phys. Rev. D **69**, 093006 (2004). arXiv:hep-ph/0312265
26. Ma, E.: Phys. Rev. D **70**, 031901 (2004). arXiv:hep-ph/0404199
27. Altarelli, G., Feruglio, F.: Nucl. Phys. B **720**, 64 (2005). arXiv:hep-ph/0504165
28. Chen, S.L., Frigerio, M., Ma, E.: Nucl. Phys. B **724**, 423 (2005). arXiv:hep-ph/0504181
29. Zee, A.: Phys. Lett. B **630**, 58 (2005). arXiv:hep-ph/0508278
30. Ma, E.: Mod. Phys. Lett. A **20**, 2601 (2005). arXiv:hep-ph/0508099
31. Ma, E.: Phys. Lett. B **632**, 352 (2006). arXiv:hep-ph/0508231
32. Ma, E.: Phys. Rev. D **73**, 057304 (2006). arXiv:hep-ph/0511133
33. Altarelli, G., Feruglio, F.: Nucl. Phys. B **741**, 215 (2006). arXiv:hep-ph/0512103
34. He, X.G., Keum, Y.Y., Volkas, R.R.: J. High Energy Phys. **0604**, 039 (2006). arXiv:hep-ph/0601001
35. Adhikary, B., Brahmachari, B., Ghosal, A., Ma, E., Parida, M.K.: Phys. Lett. B **638**, 345 (2006). arXiv:hep-ph/0603059
36. Ma, E., Sawanaka, H., Tanimoto, M.: Phys. Lett. B **641**, 301 (2006). arXiv:hep-ph/0606103
37. Valle, J.W.F.: J. Phys. Conf. Ser. **53**, 473 (2006). arXiv:hep-ph/0608101
38. Adhikary, B., Ghosal, A.: Phys. Rev. D **75**, 073020 (2007). arXiv:hep-ph/0609193
39. Lavoura, L., Kuhbock, H.: Mod. Phys. Lett. A **22**, 181 (2007). arXiv:hep-ph/0610050
40. King, S.F., Malinsky, M.: Phys. Lett. B **645**, 351 (2007). arXiv:hep-ph/0610250
41. Hirsch, M., Joshipura, A.S., Kaneko, S., Valle, J.W.F.: Phys. Rev. Lett. **99**, 151802 (2007). arXiv:hep-ph/0703046
42. Bazzocchi, F., Kaneko, S., Morisi, S.: J. High Energy Phys. **0803**, 063 (2008). arXiv:0707.3032 [hep-ph]
43. Grimus, W., Kuhbock, H.: Phys. Rev. D **77**, 055008 (2008). arXiv:0710.1585 [hep-ph]
44. Honda, M., Tanimoto, M.: Prog. Theor. Phys. **119**, 583 (2008). arXiv:0801.0181 [hep-ph]
45. Brahmachari, B., Choubey, S., Mitra, M.: Phys. Rev. D **77**, 073008 (2008) [Erratum ibid. D **77**, 119901 (2008)]. arXiv:0801.3554 [hep-ph]
46. Adhikary, B., Ghosal, A.: Phys. Rev. D **78**, 073007 (2008). arXiv:0803.3582 [hep-ph]
47. Fukuyama, T.: arXiv:0804.2107 [hep-ph]
48. Lin, Y.: Nucl. Phys. B **813**, 91 (2009). arXiv:0804.2867 [hep-ph]
49. Frampton, P.H., Matsuzaki, S.: arXiv:0806.4592 [hep-ph]

50. Feruglio, F., Hagedorn, C., Lin, Y., Merlo, L.: Nucl. Phys. B **809**, 218 (2009). arXiv:0807.3160 [hep-ph]
51. Morisi, S.: Nuovo Cimento B **123**, 886 (2008). arXiv:0807.4013 [hep-ph]
52. Ishimori, H., Kobayashi, T., Omura, Y., Tanimoto, M.: J. High Energy Phys. **0812**, 082 (2008). arXiv:0807.4625 [hep-ph]
53. Ma, E.: Phys. Lett. B **671**, 366 (2009). arXiv:0808.1729 [hep-ph]
54. Bazzocchi, F., Frigerio, M., Morisi, S.: Phys. Rev. D **78**, 116018 (2008). arXiv:0809.3573 [hep-ph]
55. Hirsch, M., Morisi, S., Valle, J.W.F.: Phys. Rev. D **79**, 016001 (2009). arXiv:0810.0121 [hep-ph]
56. Merlo, L.: arXiv:0811.3512 [hep-ph]
57. Baek, S., Oh, M.C.: arXiv:0812.2704 [hep-ph]
58. Morisi, S.: Phys. Rev. D **79**, 033008 (2009). arXiv:0901.1080 [hep-ph]
59. Ciafaloni, P., Picariello, M., Torrente-Lujan, E., Urbano, A.: Phys. Rev. D **79**, 116010 (2009). arXiv:0901.2236 [hep-ph]
60. Merlo, L.: J. Phys. Conf. Ser. **171**, 012083 (2009). arXiv:0902.3067 [hep-ph]
61. Chen, M.C., King, S.F.: J. High Energy Phys. **0906**, 072 (2009). arXiv:0903.0125 [hep-ph]
62. Branco, G.C., Gonzalez Felipe, R., Rebelo, M.N., Serodio, H.: Phys. Rev. D **79**, 093008 (2009). arXiv:0904.3076 [hep-ph]
63. Hayakawa, A., Ishimori, H., Shimizu, Y., Tanimoto, M.: Phys. Lett. B **680**, 334 (2009). arXiv:0904.3820 [hep-ph]
64. Altarelli, G., Meloni, D.: J. Phys. G **36**, 085005 (2009). arXiv:0905.0620 [hep-ph]
65. Urbano, A.: arXiv:0905.0863 [hep-ph]
66. Hirsch, M., Morisi, S., Valle, J.W.F.: Phys. Lett. B **679**, 454 (2009). arXiv:0905.3056 [hep-ph]
67. Lin, Y.: Nucl. Phys. B **824**, 95 (2010). arXiv:0905.3534 [hep-ph]
68. Grimus, W., Lavoura, L.: J. High Energy Phys. **0809**, 106 (2008). arXiv:0809.0226 [hep-ph]
69. Haba, N., Watanabe, A., Yoshioka, K.: Phys. Rev. Lett. **97**, 041601 (2006). arXiv:hep-ph/0603116
70. Ishimori, H., Shimizu, Y., Tanimoto, M., Watanabe, A.: Phys. Rev. D **83**, 033004 (2011). arXiv:1010.3805 [hep-ph]
71. Ishimori, H., Shimizu, Y., Tanimoto, M.: Prog. Theor. Phys. **121**, 769 (2009). arXiv:0812.5031 [hep-ph]
72. Froggatt, C.D., Nielsen, H.B.: Nucl. Phys. B **147**, 277 (1979)
73. Fusaoka, H., Koide, Y.: Phys. Rev. D **57**, 3986 (1998). arXiv:hep-ph/9712201
74. Jarlskog, C.: Phys. Rev. Lett. **55**, 1039 (1985)
75. Everett, L.L., Stuart, A.J.: Phys. Rev. D **79**, 085005 (2009). arXiv:0812.1057 [hep-ph]
76. Adulpravitchai, A., Blum, A., Rodejohann, W.: New J. Phys. **11**, 063026 (2009). arXiv:0903.0531 [hep-ph]
77. Kajiyama, Y., Raidal, M., Strumia, A.: Phys. Rev. D **76**, 117301 (2007). arXiv:0705.4559 [hep-ph]
78. Barger, V.D., Pakvasa, S., Weiler, T.J., Whisnant, K.: Phys. Lett. B **437**, 107 (1998). arXiv:hep-ph/9806387
79. Altarelli, G., Feruglio, F., Merlo, L.: J. High Energy Phys. **0905**, 020 (2009). arXiv:0903.1940 [hep-ph]
80. Minakata, H., Smirnov, A.Y.: Phys. Rev. D **70**, 073009 (2004). arXiv:hep-ph/0405088
81. Merlo, L.: arXiv:0909.2760 [hep-ph]
82. Kobayashi, T., Kubo, J., Terao, H.: Phys. Lett. B **568**, 83 (2003). arXiv:hep-ph/0303084
83. Ko, P., Kobayashi, T., Park, J.H., Raby, S.: Phys. Rev. D **76**, 035005 (2007) [Erratum ibid. D **76**, 059901 (2007)]. arXiv:0704.2807 [hep-ph]
84. Ishimori, H., Kobayashi, T., Ohki, H., Omura, Y., Takahashi, R., Tanimoto, M.: Phys. Rev. D **77**, 115005 (2008). arXiv:0803.0796 [hep-ph]
85. Ishimori, H., Kobayashi, T., Okada, H., Shimizu, Y., Tanimoto, M.: J. High Energy Phys. **0912**, 054 (2009). arXiv:0907.2006 [hep-ph]

86. Hirsch, M., Morisi, S., Peinado, E., Valle, J.W.F.: Phys. Rev. D **82**, 116003 (2010). arXiv:1007.0871 [hep-ph]
87. Meloni, D., Morisi, S., Peinado, E.: Phys. Lett. B **697**, 339–342 (2011). arXiv:1011.1371 [hep-ph]
88. Boucenna, M.S., Hirsch, M., Morisi, S., Peinado, E., Taoso, M., Valle, J.W.F.: J. High Energy Phys. **1105**, 037 (2011). arXiv:1101.2874 [hep-ph]
89. Morisi, S.: J. Phys. Conf. Ser. **259**, 012094 (2010)
90. Meloni, D., Morisi, S., Peinado, E.: Phys. Lett. B **703**, 281–287 (2011). arXiv:1104.0178 [hep-ph]
91. Toorop, R.d.A., Bazzocchi, F., Morisi, S.: arXiv:1104.5676 [hep-ph]
92. Batell, B., Pradler, J., Spannowsky, M.: J. High Energy Phys. **1108**, 038 (2011). arXiv:1105.1781 [hep-ph]
93. Adriani, O., et al. (PAMELA Collaboration): Nature **458**, 607 (2009). arXiv:0810.4995 [astro-ph]
94. Abdo, A.A., et al. (Fermi LAT Collaboration): Phys. Rev. Lett. **102**, 181101 (2009). arXiv:0905.0025 [astro-ph.HE]
95. Adriani, O., Barbarino, G.C., Bazilevskaya, G.A., Bellotti, R., Boezio, M., Bogomolov, E.A., Bonechi, L., Bongi, M., et al.: Phys. Rev. Lett. **102**, 051101 (2009). arXiv:0810.4994 [astro-ph]
96. Papucci, M., Strumia, A.: J. Cosmol. Astropart. Phys. **1003**, 014 (2010). arXiv:0912.0742 [hep-ph]
97. Haba, N., Kajiyama, Y., Matsumoto, S., Okada, H., Yoshioka, K.: Phys. Lett. B **695**, 476–481 (2011). arXiv:1008.4777 [hep-ph]
98. Kajiyama, Y., Okada, H.: Nucl. Phys. B **848**, 303–313 (2011). arXiv:1011.5753 [hep-ph]
99. Daikoku, Y., Okada, H., Toma, T.: arXiv:1106.4717 [hep-ph]
100. Kajiyama, Y., Okada, H., Toma, T.: arXiv:1109.2722 [hep-ph]
101. Kajiyama, Y., Okada, H., Toma, T.: Eur. Phys. J. C **71**, 1688 (2011). arXiv:1104.0367 [hep-ph]
102. Altarelli, G., Feruglio, F., Lin, Y.: Nucl. Phys. B **775**, 31 (2007). arXiv:hep-ph/0610165
103. Adulpravitchai, A., Blum, A., Lindner, M.: J. High Energy Phys. **0907**, 053 (2009). arXiv:0906.0468 [hep-ph]
104. Abe, H., Choi, K.-S., Kobayashi, T., Ohki, H., Sakai, M.: Int. J. Mod. Phys. A **26**, 4067–4082 (2011). arXiv:1009.5284 [hep-th]
105. Kobayashi, T., Raby, S., Zhang, R.J.: Nucl. Phys. B **704**, 3 (2005). arXiv:hep-ph/0409098
106. Kobayashi, T., Nilles, H.P., Ploger, F., Raby, S., Ratz, M.: Nucl. Phys. B **768**, 135 (2007). arXiv:hep-ph/0611020
107. Abe, H., Choi, K.S., Kobayashi, T., Ohki, H.: Nucl. Phys. B **820**, 317 (2009). arXiv:0904.2631 [hep-ph]
108. Abe, H., Choi, K.S., Kobayashi, T., Ohki, H.: Phys. Rev. D **80**, 126006 (2009). arXiv:0907.5274 [hep-th]
109. Abe, H., Choi, K.S., Kobayashi, T., Ohki, H.: arXiv:1001.1788 [hep-th]
110. de Medeiros Varzielas, I., King, S.F., Ross, G.G.: Phys. Lett. B **644**, 153 (2007). arXiv:hep-ph/0512313
111. Adulpravitchai, A., Blum, A., Lindner, M.: J. High Energy Phys. **0909**, 018 (2009). arXiv:0907.2332 [hep-ph]
112. Frampton, P.H., Kephart, T.W., Rohm, R.M.: Phys. Lett. B **679**, 478 (2009). arXiv:0904.0420 [hep-ph]
113. Kobayashi, T., Omura, Y., Yoshioka, K.: Phys. Rev. D **78**, 115006 (2008). arXiv:0809.3064
114. Seidl, G.: Phys. Rev. D **81**, 025004 (2010). arXiv:0811.3775 [hep-ph]
115. Adulpravitchai, A., Schmidt, M.A.: arXiv:1001.3172 [hep-ph]

Appendix A
Useful Theorems

In this appendix, we give simple proofs of useful theorems (see also, e.g., [1–4]).

Lagrange's Theorem *The order N_H of a subgroup of a finite group G is a divisor of the order N_G of G.*

Proof If $H = G$, the claim is trivial since $N_H = N_G$. Thus, we consider $H \neq G$. Let a_1 be an element of G, but not contained in H. Here, we denote all elements in H by $\{e = h_0, h_1, \ldots, h_{N_H-1}\}$. Then we consider the products of a_1 and elements of H:

$$a_1 H = \{a_1, a_1 h_1, \ldots, a_1 h_{N_h-1}\}. \tag{A.1}$$

All of $a_1 h_i$ are different from each other. None of $a_1 h_i$ are contained in H, since if $a_1 h_i = h_j$, we would find $a_1 = h_j h_i^{-1}$, that is, a_1 would be an element in H after all. Thus, the set $a_1 H$ contains N_H elements. Next let a_2 be an element of G, but contained in neither H nor $a_1 H$. If $a_2 h_i = a_1 h_j$, the element a_2 could be written as $a_2 = a_1 h_j h_i^{-1}$, that is, it would be an element of $a_1 H$. Thus, when $a_2 \notin H$ and $a_2 \notin a_1 H$, the set $a_2 H$ yields N_H new elements. We repeat this process. Then we can decompose

$$G = H + a_1 H + \cdots + a_{m-1} H. \tag{A.2}$$

This implies $N_G = m N_H$. □

Theorem *For a finite group, every representation is equivalent to a unitary representation.*

Proof Every group element a is represented by a matrix $D(a)$, which acts on the vector space. We denote the basis of the representation vector space by $\{e_1, \ldots, e_d\}$. We consider two vectors v and w:

$$v = \sum_{i=1}^{d} v_i e_i, \qquad w = \sum_{i=1}^{d} w_i e_i. \tag{A.3}$$

We define the scalar product between v and w as

$$(v, w) = \sum_{i=1}^{d} v_i^* w_i. \tag{A.4}$$

Then we define another scalar product by

$$\langle v, w \rangle = \frac{1}{N_G} \sum_{a \in G} (D(a)v, D(a)w). \tag{A.5}$$

It follows that

$$\langle D(b)v, D(b)w \rangle = \frac{1}{N_G} \sum_{a \in G} (D(b)D(a)v, D(b)D(a)w)$$

$$= \frac{1}{N_G} \sum_{a \in G} (D(ba)v, D(ba)w)$$

$$= \frac{1}{N_G} \sum_{c \in G} (D(c)v, D(c)w)$$

$$= \langle v, w \rangle. \tag{A.6}$$

This implies that $D(b)$ is unitary with respect to the scalar product $\langle v, w \rangle$. The orthogonal bases $\{e_i\}$ and $\{e'_i\}$ for the two scalar products (v, w) and $\langle v, w \rangle$ can be related by the linear transformation T such that $e'_i = T e_i$, i.e., $(v, w) = \langle Tv, Tw \rangle$. We define $D'(g) = T^{-1} D(g) T$. Then it follows that

$$(T^{-1} D(a) T v, T^{-1} D(a) T w) = \langle D(a) T v, D(a) T w \rangle$$

$$= \langle T v, T w \rangle$$

$$= (v, w). \tag{A.7}$$

That is, the matrix $D'(g)$ is unitary and is equivalent to $D(g)$. □

Schur's Lemma

(I) Let $D_1(g)$ and $D_2(g)$ be irreducible representations of G, which are inequivalent to each other. If

$$A D_1(g) = D_2(g) A, \quad \forall g \in G, \tag{A.8}$$

the matrix A must vanish, i.e., $A = 0$.

(II) If

$$D(g) A = A D(g), \quad \forall g \in G, \tag{A.9}$$

the matrix A must be proportional to the identity matrix I, i.e., $A = \lambda I$.

A Useful Theorems

Proof (I) We denote the representation vector spaces for $D_1(g)$ and $D_2(g)$ by V and W, respectively. Consider a map $A: V \to W$ satisfying (A.8). We consider the kernel of A, viz.,

$$\mathrm{Ker}(A) = \{v \in V | Av = 0\}. \tag{A.10}$$

Let $v \in \mathrm{Ker}(A)$. Then we have

$$AD_1(g)v = D_2(g)Av = 0. \tag{A.11}$$

It follows that $D_1(g)\mathrm{Ker}(A) \subset \mathrm{Ker}(A)$, that is, $\mathrm{Ker}(A)$ is invariant under $D_1(g)$. Because $D_1(g)$ is irreducible, this implies that

$$\mathrm{Ker}(A) = \{0\}, \quad \text{or} \quad \mathrm{Ker}(A) = V. \tag{A.12}$$

But $\mathrm{Ker}(A) = V$ implies that $A = 0$. On the other hand, if we consider the image

$$\mathrm{Im}(A) = \{Av | v \in V\}, \tag{A.13}$$

we find

$$D_2(g)Av = AD_1(g)v \in \mathrm{Im}(A). \tag{A.14}$$

That is, $\mathrm{Im}(A)$ is invariant under $D_2(g)$. Once again, since $D_2(g)$ is irreducible, this implies that

$$\mathrm{Im}(A) = \{0\}, \quad \text{or} \quad \mathrm{Im}(A) = W. \tag{A.15}$$

However, $\mathrm{Im}(A) = \{0\}$ means that $A = 0$. As a result, we find that A must satisfy

$$A = 0, \quad \text{or} \quad AD_1(g)A^{-1} = D_2(g). \tag{A.16}$$

The latter would mean that the representations $D_1(g)$ and $D_2(g)$ were equivalent to each other. Therefore, A must vanish, i.e., $A = 0$, if $D_1(g)$ and $D_2(g)$ are not equivalent.

(II) Now we consider the case with $D(g) = D_1(g) = D_2(g)$ and $V = W$. Hence, A is a linear operator on V. The finite-dimensional matrix A has at least one eigenvalue, because the characteristic equation $\det(A - \lambda I) = 0$ has at least one root λ, and this λ is an eigenvalue. Then (A.9) leads to

$$D(g)(A - \lambda I) = (A - \lambda I)D(g), \quad \forall g \in G. \tag{A.17}$$

Using the above proof of Schur's lemma (I) and the fact that $\mathrm{Ker}(A - \lambda I) \neq \{0\}$, we find $\mathrm{Ker}(A - \lambda I) = V$, that is, $A - \lambda I = 0$. □

Theorem *Let $D_\alpha(g)$ and $D_\beta(g)$ be irreducible representations of a group G on the*

d_α and d_β dimensional vector spaces. Then they satisfy the orthogonality relation

$$\sum_{a \in G} D_\alpha(a)_{i\ell} D_\beta(a^{-1})_{mj} = \frac{N_G}{d_\alpha} \delta_{\alpha\beta} \delta_{ij} \delta_{\ell m}. \tag{A.18}$$

Proof We define

$$A = \sum_{a \in G} D_\alpha(a) B D_\alpha(a^{-1}), \tag{A.19}$$

where B is an arbitrary $(d_\alpha \times d_\alpha)$ matrix. We find $D(b)A = AD(b)$, since

$$D_\alpha(b) A = \sum_{a \in G} D_\alpha(b) D_\alpha(a) B D_\alpha(a^{-1})$$

$$= \sum_{a \in G} D_\alpha(ba) B D_\alpha((ba)^{-1}) D_\alpha(b)$$

$$= \sum_{c \in G} D_\alpha(c) B D_\alpha(c^{-1}) D_\alpha(b). \tag{A.20}$$

That is, by Schur's lemma (II), we find that the matrix A must be proportional to the $(d_\alpha \times d_\alpha)$ identity matrix. We choose $B_{ij} = \delta_{i\ell} \delta_{jm}$. Then we obtain

$$A_{ij} = \sum_{a \in G} D_\alpha(a)_{i\ell} D_\alpha(a^{-1})_{mj}, \tag{A.21}$$

and the right-hand side (RHS) can be written $\lambda(\ell, m) \delta_{ij}$, that is,

$$\sum_{a \in G} D_\alpha(a)_{i\ell} D_\alpha(a^{-1})_{mj} = \lambda(\ell, m) \delta_{ij}. \tag{A.22}$$

Furthermore, we compute the trace of both sides. The trace of the RHS is computed to be

$$\lambda(\ell, m) \operatorname{tr} \delta_{ij} = d_\alpha \lambda(\ell, m), \tag{A.23}$$

while the trace of left-hand side (LHS) is obtained as

$$\sum_{i=1}^{d} \sum_{a \in G} D_\alpha(a)_{i\ell} D_\alpha(a^{-1})_{mi} = \sum_{a \in G} D_\alpha(aa^{-1})_{\ell m}$$

$$= N_G \delta_{\ell m}. \tag{A.24}$$

By comparing these results, we obtain

$$\lambda(\ell, m) = \frac{N_G}{d_\alpha} \delta_{\ell m}.$$

A Useful Theorems

Hence, we find

$$\sum_{a \in G} D_\alpha(a)_{i\ell} D_\alpha(a^{-1})_{mj} = \frac{N_G}{d_\alpha} \delta_{ij} \delta_{\ell m}. \tag{A.25}$$

Similarly, we define

$$A^{(\alpha\beta)} = \sum_{a \in G} D_\alpha(a) B D_\beta(a^{-1}), \tag{A.26}$$

where $D_\alpha(a)$ and $D_\beta(a)$ are inequivalent. Then we find $D_\alpha(a)A = AD_\beta(a)$. Similarly to the previous analysis, using Schur's lemma (I), we obtain

$$\sum_{a \in G} D_\alpha(a)_{i\ell} D_\beta(a^{-1})_{mj} = 0. \tag{A.27}$$

Thus, we arrive at (A.18). Furthermore, if the representation is unitary, (A.18) becomes

$$\sum_{a \in G} D_\alpha(a)_{i\ell} D_\beta^*(a)_{jm} = \frac{N_G}{d_\alpha} \delta_{\alpha\beta} \delta_{ij} \delta_{\ell m}. \tag{A.28}$$

□

Because of this orthogonality, we can expand an arbitrary function $F(a)$ of a in terms of the matrix elements of irreducible representations:

$$F(a) = \sum_{\alpha, j, k} c_{j,k}^\alpha D_\alpha(a)_{jk}. \tag{A.29}$$

Theorem *The characters $\chi_\alpha(g)$ and $\chi_\beta(g)$ of representations of $D_\alpha(g)$ and $D_\beta(g)$ satisfy the orthogonality relation*

$$\sum_{g \in G} \chi_{D_\alpha}(g)^* \chi_{D_\beta}(g) = N_G \delta_{\alpha\beta}. \tag{A.30}$$

Proof From (A.28), we obtain

$$\sum_{g \in G} D_\alpha(g)_{ii} D_\beta^*(g)_{jj} = \frac{N_G}{d_\alpha} \delta_{\alpha\beta} \delta_{ij}. \tag{A.31}$$

Thus, by summing over all i and j, we obtain (A.30). □

A *class function* is defined as any function $F(a)$ of a which satisfies

$$F(g^{-1}ag) = F(a), \quad \forall g \in G. \tag{A.32}$$

Theorem *The number of irreducible representations is equal to the number of conjugacy classes.*

Proof The class function can also be expanded in terms of the matrix elements of the irreducible representations as in (A.29). It then follows that

$$F(a) = \frac{1}{N_G} \sum_{g \in G} F(g^{-1}ag)$$

$$= \frac{1}{N_G} \sum_{g \in G, \alpha, j, k} c^{\alpha}_{j,k} D_{\alpha}(g^{-1}ag)_{jk}$$

$$= \frac{1}{N_G} \sum_{g \in G, \alpha, j, k} c^{\alpha}_{j,k} \left(D_{\alpha}(g^{-1}) D_{\alpha}(a) D_{\alpha}(g) \right)_{jk}. \quad \text{(A.33)}$$

Using the orthogonality relation (A.28), we obtain

$$F(a) = \sum_{\alpha, j, \ell} \frac{1}{d_{\alpha}} c^{\alpha}_{j,j} D_{\alpha}(a)_{\ell\ell}$$

$$= \sum_{\alpha, j} \frac{1}{d_{\alpha}} c^{\alpha}_{j,j} \chi_{\alpha}(a). \quad \text{(A.34)}$$

That is, any class function $F(a)$, which is constant on conjugacy classes, can be expanded in terms of the characters $\chi_{\alpha}(a)$. This implies that the number of irreducible representations is equal to the number of conjugacy classes. □

Theorem *The characters satisfy the orthogonality relation*

$$\sum_{\alpha} \chi_{D_{\alpha}}(C_i)^* \chi_{D_{\alpha}}(C_j) = \frac{N_G}{n_i} \delta_{C_i C_j}, \quad \text{(A.35)}$$

where C_i and C_j denote the conjugacy classes and n_i is the number of elements in the conjugacy class C_i.

Proof We define the matrix $V_{i\alpha}$ by

$$V_{i\alpha} = \sqrt{\frac{n_i}{N_G}} \chi_{\alpha}(C_i), \quad \text{(A.36)}$$

where n_i is the number of elements in the conjugacy class C_i. Note that i and α label the conjugacy class C_i and the irreducible representation, respectively. The matrix $V_{i\alpha}$ is a square matrix because the number of irreducible representations is equal to the number of conjugacy classes. Using $V_{i\alpha}$, the orthogonality relation (A.30) can be rewritten as $V^{\dagger} V = 1$, that is, V is unitary. Thus, we also obtain $V V^{\dagger} = 1$. This is the content of (A.35). □

References

1. Hamermesh, M.: Group Theory and Its Application to Physical Problems. Addison-Wesley, Reading (1962)
2. Georgi, H.: Front. Phys. **54**, 1 (1982)
3. Ludl, P.O.: arXiv:0907.5587 [hep-ph]
4. Ramond, P.: Group Theory: A Physicist's Survey. Cambridge University Press, Cambridge (2010)

Appendix B
Representations of S_4 in Different Bases

For the group S_4, several bases of representations have been used in the literature. Most group-theoretical aspects such as conjugacy classes and characters are independent of the representation basis. Tensor products are also independent of the basis. For example, we always have

$$2 \otimes 2 = 1_1 \oplus 1_2 \oplus 2, \tag{B.1}$$

in any basis. However, the component form of this equation is basis-dependent. For example, the singlets 1_1 and 1_2 on the RHS are represented by components of 2 on the LHS, but their forms depend on the representation basis, as we shall see below. For applications, it is useful to show the transformation of bases and tensor products explicitly for several bases. This is what we shall do below.

B.1 Basis I

First, we examine the basis used in Sect. 3.2. All the elements of S_4 can be written as products of the generators b_1 and d_4, which satisfy

$$(b_1)^3 = (d_4)^4 = e, \qquad d_4 (b_1)^2 d_4 = b_1, \qquad d_4 b_1 d_4 = b_1 (d_4)^2 b_1. \tag{B.2}$$

These generators are represented on 2, 3, and $3'$ as follows:

$$b_1 = \begin{pmatrix} \omega & 0 \\ 0 & \omega^2 \end{pmatrix}, \quad d_4 = \begin{pmatrix} 0 & 1 \\ 1 & 0 \end{pmatrix}, \quad \text{on } 2, \tag{B.3}$$

$$b_1 = \begin{pmatrix} 0 & 0 & 1 \\ 1 & 0 & 0 \\ 0 & 1 & 0 \end{pmatrix}, \quad d_4 = \begin{pmatrix} -1 & 0 & 0 \\ 0 & 0 & -1 \\ 0 & 1 & 0 \end{pmatrix}, \quad \text{on } 3, \tag{B.4}$$

$$b_1 = \begin{pmatrix} 0 & 0 & 1 \\ 1 & 0 & 0 \\ 0 & 1 & 0 \end{pmatrix}, \quad d_4 = \begin{pmatrix} 1 & 0 & 0 \\ 0 & 0 & 1 \\ 0 & -1 & 0 \end{pmatrix}, \quad \text{on } 3'. \tag{B.5}$$

The multiplication rules are then obtained as follows:

$$\begin{pmatrix} a_1 \\ a_2 \end{pmatrix}_2 \otimes \begin{pmatrix} b_1 \\ b_2 \end{pmatrix}_2 = (a_1 b_2 + a_2 b_1)_1 \oplus (a_1 b_2 - a_2 b_1)_{1'} \oplus \begin{pmatrix} a_2 b_2 \\ a_1 b_1 \end{pmatrix}_2, \qquad (B.6)$$

$$\begin{pmatrix} a_1 \\ a_2 \end{pmatrix}_2 \otimes \begin{pmatrix} b_1 \\ b_2 \\ b_3 \end{pmatrix}_3 = \begin{pmatrix} a_1 b_1 + a_2 b_1 \\ \omega^2 a_1 b_2 + \omega a_2 b_2 \\ \omega a_1 b_3 + \omega^2 a_2 b_3 \end{pmatrix}_3 \oplus \begin{pmatrix} a_1 b_1 - a_2 b_1 \\ \omega^2 a_1 b_2 - \omega a_2 b_2 \\ \omega a_1 b_3 - \omega^2 a_2 b_3 \end{pmatrix}_{3'}, \qquad (B.7)$$

$$\begin{pmatrix} a_1 \\ a_2 \end{pmatrix}_2 \otimes \begin{pmatrix} b_1 \\ b_2 \\ b_3 \end{pmatrix}_{3'} = \begin{pmatrix} a_1 b_1 - a_2 b_1 \\ \omega^2 a_1 b_2 - \omega a_2 b_2 \\ \omega a_1 b_3 - \omega^2 a_2 b_3 \end{pmatrix}_3 \oplus \begin{pmatrix} a_1 b_1 + a_2 b_1 \\ \omega^2 a_1 b_2 + \omega a_2 b_2 \\ \omega a_1 b_3 + \omega^2 a_2 b_3 \end{pmatrix}_{3'}, \qquad (B.8)$$

$$\begin{pmatrix} a_1 \\ a_2 \\ a_3 \end{pmatrix}_3 \otimes \begin{pmatrix} b_1 \\ b_2 \\ b_3 \end{pmatrix}_3 = (a_1 b_1 + a_2 b_2 + a_3 b_3)_1 \oplus \begin{pmatrix} a_1 b_1 + \omega a_2 b_2 + \omega^2 a_3 b_3 \\ a_1 b_1 + \omega^2 a_2 b_2 + \omega a_3 b_3 \end{pmatrix}_2$$

$$\oplus \begin{pmatrix} a_2 b_3 + a_3 b_2 \\ a_3 b_1 + a_1 b_3 \\ a_1 b_2 + a_2 b_1 \end{pmatrix}_3 \oplus \begin{pmatrix} a_2 b_3 - a_3 b_2 \\ a_3 b_1 - a_1 b_3 \\ a_1 b_2 - a_2 b_1 \end{pmatrix}_{3'}, \qquad (B.9)$$

$$\begin{pmatrix} a_1 \\ a_2 \\ a_3 \end{pmatrix}_{3'} \otimes \begin{pmatrix} b_1 \\ b_2 \\ b_3 \end{pmatrix}_{3'} = (a_1 b_1 + a_2 b_2 + a_3 b_3)_1 \oplus \begin{pmatrix} a_1 b_1 + \omega a_2 b_2 + \omega^2 a_3 b_3 \\ a_1 b_1 + \omega^2 a_2 b_2 + \omega a_3 b_3 \end{pmatrix}_2$$

$$\oplus \begin{pmatrix} a_2 b_3 + a_3 b_2 \\ a_3 b_1 + a_1 b_3 \\ a_1 b_2 + a_2 b_1 \end{pmatrix}_3 \oplus \begin{pmatrix} a_2 b_3 - a_3 b_2 \\ a_3 b_1 - a_1 b_3 \\ a_1 b_2 - a_2 b_1 \end{pmatrix}_{3'}, \qquad (B.10)$$

$$\begin{pmatrix} a_1 \\ a_2 \\ a_3 \end{pmatrix}_3 \otimes \begin{pmatrix} b_1 \\ b_2 \\ b_3 \end{pmatrix}_{3'} = (a_1 b_1 + a_2 b_2 + a_3 b_3)_{1'} \oplus \begin{pmatrix} a_1 b_1 + \omega a_2 b_2 + \omega^2 a_3 b_3 \\ -a_1 b_1 - \omega^2 a_2 b_2 - \omega a_3 b_3 \end{pmatrix}_2$$

$$\oplus \begin{pmatrix} a_2 b_3 - a_3 b_2 \\ a_3 b_1 - a_1 b_3 \\ a_1 b_2 - a_2 b_1 \end{pmatrix}_3 \oplus \begin{pmatrix} a_2 b_3 + a_3 b_2 \\ a_3 b_1 + a_1 b_3 \\ a_1 b_2 + a_2 b_1 \end{pmatrix}_{3'}. \qquad (B.11)$$

B.2 Basis II

Next we consider another basis, which is used, e.g., in [1]. Following this reference, we define the generators b_1 and d_4 by $b = b_1$ and $a = d_4$. In this basis, the generators a and b are represented by

$$a = \begin{pmatrix} -1 & 0 \\ 0 & 1 \end{pmatrix}, \qquad b = -\frac{1}{2} \begin{pmatrix} 1 & \sqrt{3} \\ -\sqrt{3} & 1 \end{pmatrix}, \qquad \text{on } \mathbf{2}, \qquad (B.12)$$

B.2 Basis II

$$a = \begin{pmatrix} -1 & 0 & 0 \\ 0 & 0 & -1 \\ 0 & 1 & 0 \end{pmatrix}, \quad b = \begin{pmatrix} 0 & 0 & 1 \\ 1 & 0 & 0 \\ 0 & 1 & 0 \end{pmatrix}, \quad \text{on } 3_1, \quad (B.13)$$

$$a = \begin{pmatrix} 1 & 0 & 0 \\ 0 & 0 & 1 \\ 0 & -1 & 0 \end{pmatrix}, \quad b = \begin{pmatrix} 0 & 0 & 1 \\ 1 & 0 & 0 \\ 0 & 1 & 0 \end{pmatrix}, \quad \text{on } 3_2, \quad (B.14)$$

where we define $3_1 \equiv 3$ and $3_2 \equiv 3'$ hereafter. The generators a and b are represented in the real basis. On the other hand, the above generators b_1 and d_4 are represented in the complex basis. The bases for **2** transform by the unitary transformation $U^\dagger g U$, where

$$U = \frac{1}{\sqrt{2}} \begin{pmatrix} 1 & i \\ -1 & i \end{pmatrix}. \quad (B.15)$$

That is, the elements a and b are written in terms of b_1 and d_4 as

$$b = U^\dagger b_1 U = -\frac{1}{2} \begin{pmatrix} 1 & \sqrt{3} \\ -\sqrt{3} & 1 \end{pmatrix}, \quad a = U^\dagger d_4 U = \begin{pmatrix} -1 & 0 \\ 0 & 1 \end{pmatrix}, \quad (B.16)$$

in the real basis. For the triplets, the (b_1, d_4) basis is the same as the (b, a) basis.
The multiplication rules are thus obtained as follows:

$$\begin{pmatrix} a_1 \\ a_2 \end{pmatrix}_2 \otimes \begin{pmatrix} b_1 \\ b_2 \end{pmatrix}_2 = (a_1 b_1 + a_2 b_2)_{1_1} \oplus (-a_1 b_2 + a_2 b_1)_{1_2} \oplus \begin{pmatrix} a_1 b_2 + a_2 b_1 \\ a_1 b_1 - a_2 b_2 \end{pmatrix}_2,$$
$$(B.17)$$

$$\begin{pmatrix} a_1 \\ a_2 \end{pmatrix}_2 \otimes \begin{pmatrix} b_1 \\ b_2 \\ b_3 \end{pmatrix}_{3_1} = \begin{pmatrix} a_2 b_1 \\ -\frac{1}{2}(\sqrt{3}a_1 b_2 + a_2 b_2) \\ \frac{1}{2}(\sqrt{3}a_1 b_3 - a_2 b_3) \end{pmatrix}_{3_1} \oplus \begin{pmatrix} a_1 b_1 \\ \frac{1}{2}(\sqrt{3}a_2 b_2 - a_1 b_2) \\ -\frac{1}{2}(\sqrt{3}a_2 b_3 + a_1 b_3) \end{pmatrix}_{3_2},$$
$$(B.18)$$

$$\begin{pmatrix} a_1 \\ a_2 \end{pmatrix}_2 \otimes \begin{pmatrix} b_1 \\ b_2 \\ b_3 \end{pmatrix}_{3_2} = \begin{pmatrix} a_1 b_1 \\ \frac{1}{2}(\sqrt{3}a_2 b_2 - a_1 b_2) \\ -\frac{1}{2}(\sqrt{3}a_2 b_3 + a_1 b_3) \end{pmatrix}_{3_1} \oplus \begin{pmatrix} a_2 b_1 \\ -\frac{1}{2}(\sqrt{3}a_1 b_2 + a_2 b_2) \\ \frac{1}{2}(\sqrt{3}a_1 b_3 - a_2 b_3) \end{pmatrix}_{3_2},$$
$$(B.19)$$

$$\begin{pmatrix} a_1 \\ a_2 \\ a_3 \end{pmatrix}_{3_1} \otimes \begin{pmatrix} b_1 \\ b_2 \\ b_3 \end{pmatrix}_{3_1} = (a_1 b_1 + a_2 b_2 + a_3 b_3)_{1_1} \oplus \begin{pmatrix} \frac{1}{\sqrt{2}}(a_2 b_2 - a_3 b_3) \\ \frac{1}{\sqrt{6}}(-2a_1 b_1 + a_2 b_2 + a_3 b_3) \end{pmatrix}_2$$

$$\oplus \begin{pmatrix} a_2 b_3 + a_3 b_2 \\ a_1 b_3 + a_3 b_1 \\ a_1 b_2 + a_2 b_1 \end{pmatrix}_{3_1} \oplus \begin{pmatrix} a_3 b_2 - a_2 b_3 \\ a_1 b_3 - a_3 b_1 \\ a_2 b_1 - a_1 b_2 \end{pmatrix}_{3_2}, \quad (B.20)$$

$$\begin{pmatrix} a_1 \\ a_2 \\ a_3 \end{pmatrix}_{3_2} \otimes \begin{pmatrix} b_1 \\ b_2 \\ b_3 \end{pmatrix}_{3_2} = (a_1b_1 + a_2b_2 + a_3b_3)_{1_1} \oplus \begin{pmatrix} \frac{1}{\sqrt{2}}(a_2b_2 - a_3b_3) \\ \frac{1}{\sqrt{6}}(-2a_1b_1 + a_2b_2 + a_3b_3) \end{pmatrix}_{2}$$

$$\oplus \begin{pmatrix} a_2b_3 + a_3b_2 \\ a_1b_3 + a_3b_1 \\ a_1b_2 + a_2b_1 \end{pmatrix}_{3_1} \oplus \begin{pmatrix} a_3b_2 - a_2b_3 \\ a_1b_3 - a_3b_1 \\ a_2b_1 - a_1b_2 \end{pmatrix}_{3_2}, \quad \text{(B.21)}$$

$$\begin{pmatrix} a_1 \\ a_2 \\ a_3 \end{pmatrix}_{3_1} \otimes \begin{pmatrix} b_1 \\ b_2 \\ b_3 \end{pmatrix}_{3_2} = (a_1b_1 + a_2b_2 + a_3b_3)_{1_2} \oplus \begin{pmatrix} \frac{1}{\sqrt{6}}(2a_1b_1 - a_2b_2 - a_3b_3) \\ \frac{1}{\sqrt{2}}(a_2b_2 - a_3b_3) \end{pmatrix}_{2}$$

$$\oplus \begin{pmatrix} a_3b_2 - a_2b_3 \\ a_1b_3 - a_3b_1 \\ a_2b_1 - a_1b_2 \end{pmatrix}_{3_1} \oplus \begin{pmatrix} a_2b_3 + a_3b_2 \\ a_1b_3 + a_3b_1 \\ a_1b_2 + a_2b_1 \end{pmatrix}_{3_2}. \quad \text{(B.22)}$$

B.3 Basis III

Next, we consider a different basis, which is used, e.g., in [2], with the generators s and t corresponding to d_4 and b_1, respectively. These generators are represented as follows:

$$s = \begin{pmatrix} 0 & 1 \\ 1 & 0 \end{pmatrix}, \quad t = \begin{pmatrix} \omega & 0 \\ 0 & \omega^2 \end{pmatrix}, \quad \text{on } \mathbf{2}, \quad \text{(B.23)}$$

$$s = \frac{1}{3}\begin{pmatrix} -1 & 2\omega & 2\omega^2 \\ 2\omega & 2\omega^2 & -1 \\ 2\omega^2 & -1 & 2\omega \end{pmatrix}, \quad t = \begin{pmatrix} 1 & 0 & 0 \\ 0 & \omega^2 & 0 \\ 0 & 0 & \omega \end{pmatrix}, \quad \text{on } \mathbf{3}_1, \quad \text{(B.24)}$$

$$s = \frac{1}{3}\begin{pmatrix} 1 & -2\omega & -2\omega^2 \\ -2\omega & -2\omega^2 & 1 \\ -2\omega^2 & 1 & -2\omega \end{pmatrix}, \quad t = \begin{pmatrix} 1 & 0 & 0 \\ 0 & \omega^2 & 0 \\ 0 & 0 & \omega \end{pmatrix}, \quad \text{on } \mathbf{3}_2.$$

$$\text{(B.25)}$$

The doublet of this basis [2] is the same as the (d_4, b_1) basis. In the representations $\mathbf{3}_1$ and $\mathbf{3}_2$, the (s, t) basis and (d_4, b_1) basis are transformed by the unitary matrix

$$U_\omega = \frac{1}{\sqrt{3}}\begin{pmatrix} 1 & 1 & 1 \\ 1 & \omega & \omega^2 \\ 1 & \omega^2 & \omega \end{pmatrix}, \quad \text{(B.26)}$$

B.3 Basis III

which is the so-called magic matrix. That is, the elements s and t are obtained from d_4 and b_1 as follows:

$$s = U_\omega^\dagger d_4 U_\omega = \frac{1}{3} \begin{pmatrix} -1 & 2\omega & 2\omega^2 \\ 2\omega & 2\omega^2 & -1 \\ 2\omega^2 & -1 & 2\omega^2 \end{pmatrix}, \quad t = U_\omega^\dagger b_1 U_\omega = \begin{pmatrix} 1 & 0 & 0 \\ 0 & \omega^2 & 0 \\ 0 & 0 & \omega \end{pmatrix}. \tag{B.27}$$

For 3_2, we also find s and t in the same way.

The multiplication rules are then obtained as follows:

$$\begin{pmatrix} a_1 \\ a_2 \end{pmatrix}_2 \otimes \begin{pmatrix} b_1 \\ b_2 \end{pmatrix}_2 = (a_1 b_2 + a_2 b_1)_{1_1} \oplus (a_1 b_2 - a_2 b_1)_{1_2} \oplus \begin{pmatrix} a_2 b_2 \\ a_1 b_1 \end{pmatrix}_2, \tag{B.28}$$

$$\begin{pmatrix} a_1 \\ a_2 \end{pmatrix}_2 \otimes \begin{pmatrix} b_1 \\ b_2 \\ b_3 \end{pmatrix}_{3_1} = \begin{pmatrix} a_1 b_2 + a_2 b_3 \\ a_1 b_3 + a_2 b_1 \\ a_1 b_1 + a_2 b_2 \end{pmatrix}_{3_1} \oplus \begin{pmatrix} a_1 b_2 - a_2 b_3 \\ a_1 b_3 - a_2 b_1 \\ a_1 b_1 - a_2 b_2 \end{pmatrix}_{3_2}, \tag{B.29}$$

$$\begin{pmatrix} a_1 \\ a_2 \end{pmatrix}_2 \otimes \begin{pmatrix} b_1 \\ b_2 \\ b_3 \end{pmatrix}_{3_2} = \begin{pmatrix} a_1 b_2 - a_2 b_3 \\ a_1 b_3 - a_2 b_1 \\ a_1 b_1 - a_2 b_2 \end{pmatrix}_{3_1} \oplus \begin{pmatrix} a_1 b_2 + a_2 b_3 \\ a_1 b_3 + a_2 b_1 \\ a_1 b_1 + a_2 b_2 \end{pmatrix}_{3_2}, \tag{B.30}$$

$$\begin{pmatrix} a_1 \\ a_2 \\ a_3 \end{pmatrix}_{3_1} \otimes \begin{pmatrix} b_1 \\ b_2 \\ b_3 \end{pmatrix}_{3_1} = (a_1 b_1 + a_2 b_3 + a_3 b_2)_{1_1} \oplus \begin{pmatrix} a_2 b_2 + a_1 b_3 + a_3 b_1 \\ a_3 b_3 + a_1 b_2 + a_2 b_1 \end{pmatrix}_2$$

$$\oplus \begin{pmatrix} 2a_1 b_1 - a_2 b_3 - a_3 b_2 \\ 2a_3 b_3 - a_1 b_2 - a_2 b_1 \\ 2a_2 b_2 - a_1 b_3 - a_3 b_1 \end{pmatrix}_{3_1} \oplus \begin{pmatrix} a_2 b_3 - a_3 b_2 \\ a_1 b_2 - a_2 b_1 \\ a_3 b_1 - a_1 b_3 \end{pmatrix}_{3_2}, \tag{B.31}$$

$$\begin{pmatrix} a_1 \\ a_2 \\ a_3 \end{pmatrix}_{3_2} \otimes \begin{pmatrix} b_1 \\ b_2 \\ b_3 \end{pmatrix}_{3_2} = (a_1 b_1 + a_2 b_3 + a_3 b_2)_{1_1} \oplus \begin{pmatrix} a_2 b_2 + a_1 b_3 + a_3 b_1 \\ a_3 b_3 + a_1 b_2 + a_2 b_1 \end{pmatrix}_2$$

$$\oplus \begin{pmatrix} 2a_1 b_1 - a_2 b_3 - a_3 b_2 \\ 2a_3 b_3 - a_1 b_2 - a_2 b_1 \\ 2a_2 b_2 - a_1 b_3 - a_3 b_1 \end{pmatrix}_{3_1} \oplus \begin{pmatrix} a_2 b_3 - a_3 b_2 \\ a_1 b_2 - a_2 b_1 \\ a_3 b_1 - a_1 b_3 \end{pmatrix}_{3_2}, \tag{B.32}$$

$$\begin{pmatrix} a_1 \\ a_2 \\ a_3 \end{pmatrix}_{3_1} \otimes \begin{pmatrix} b_1 \\ b_2 \\ b_3 \end{pmatrix}_{3_2} = (a_1 b_1 + a_2 b_3 + a_3 b_2)_{1_2} \oplus \begin{pmatrix} a_2 b_2 + a_1 b_3 + a_3 b_1 \\ -a_3 b_3 - a_1 b_2 - a_2 b_1 \end{pmatrix}_2$$

$$\oplus \begin{pmatrix} a_2 b_3 - a_3 b_2 \\ a_1 b_2 - a_2 b_1 \\ a_3 b_1 - a_1 b_3 \end{pmatrix}_{3_1} \oplus \begin{pmatrix} 2a_1 b_1 - a_2 b_3 - a_3 b_2 \\ 2a_3 b_3 - a_1 b_2 - a_2 b_1 \\ 2a_2 b_2 - a_1 b_3 - a_3 b_1 \end{pmatrix}_{3_2}. \tag{B.33}$$

B.4 Basis IV

Here we consider another basis, which is used, e.g., in [3], with the generators $\tilde{t}$ and $\tilde{s}$ satisfying

$$\tilde{t}^4 = \tilde{s}^2 = (\tilde{s}\tilde{t})^3 = (\tilde{t}\tilde{s})^3 = e. \tag{B.34}$$

These generators are represented by

$$\tilde{t} = \begin{pmatrix} 1 & 0 \\ 0 & -1 \end{pmatrix}, \quad \tilde{s} = \frac{1}{2}\begin{pmatrix} -1 & \sqrt{3} \\ \sqrt{3} & 1 \end{pmatrix}, \quad \tilde{s}\tilde{t} = \frac{1}{2}\begin{pmatrix} -1 & -\sqrt{3} \\ \sqrt{3} & -1 \end{pmatrix}, \quad \text{on } \mathbf{2}, \tag{B.35}$$

$$\tilde{t} = \begin{pmatrix} -1 & 0 & 0 \\ 0 & -i & 0 \\ 0 & 0 & i \end{pmatrix}, \quad \tilde{s} = \begin{pmatrix} 0 & -\frac{1}{\sqrt{2}} & -\frac{1}{\sqrt{2}} \\ -\frac{1}{\sqrt{2}} & \frac{1}{2} & -\frac{1}{2} \\ -\frac{1}{\sqrt{2}} & -\frac{1}{2} & \frac{1}{2} \end{pmatrix},$$

$$\tilde{s}\tilde{t} = \begin{pmatrix} 0 & \frac{i}{\sqrt{2}} & -\frac{i}{\sqrt{2}} \\ \frac{1}{\sqrt{2}} & -\frac{i}{2} & -\frac{i}{2} \\ \frac{1}{\sqrt{2}} & \frac{i}{2} & \frac{i}{2} \end{pmatrix}, \tag{B.36}$$

on $\mathbf{3}_1$, and

$$\tilde{t} = \begin{pmatrix} 1 & 0 & 0 \\ 0 & i & 0 \\ 0 & 0 & -i \end{pmatrix}, \quad \tilde{s} = \begin{pmatrix} 0 & \frac{1}{\sqrt{2}} & \frac{1}{\sqrt{2}} \\ \frac{1}{\sqrt{2}} & -\frac{1}{2} & \frac{1}{2} \\ \frac{1}{\sqrt{2}} & \frac{1}{2} & -\frac{1}{2} \end{pmatrix}, \quad \tilde{s}\tilde{t} = \begin{pmatrix} 0 & \frac{i}{\sqrt{2}} & -\frac{i}{\sqrt{2}} \\ \frac{1}{\sqrt{2}} & -\frac{i}{2} & -\frac{i}{2} \\ \frac{1}{\sqrt{2}} & \frac{i}{2} & \frac{i}{2} \end{pmatrix}, \tag{B.37}$$

on $\mathbf{3}_2$.

For the representation $\mathbf{2}$, the unitary transformation matrix U_{doublet} given by

$$U_{\text{doublet}} = \frac{1}{\sqrt{2}}\begin{pmatrix} 1 & i \\ 1 & -i \end{pmatrix} \tag{B.38}$$

is used and the elements $\tilde{t}$ and $\tilde{s}\tilde{t}$ are given in terms of d_1 and b_1 by

$$\tilde{t} = U_{\text{doublet}}^\dagger d_4 U_{\text{doublet}} = \begin{pmatrix} 1 & 0 \\ 0 & -1 \end{pmatrix},$$

$$\tilde{s}\tilde{t} = U_{\text{doublet}}^\dagger b_1 U_{\text{doublet}} = \frac{1}{2}\begin{pmatrix} -1 & -\sqrt{3} \\ \sqrt{3} & -1 \end{pmatrix}. \tag{B.39}$$

On the other hand, for the representations $\mathbf{3}_1$ and $\mathbf{3}_2$, the unitary transformation matrix U_{triplet} given by

$$U_{\text{triplet}} = \begin{pmatrix} 1 & 0 & 0 \\ 0 & \frac{1}{\sqrt{2}} & \frac{1}{\sqrt{2}} \\ 0 & \frac{i}{\sqrt{2}} & -\frac{i}{\sqrt{2}} \end{pmatrix} \tag{B.40}$$

is used.

B.4 Basis IV

For 3_1, the elements $\tilde{t}$ and $\tilde{st}$ are given in terms of d_4 and b_1 by

$$\tilde{t} = U^\dagger_{\text{triplet}} d_4 U_{\text{triplet}} = \begin{pmatrix} -1 & 0 & 0 \\ 0 & -i & 0 \\ 0 & 0 & i \end{pmatrix},$$

$$\tilde{st} = U^\dagger_{\text{triplet}} b_1 U_{\text{triplet}} = \begin{pmatrix} 0 & \frac{i}{\sqrt{2}} & -\frac{i}{\sqrt{2}} \\ \frac{1}{\sqrt{2}} & -\frac{i}{2} & -\frac{i}{2} \\ \frac{1}{\sqrt{2}} & \frac{i}{2} & \frac{i}{2} \end{pmatrix}.$$
(B.41)

For 3_2, we find the same transformations.
The multiplication rules are as follows:

$$\begin{pmatrix} a_1 \\ a_2 \end{pmatrix}_2 \otimes \begin{pmatrix} b_1 \\ b_2 \end{pmatrix}_2 = (a_1 b_1 + a_2 b_2)_{1_1} \oplus (a_1 b_2 - a_2 b_1)_{1_2} \oplus \begin{pmatrix} a_2 b_2 - a_1 b_1 \\ a_1 b_2 + a_2 b_1 \end{pmatrix}_2,$$
(B.42)

$$\begin{pmatrix} a_1 \\ a_2 \end{pmatrix}_2 \otimes \begin{pmatrix} b_1 \\ b_2 \\ b_3 \end{pmatrix}_{3_1} = \begin{pmatrix} a_1 b_1 \\ \frac{\sqrt{3}}{2} a_2 b_3 - \frac{1}{2} a_1 b_2 \\ \frac{\sqrt{3}}{2} a_2 b_2 - \frac{1}{2} a_1 b_3 \end{pmatrix}_{3_1} \oplus \begin{pmatrix} -a_2 b_1 \\ \frac{\sqrt{3}}{2} a_1 b_3 + \frac{1}{2} a_2 b_2 \\ \frac{\sqrt{3}}{2} a_1 b_2 + \frac{1}{2} a_2 b_3 \end{pmatrix}_{3_2},$$ (B.43)

$$\begin{pmatrix} a_1 \\ a_2 \end{pmatrix}_2 \otimes \begin{pmatrix} b_1 \\ b_2 \\ b_3 \end{pmatrix}_{3_2} = \begin{pmatrix} -a_2 b_1 \\ \frac{\sqrt{3}}{2} a_1 b_3 + \frac{1}{2} a_2 b_2 \\ \frac{\sqrt{3}}{2} a_1 b_2 + \frac{1}{2} a_2 b_3 \end{pmatrix}_{3_1} \oplus \begin{pmatrix} a_1 b_1 \\ \frac{\sqrt{3}}{2} a_2 b_3 - \frac{1}{2} a_1 b_2 \\ \frac{\sqrt{3}}{2} a_2 b_2 - \frac{1}{2} a_1 b_3 \end{pmatrix}_{3_2},$$ (B.44)

$$\begin{pmatrix} a_1 \\ a_2 \\ a_3 \end{pmatrix}_{3_1} \otimes \begin{pmatrix} b_1 \\ b_2 \\ b_3 \end{pmatrix}_{3_1} = (a_1 b_1 + a_2 b_3 + a_3 b_2)_{1_1} \oplus \begin{pmatrix} a_1 b_1 - \frac{1}{2}(a_2 b_3 + a_3 b_2) \\ \frac{\sqrt{3}}{2}(a_2 b_2 + a_3 b_3) \end{pmatrix}_2$$

$$\oplus \begin{pmatrix} a_3 b_3 - a_2 b_2 \\ a_1 b_3 + a_3 b_1 \\ -a_1 b_2 - a_2 b_1 \end{pmatrix}_{3_1} \oplus \begin{pmatrix} a_3 b_2 - a_2 b_3 \\ a_2 b_1 - a_1 b_2 \\ a_1 b_3 - a_3 b_1 \end{pmatrix}_{3_2},$$ (B.45)

$$\begin{pmatrix} a_1 \\ a_2 \\ a_3 \end{pmatrix}_{3_2} \otimes \begin{pmatrix} b_1 \\ b_2 \\ b_3 \end{pmatrix}_{3_2} = (a_1 b_1 + a_2 b_3 + a_3 b_2)_{1_1} \oplus \begin{pmatrix} a_1 b_1 - \frac{1}{2}(a_2 b_3 + a_3 b_2) \\ \frac{\sqrt{3}}{2}(a_2 b_2 + a_3 b_3) \end{pmatrix}_2$$

$$\oplus \begin{pmatrix} a_3 b_3 - a_2 b_2 \\ a_1 b_3 + a_3 b_1 \\ -a_1 b_2 - a_2 b_1 \end{pmatrix}_{3_1} \oplus \begin{pmatrix} a_3 b_2 - a_2 b_3 \\ a_2 b_1 - a_1 b_2 \\ a_1 b_3 - a_3 b_1 \end{pmatrix}_{3_2},$$ (B.46)

$$\begin{pmatrix} a_1 \\ a_2 \\ a_3 \end{pmatrix}_{3_1} \otimes \begin{pmatrix} b_1 \\ b_2 \\ b_3 \end{pmatrix}_{3_2} = (a_1 b_1 + a_2 b_3 + a_3 b_2)_{1_2}$$

$$\oplus \begin{pmatrix} \frac{\sqrt{3}}{2}(a_2 b_2 + a_3 b_3) \\ -a_1 b_1 + \frac{1}{2}(a_2 b_3 + a_3 b_2) \end{pmatrix}_2$$

$$\oplus \begin{pmatrix} a_3b_2 - a_2b_3 \\ a_2b_1 - a_1b_2 \\ a_1b_3 - a_3b_1 \end{pmatrix}_{3_1} \oplus \begin{pmatrix} a_3b_3 - a_2b_2 \\ a_1b_3 + a_3b_1 \\ -a_1b_2 - a_2b_1 \end{pmatrix}_{3_2} . \tag{B.47}$$

References

1. Hagedorn, C., Lindner, M., Mohapatra, R.N.: J. High Energy Phys. **0606**, 042 (2006). arXiv:hep-ph/0602244
2. Bazzocchi, F., Merlo, L., Morisi, S.: Nucl. Phys. B **816**, 204 (2009). arXiv:0901.2086 [hep-ph]
3. Altarelli, G., Feruglio, F., Merlo, L.: J. High Energy Phys. **0905**, 020 (2009). arXiv:0903.1940 [hep-ph]

Appendix C
Representations of A_4 in Different Bases

C.1 Basis I

In Sect. C.2, we discuss another basis for representation of the A_4 group. But first consider the basis used in Sect. 4.1. All elements of A_4 can be written as products of the generators s and t, which satisfy

$$s^2 = t^3 = (st)^3 = e. \tag{C.1}$$

On the representation **3**, these generators are represented by

$$s = a_2 = \begin{pmatrix} 1 & 0 & 0 \\ 0 & -1 & 0 \\ 0 & 0 & -1 \end{pmatrix}, \quad t = b_1 = \begin{pmatrix} 0 & 0 & 1 \\ 1 & 0 & 0 \\ 0 & 1 & 0 \end{pmatrix}. \tag{C.2}$$

The multiplication rule of the triplet is thus

$$\begin{pmatrix} a_1 \\ a_2 \\ a_3 \end{pmatrix}_3 \otimes \begin{pmatrix} b_1 \\ b_2 \\ b_3 \end{pmatrix}_3 = (a_1 b_1 + a_2 b_2 + a_3 b_3)_1 \oplus (a_1 b_1 + \omega a_2 b_2 + \omega^2 a_3 b_3)_{1'}$$

$$\oplus (a_1 b_1 + \omega^2 a_2 b_2 + \omega a_3 b_3)_{1''}$$

$$\oplus \begin{pmatrix} a_2 b_3 + a_3 b_2 \\ a_3 b_1 + a_1 b_3 \\ a_1 b_2 + a_2 b_1 \end{pmatrix}_3 \oplus \begin{pmatrix} a_2 b_3 - a_3 b_2 \\ a_3 b_1 - a_1 b_3 \\ a_1 b_2 - a_2 b_1 \end{pmatrix}_3. \tag{C.3}$$

C.2 Basis II

Next we consider another basis, used, e.g., in [1]. In this basis, we denote the generators by a and b, which correspond to s and t, respectively, and these generators are represented as follows:

$$a = \frac{1}{3}\begin{pmatrix} -1 & 2 & 2 \\ 2 & -1 & 2 \\ 2 & 2 & -1 \end{pmatrix}, \quad b = \begin{pmatrix} 1 & 0 & 0 \\ 0 & \omega^2 & 0 \\ 0 & 0 & \omega \end{pmatrix}, \quad (C.4)$$

on the representation **3**. These bases are transformed by the unitary transformation matrix U_ω given by

$$U_\omega = \frac{1}{\sqrt{3}}\begin{pmatrix} 1 & 1 & 1 \\ 1 & \omega & \omega^2 \\ 1 & \omega^2 & \omega \end{pmatrix}, \quad (C.5)$$

and the elements a and b are written as

$$a = U_\omega^\dagger s U_\omega = \frac{1}{3}\begin{pmatrix} -1 & 2 & 2 \\ 2 & -1 & 2 \\ 2 & 2 & -1 \end{pmatrix}, \quad b = U_\omega^\dagger t U_\omega = \begin{pmatrix} 1 & 0 & 0 \\ 0 & \omega^2 & 0 \\ 0 & 0 & \omega \end{pmatrix}. \quad (C.6)$$

Therefore, the multiplication rule of the triplet is

$$\begin{pmatrix} a_1 \\ a_2 \\ a_3 \end{pmatrix}_3 \otimes \begin{pmatrix} b_1 \\ b_2 \\ b_3 \end{pmatrix}_3 = (a_1 b_1 + a_2 b_3 + a_3 b_2)_1 \oplus (a_3 b_3 + a_1 b_2 + a_2 b_1)_{1'}$$

$$\oplus (a_2 b_2 + a_1 b_3 + a_3 b_1)_{1''}$$

$$\oplus \frac{1}{3}\begin{pmatrix} 2a_1 b_1 - a_2 b_3 - a_3 b_2 \\ 2a_3 b_3 - a_1 b_2 - a_2 b_1 \\ 2a_2 b_2 - a_1 b_3 - a_3 b_1 \end{pmatrix}_3 \oplus \frac{1}{2}\begin{pmatrix} a_2 b_3 - a_3 b_2 \\ a_1 b_2 - a_2 b_1 \\ a_1 b_3 - a_3 b_1 \end{pmatrix}_3.$$

(C.7)

References

1. Altarelli, G., Feruglio, F.: Nucl. Phys. B **741**, 215 (2006). arXiv:hep-ph/0512103

Appendix D
Representations of A_5 in Different Bases

D.1 Basis I

In Sect. D.2, we show another basis for representations of the A_5 group, but first, let us discuss the basis used in Sect. 4.2. All the elements of A_5 can be written as products of the generators s and t, which satisfy

$$s^2 = t^5 = \left(t^2 st^3 st^{-1} stst^{-1}\right)^3 = e. \tag{D.1}$$

The generators s and t are represented as follows [1]:

$$s = \frac{1}{2}\begin{pmatrix} -1 & \phi & \frac{1}{\phi} \\ \phi & \frac{1}{\phi} & 1 \\ \frac{1}{\phi} & 1 & -\phi \end{pmatrix}, \quad t = \frac{1}{2}\begin{pmatrix} 1 & \phi & \frac{1}{\phi} \\ -\phi & \frac{1}{\phi} & 1 \\ \frac{1}{\phi} & -1 & \phi \end{pmatrix}, \quad \text{on } \mathbf{3}, \tag{D.2}$$

$$s = \frac{1}{2}\begin{pmatrix} -\phi & \frac{1}{\phi} & 1 \\ \frac{1}{\phi} & -1 & \phi \\ 1 & \phi & \frac{1}{\phi} \end{pmatrix}, \quad t = \frac{1}{2}\begin{pmatrix} -\phi & -\frac{1}{\phi} & 1 \\ \frac{1}{\phi} & 1 & \phi \\ -1 & \phi & -\frac{1}{\phi} \end{pmatrix}, \quad \text{on } \mathbf{3'}, \tag{D.3}$$

$$s = \frac{1}{4}\begin{pmatrix} -1 & -1 & -3 & -\sqrt{5} \\ -1 & 3 & 1 & -\sqrt{5} \\ -3 & 1 & -1 & \sqrt{5} \\ -\sqrt{5} & -\sqrt{5} & \sqrt{5} & -1 \end{pmatrix},$$

$$t = \frac{1}{4}\begin{pmatrix} -1 & 1 & -3 & \sqrt{5} \\ -1 & -3 & 1 & \sqrt{5} \\ 3 & 1 & 1 & \sqrt{5} \\ \sqrt{5} & -\sqrt{5} & -\sqrt{5} & -1 \end{pmatrix}, \quad \text{on } \mathbf{4}, \tag{D.4}$$

$$s = \frac{1}{2}\begin{pmatrix} \frac{1-3\phi}{4} & \frac{\phi^2}{2} & -\frac{1}{2\phi^2} & \frac{\sqrt{5}}{2} & \frac{\sqrt{3}}{4\phi} \\ \frac{\phi^2}{2} & 1 & 1 & 0 & \frac{\sqrt{3}}{2\phi} \\ -\frac{1}{2\phi^2} & 1 & 0 & -1 & -\frac{\sqrt{3}\phi}{2} \\ \frac{\sqrt{5}}{2} & 0 & -1 & 1 & -\frac{\sqrt{3}}{2} \\ \frac{\sqrt{3}}{4\phi} & \frac{\sqrt{3}}{2\phi} & -\frac{\sqrt{3}\phi}{2} & -\frac{\sqrt{3}}{2} & \frac{3\phi-1}{4} \end{pmatrix},$$

(D.5)

$$t = \frac{1}{2}\begin{pmatrix} \frac{1-3\phi}{4} & -\frac{\phi^2}{2} & -\frac{1}{2\phi^2} & -\frac{\sqrt{5}}{2} & \frac{\sqrt{3}}{4\phi} \\ \frac{\phi^2}{2} & -1 & 1 & 0 & \frac{\sqrt{3}}{2\phi} \\ \frac{1}{2\phi^2} & 1 & 0 & -1 & \frac{\sqrt{3}\phi}{2} \\ -\frac{\sqrt{5}}{2} & 0 & 1 & 1 & \frac{\sqrt{3}}{2} \\ \frac{\sqrt{3}}{4\phi} & -\frac{\sqrt{3}}{2\phi} & -\frac{\sqrt{3}\phi}{2} & \frac{\sqrt{3}}{2} & \frac{3\phi-1}{4} \end{pmatrix},$$

on **5**, where $\phi = (1+\sqrt{5})/2$.

The tensor products decompose as follows:

$$\begin{pmatrix} x_1 \\ x_2 \\ x_3 \end{pmatrix}_3 \otimes \begin{pmatrix} y_1 \\ y_2 \\ y_3 \end{pmatrix}_3 = (x_1 y_1 + x_2 y_2 + x_3 y_3)_1 \oplus \begin{pmatrix} x_3 y_2 - x_2 y_3 \\ x_1 y_3 - x_3 y_1 \\ x_2 y_1 - x_1 y_2 \end{pmatrix}_3$$

$$\oplus \begin{pmatrix} x_2 y_2 - x_1 y_1 \\ x_2 y_1 + x_1 y_2 \\ x_3 y_2 + x_2 y_3 \\ x_1 y_3 + x_3 y_1 \\ -\frac{1}{\sqrt{3}}(x_1 y_1 + x_2 y_2 - 2 x_3 y_3) \end{pmatrix}_5,$$

(D.6)

$$\begin{pmatrix} x_1 \\ x_2 \\ x_3 \end{pmatrix}_{3'} \otimes \begin{pmatrix} y_1 \\ y_2 \\ y_3 \end{pmatrix}_{3'} = (x_1 y_1 + x_2 y_2 + x_3 y_3)_1 \oplus \begin{pmatrix} x_3 y_2 - x_2 y_3 \\ x_1 y_3 - x_3 y_1 \\ x_2 y_1 - x_1 y_2 \end{pmatrix}_{3'}$$

$$\oplus \begin{pmatrix} \frac{1}{2}(-\frac{1}{\phi}x_1 y_1 - \phi x_2 y_2 + \sqrt{5} x_3 y_3) \\ x_2 y_1 + x_1 y_2 \\ -(x_3 y_1 + x_1 y_3) \\ x_2 y_3 + x_3 y_2 \\ \frac{1}{2\sqrt{3}}[(1-3\phi)x_1 y_1 + (3\phi-2)x_2 y_2 + x_3 y_3] \end{pmatrix}_5,$$

(D.7)

$$\begin{pmatrix} x_1 \\ x_2 \\ x_3 \end{pmatrix}_3 \otimes \begin{pmatrix} y_1 \\ y_2 \\ y_3 \end{pmatrix}_{3'} = \begin{pmatrix} \frac{1}{\phi} x_3 y_2 - \phi x_1 y_3 \\ \phi x_3 y_1 + \frac{1}{\phi} x_2 y_3 \\ -\frac{1}{\phi} x_1 y_1 + \phi x_2 y_2 \\ x_2 y_1 - x_1 y_2 + x_3 y_3 \end{pmatrix}_4,$$

D.1 Basis I

$$\oplus \begin{pmatrix} \frac{1}{2}\left(\phi^2 x_2 y_1 + \frac{1}{\phi^2} x_1 y_2 - \sqrt{5} x_3 y_3\right) \\ -\left(\phi x_1 y_1 + \frac{1}{\phi} x_2 y_2\right) \\ \frac{1}{\phi} x_3 y_1 - \phi x_2 y_3 \\ \phi x_3 y_2 + \frac{1}{\phi} x_1 y_3 \\ \frac{\sqrt{3}}{2}\left(\frac{1}{\phi} x_2 y_1 + \phi x_1 y_2 + x_3 y_3\right) \end{pmatrix}_5, \quad (D.8)$$

$$\begin{pmatrix} x_1 \\ x_2 \\ x_3 \end{pmatrix}_3 \otimes \begin{pmatrix} y_1 \\ y_2 \\ y_3 \\ y_4 \end{pmatrix}_4 = \begin{pmatrix} -\frac{1}{\phi^2} x_1 y_3 + \frac{1}{\phi} x_2 y_4 + x_3 y_2 \\ -\frac{1}{\phi} x_1 y_4 + x_2 y_3 + \frac{1}{\phi^2} x_3 y_1 \\ -x_1 y_1 + \frac{1}{\phi^2} x_2 y_2 + \frac{1}{\phi} x_3 y_4 \end{pmatrix}_{3'}$$

$$\oplus \begin{pmatrix} -x_1 y_3 + x_2 y_4 - x_3 y_2 \\ -x_1 y_4 - x_2 y_3 + x_3 y_1 \\ x_1 y_1 + x_2 y_2 + x_3 y_4 \\ x_1 y_2 - x_2 y_1 - x_3 y_3 \end{pmatrix}_4$$

$$\oplus \begin{pmatrix} \frac{1}{2}\left[(6\phi + 5) x_1 y_2 + (3\phi + 4) x_2 y_1 + (3\phi + 1) x_3 y_3\right] \\ -x_1 y_1 + (3\phi + 2) x_2 y_2 - (3\phi + 1) x_3 y_4 \\ -(3\phi + 1) x_1 y_4 - x_2 y_3 - (3\phi + 2) x_3 y_1 \\ -(3\phi + 2) x_1 y_3 - (3\phi + 1) x_2 y_4 + x_3 y_2 \\ \frac{\sqrt{3}}{2}\left[x_1 y_2 - (3\phi + 2) x_2 y_1 + 3(\phi + 1) x_3 y_3\right] \end{pmatrix}_5, \quad (D.9)$$

$$\begin{pmatrix} x_1 \\ x_2 \\ x_3 \end{pmatrix}_{3'} \otimes \begin{pmatrix} y_1 \\ y_2 \\ y_3 \\ y_4 \end{pmatrix}_4 = \begin{pmatrix} x_1 y_3 + \phi x_2 y_4 + \phi^2 x_3 y_1 \\ -\phi x_1 y_4 - \phi^2 x_2 y_3 - x_3 y_2 \\ -\phi^2 x_1 y_2 - x_2 y_1 - \phi x_3 y_4 \end{pmatrix}_3$$

$$\oplus \begin{pmatrix} x_1 y_4 - x_2 y_3 + x_3 y_2 \\ x_1 y_3 + x_2 y_4 - x_3 y_1 \\ -x_1 y_2 + x_2 y_1 + x_3 y_4 \\ -(x_1 y_1 + x_2 y_2 + x_3 y_3) \end{pmatrix}_4$$

$$\oplus \begin{pmatrix} x_1 y_1 - \phi^4 x_2 y_2 + \phi^2 (2\phi - 1) x_3 y_3 \\ x_1 y_2 - \phi^4 x_2 y_1 + \phi^2 (2\phi - 1) x_3 y_4 \\ \phi^4 x_1 y_3 - \phi^2 (2\phi - 1) x_2 y_4 + x_3 y_1 \\ \phi^2 (2\phi - 1) x_1 y_4 - x_2 y_3 - \phi^4 x_3 y_2 \\ -\sqrt{3}\phi\left(\phi^2 x_1 y_1 - x_2 y_2 - \phi x_3 y_3\right) \end{pmatrix}_5, \quad (D.10)$$

$$\begin{pmatrix} x_1 \\ x_2 \\ x_3 \end{pmatrix}_3 \otimes \begin{pmatrix} y_1 \\ y_2 \\ y_3 \\ y_4 \\ y_5 \end{pmatrix}_5 = \begin{pmatrix} x_1\left(y_1 + \frac{1}{\sqrt{3}} y_5\right) - x_2 y_2 - x_3 y_4 \\ -x_1 y_2 - x_2\left(y_1 - \frac{1}{\sqrt{3}} y_5\right) - x_3 y_3 \\ -x_1 y_4 - x_2 y_3 - \frac{2}{\sqrt{3}} x_3 y_5 \end{pmatrix}_3$$

$$\oplus \begin{pmatrix} x_1 y_2 - \frac{\phi}{2} x_2 y_1 - \frac{\sqrt{3}}{2\phi^2} x_2 y_5 - \frac{1}{\phi^2} x_3 y_3 \\ -\frac{\sqrt{3}}{2} x_1 y_5 - \frac{1}{2\phi^3} x_1 y_1 + \frac{1}{\phi^2} x_2 y_2 - x_3 y_4 \\ -\frac{1}{\phi^2} x_1 y_4 + x_2 y_3 + \frac{\sqrt{5}}{2\phi} x_3 y_1 - \frac{\sqrt{3}}{2\phi} x_3 y_5 \end{pmatrix}_{3'}$$

$$\oplus \begin{pmatrix} \frac{1}{\phi^2} x_1 y_2 + \frac{\phi^2-6}{2} x_2 y_1 + \frac{\sqrt{3}}{2} \phi^2 x_2 y_5 + \phi^2 x_3 y_3 \\ -\frac{\phi+4}{2} x_1 y_1 - \frac{\sqrt{3}}{2\phi^2} x_1 y_5 - \phi^2 x_2 y_2 - \frac{1}{\phi^2} x_3 y_4 \\ \phi^2 x_1 y_4 + \frac{1}{\phi^2} x_2 y_3 - \frac{\sqrt{5}}{2} x_3 y_1 - \frac{3\sqrt{3}}{2} x_3 y_5 \\ \sqrt{5}(x_1 y_3 + x_2 y_4 + x_3 y_2) \end{pmatrix}_4$$

$$\oplus \begin{pmatrix} x_1 y_3 + x_2 y_4 - 2 x_3 y_2 \\ x_1 y_4 - x_2 y_3 + 2 x_3 y_1 \\ -x_1 y_1 + x_2 y_2 - x_3 y_4 + \sqrt{3} x_1 y_5 \\ -x_1 y_2 - x_2 y_1 + x_3 y_3 - \sqrt{3} x_2 y_5 \\ -\sqrt{3}(x_1 y_3 - x_2 y_4) \end{pmatrix}_5 , \qquad (D.11)$$

$$\begin{pmatrix} x_1 \\ x_2 \\ x_3 \end{pmatrix}_{3'} \otimes \begin{pmatrix} y_1 \\ y_2 \\ y_3 \\ y_4 \\ y_5 \end{pmatrix}_5 = \begin{pmatrix} -\phi^2 x_1 y_2 + \frac{1}{2\phi} x_2 y_1 + \frac{\sqrt{3}}{2} \phi^2 x_2 y_5 + x_3 y_4 \\ \frac{2\phi+1}{2} x_1 y_1 + \frac{\sqrt{3}}{2} x_1 y_5 - x_2 y_2 - \phi^2 x_3 y_3 \\ x_1 y_3 + \phi^2 x_2 y_4 - \frac{\sqrt{5}}{2} \phi x_3 y_1 + \frac{\sqrt{3}}{2} \phi x_3 y_5 \end{pmatrix}_3$$

$$\oplus \begin{pmatrix} \frac{1}{2\phi} x_1 y_1 - x_2 y_2 + x_3 y_3 + \frac{3\phi-1}{2\sqrt{3}} x_1 y_5 \\ -x_1 y_2 + \frac{\phi}{2} x_2 y_1 - x_3 y_4 - \frac{3\phi-2}{2\sqrt{3}} x_2 y_5 \\ x_1 y_3 - x_2 y_4 - \frac{\sqrt{5}}{2} x_3 y_1 - \frac{1}{2\sqrt{3}} x_3 y_5 \end{pmatrix}_{3'}$$

$$\oplus \begin{pmatrix} \frac{1}{\sqrt{5}} \left(\frac{1}{\phi^2} x_1 y_1 + \phi^2 x_2 y_2 + \frac{1}{\phi^2} x_3 y_3 - \sqrt{3} \phi x_1 y_5 \right) \\ \frac{1}{\sqrt{5}} \left(-\frac{1}{\phi^2} x_1 y_2 - \phi^2 x_2 y_1 - \frac{\sqrt{3}(\phi-3)}{\sqrt{5}} x_2 y_5 - \phi^2 x_3 y_4 \right) \\ \frac{1}{\sqrt{5}} \left(\phi^2 x_1 y_3 - \frac{1}{\phi^2} x_2 y_4 + \sqrt{5} x_3 y_1 + \sqrt{3} x_3 y_5 \right) \\ x_1 y_4 - x_2 y_3 + x_3 y_2 \end{pmatrix}_4$$

$$\oplus \begin{pmatrix} -(3\phi-1) x_1 y_4 + (2-3\phi) x_2 y_3 + x_3 y_2 \\ -2 x_1 y_3 - 2 x_2 y_4 - x_3 y_1 + \sqrt{15} x_3 y_5 \\ 2 x_1 y_2 - (2-3\phi) x_2 y_1 - 2 x_3 y_4 + \sqrt{3} \phi x_2 y_5 \\ (3\phi-1) x_1 y_1 + 2 x_2 y_2 + 2 x_3 y_3 - \frac{\sqrt{3}}{\phi} x_1 y_5 \\ \frac{\sqrt{3}}{\phi} x_1 y_4 - \phi \sqrt{3} x_2 y_3 - \sqrt{15} x_3 y_2 \end{pmatrix}_5 , (D.12)$$

$$\begin{pmatrix} x_1 \\ x_2 \\ x_3 \\ x_4 \end{pmatrix}_4 \otimes \begin{pmatrix} y_1 \\ y_2 \\ y_3 \\ y_4 \end{pmatrix}_4 = (x_1 y_1 + x_2 y_2 + x_3 y_3 + x_4 y_4)_{\mathbf{1}}$$

D.1 Basis I   251

$$\oplus \begin{pmatrix} x_1y_3 + x_2y_4 - x_3y_1 - x_4y_2 \\ -x_1y_4 + x_2y_3 - x_3y_2 + x_4y_1 \\ x_1y_2 - x_2y_1 - x_3y_4 + x_4y_3 \end{pmatrix}_3$$

$$\oplus \begin{pmatrix} x_1y_4 + x_2y_3 - x_3y_2 - x_4y_1 \\ -x_1y_3 + x_2y_4 + x_3y_1 - x_4y_2 \\ x_1y_2 - x_2y_1 + x_3y_4 - x_4y_3 \end{pmatrix}_{3'}$$

$$\oplus \begin{pmatrix} x_1y_4 - \sqrt{5}x_2y_3 - \sqrt{5}x_3y_2 + x_4y_1 \\ -\sqrt{5}x_1y_3 + x_2y_4 - \sqrt{5}x_3y_1 + x_4y_2 \\ -\sqrt{5}x_1y_2 - \sqrt{5}x_2y_1 + x_3y_4 + x_4y_3 \\ x_1y_1 + x_2y_2 + x_3y_3 - 3x_4y_4 \end{pmatrix}_4$$

$$\oplus \begin{pmatrix} -\tfrac{\phi^2}{\sqrt{5}}x_1y_1 + \tfrac{1}{\sqrt{5}\phi^2}x_2y_2 + x_3y_3 \\ -\tfrac{1}{\sqrt{5}}x_1y_2 - \tfrac{1}{\sqrt{5}}x_2y_1 - x_3y_4 - x_4y_3 \\ \tfrac{1}{\sqrt{5}}x_1y_3 + x_2y_4 + \tfrac{1}{\sqrt{5}}x_3y_1 + x_4y_2 \\ -x_1y_4 - \tfrac{1}{\sqrt{5}}x_2y_3 - \tfrac{1}{\sqrt{5}}x_3y_2 - x_4y_1 \\ -\sqrt{\tfrac{3}{5}}\left(\tfrac{1}{\phi}x_1y_1 - \phi x_2y_2 + x_3y_3\right) \end{pmatrix}_5 , \quad (D.13)$$

$$\begin{pmatrix} x_1 \\ x_2 \\ x_3 \\ x_4 \end{pmatrix}_4 \otimes \begin{pmatrix} y_1 \\ y_2 \\ y_3 \\ y_4 \\ y_5 \end{pmatrix}_5$$

$$= \begin{pmatrix} \tfrac{2}{\phi^2}x_1y_2 - (\phi+4)x_2y_1 + 2\phi^2 x_3y_4 + 2\sqrt{5}x_4y_3 - \tfrac{\sqrt{3}}{\phi^2}x_2y_5 \\ (\phi-5)x_1y_1 + \sqrt{3}\phi^2 x_1y_5 - 2\phi^2 x_2y_2 + \tfrac{2}{\phi^2}x_3y_3 + 2\sqrt{5}x_4y_4 \\ 2\phi^2 x_1y_3 - \tfrac{2}{\phi^2}x_2y_4 - \sqrt{5}x_3y_1 - 3\sqrt{3}x_3y_5 + 2\sqrt{5}x_4y_2 \end{pmatrix}_3$$

$$\oplus \begin{pmatrix} \tfrac{1}{\phi^2}x_1y_1 - \sqrt{3}\phi x_1y_5 - \tfrac{1}{\phi^2}x_2y_2 + \phi^2 x_3y_3 + \sqrt{5}x_4y_4 \\ -\phi^2 x_1y_2 - \phi^2 x_2y_1 + \tfrac{\sqrt{3}}{\phi}x_2y_5 - \tfrac{1}{\phi^2}x_3y_4 - \sqrt{5}x_4y_3 \\ \tfrac{1}{\phi^2}x_1y_3 - \phi^2 x_2y_4 + \sqrt{5}x_3y_1 + \sqrt{3}x_3y_5 + \sqrt{5}x_4y_2 \end{pmatrix}_{3'}$$

$$\oplus \begin{pmatrix} -\tfrac{\phi^2}{\sqrt{5}}x_1y_1 - \tfrac{1}{\sqrt{5}}x_2y_2 + \tfrac{1}{\sqrt{5}}x_3y_3 - x_4y_4 - \tfrac{\sqrt{3}}{\sqrt{5}\phi}x_1y_5 \\ -\tfrac{1}{\sqrt{5}}x_1y_2 + \tfrac{1}{\sqrt{5}\phi^2}x_2y_1 + \sqrt{\tfrac{3}{5}}\phi x_2y_5 - \tfrac{1}{\sqrt{5}}x_3y_4 + x_4y_3 \\ \tfrac{1}{\sqrt{5}}x_1y_3 - \tfrac{1}{\sqrt{5}}x_2y_4 + x_3y_1 - \sqrt{\tfrac{3}{5}}x_3y_5 - x_4y_2 \\ -x_1y_4 + x_2y_3 - x_3y_2 \end{pmatrix}_4$$

$$\oplus \begin{pmatrix} \frac{1}{2}(\phi^2 x_1 y_4 - \frac{1}{\phi^2} x_2 y_3 - 3 x_3 y_2 + 3 x_4 y_1 + \sqrt{\frac{5}{3}} x_4 y_5) \\ \phi x_1 y_3 - \frac{1}{\phi} x_2 y_4 - x_3 y_1 - x_4 y_2 + \sqrt{\frac{5}{3}} x_3 y_5 \\ \frac{1}{\phi} x_1 y_2 + \frac{1}{\phi} x_2 y_1 + \frac{1}{\sqrt{3}} \phi^2 x_2 y_5 + \phi x_3 y_4 - x_4 y_3 \\ \phi x_1 y_1 + \frac{1}{\sqrt{3}\phi^2} x_1 y_5 - \phi x_2 y_2 + \frac{1}{\phi} x_3 y_3 - x_4 y_4 \\ \frac{1}{2\sqrt{3}}[-(\phi-5)x_1 y_4 + (\phi+4)x_2 y_3 + \sqrt{5} x_3 y_2 - \sqrt{5} x_4 y_1] + \frac{3}{2} x_4 y_5 \end{pmatrix}_5 ,$$

$$\oplus \begin{pmatrix} x_1 y_4 - x_2 y_3 - 2 x_3 y_2 + \frac{3}{2} x_4 y_1 + \frac{\sqrt{15}}{2} x_4 y_5 \\ \phi^2 x_1 y_3 + \frac{1}{\phi^2} x_2 y_4 - \frac{1}{2} x_3 y_1 + \frac{\sqrt{15}}{2} x_3 y_5 - x_4 y_2 \\ -\frac{1}{\phi^2} x_1 y_2 + \frac{3\phi-2}{2} x_2 y_1 + \frac{\sqrt{3}}{2} \phi x_2 y_5 + \phi^2 x_3 y_4 - x_4 y_3 \\ \frac{3\phi-1}{2} x_1 y_1 - \phi^2 x_2 y_2 - \frac{1}{\phi^2} x_3 y_3 - x_4 y_4 - \frac{\sqrt{3}}{2\phi} x_1 y_5 \\ \sqrt{3} x_1 y_4 + \sqrt{3} x_2 y_3 + \frac{3}{2} x_4 y_5 - \frac{\sqrt{15}}{2} x_4 y_1 \end{pmatrix}_5 , \qquad (D.14)$$

$$\begin{pmatrix} x_1 \\ x_2 \\ x_3 \\ x_4 \\ x_5 \end{pmatrix}_5 \otimes \begin{pmatrix} y_1 \\ y_2 \\ y_3 \\ y_4 \\ y_5 \end{pmatrix}_5$$

$$= (x_1 y_1 + x_2 y_2 + x_3 y_3 + x_4 y_4 + x_5 y_5)_1$$

$$\oplus \begin{pmatrix} x_1 y_3 - x_3 y_1 + x_2 y_4 - x_4 y_2 + \sqrt{3}(x_3 y_5 - x_5 y_3) \\ x_1 y_4 - x_4 y_1 - (x_2 y_3 - x_3 y_2) - \sqrt{3}(x_4 y_5 - x_5 y_4) \\ -2(x_1 y_2 - x_2 y_1) - (x_3 y_4 - x_4 y_3) \end{pmatrix}_3$$

$$\oplus \begin{pmatrix} (2\phi+3)(x_1 y_4 - x_4 y_1) + 2\phi(x_2 y_3 - x_3 y_2) + \sqrt{3}(x_4 y_5 - x_5 y_4) \\ (\phi+3)(x_1 y_3 - x_3 y_1) + 2\phi(x_2 y_4 - x_4 y_2) - \sqrt{3}\phi^2(x_3 y_5 - x_5 y_3) \\ -\phi(x_1 y_2 - x_2 y_1) - \sqrt{15}\phi(x_2 y_5 - x_5 y_2) + 2\phi(x_3 y_4 - x_4 y_3) \end{pmatrix}_{3'}$$

$$\oplus \begin{pmatrix} \sqrt{5}[\frac{\phi^2}{2}(x_1 y_4 + x_4 y_1) + x_2 y_3 + x_3 y_2 + \frac{\sqrt{3}}{2\phi}(x_4 y_5 + x_5 y_4)] \\ \sqrt{5}[\frac{1}{2\phi^2}(x_1 y_3 + x_3 y_1) - (x_2 y_4 + x_4 y_2) + \frac{\sqrt{3}}{2}\phi(x_3 y_5 + x_5 y_3)] \\ \frac{\sqrt{5}}{2}[-\sqrt{5}(x_1 y_2 + x_2 y_1) + \sqrt{3}(x_2 y_5 + x_5 y_2) + 2(x_3 y_4 + x_4 y_3)] \\ 3 x_1 y_1 - 2(x_2 y_2 + x_3 y_3 + x_4 y_4) + 3 x_5 y_5 \end{pmatrix}_4$$

$$\oplus \begin{pmatrix} \frac{1}{\sqrt{5}}[\frac{1}{2\phi}(x_1 y_4 - x_4 y_1) - (x_2 y_3 - x_3 y_2) + \frac{\phi^2}{2\sqrt{3}}(x_4 y_5 - x_5 y_4)] \\ \frac{1}{\sqrt{5}}[\frac{\phi}{2}(x_1 y_3 - x_3 y_1) - (x_2 y_4 - x_4 y_2) + \frac{1}{2\sqrt{3}\phi^2}(x_3 y_5 - x_5 y_3)] \\ \frac{1}{2\sqrt{5}}(x_1 y_2 - x_2 y_1) - \frac{1}{2\sqrt{3}}(x_2 y_5 - x_5 y_2) - \frac{1}{\sqrt{5}}(x_3 y_4 - x_4 y_3) \\ -\frac{1}{\sqrt{3}}(x_1 y_5 - x_5 y_1) \end{pmatrix}_4$$

D.2 Basis II

$$\oplus \begin{pmatrix} -x_1y_1 - \frac{11}{3\sqrt{15}}(x_1y_5 + x_5y_1) + \frac{4}{3}x_2y_2 - \frac{4\sqrt{5}}{15}\left(\phi x_3y_3 + \frac{1}{\phi}x_4y_4\right) + x_5y_5 \\ \frac{4}{3}[x_1y_2 + x_2y_1 + \frac{1}{\sqrt{15}}(x_2y_5 + x_5y_2) + \frac{2}{\sqrt{5}}(x_3y_4 + x_4y_3)] \\ \frac{4}{3\sqrt{5}}[-\phi(x_1y_3 + x_3y_1) + 2(x_2y_4 + x_4y_2) - \frac{2-3\phi}{\sqrt{3}}(x_3y_5 + x_5y_3)] \\ \frac{4\sqrt{5}}{15}[-\frac{1}{\phi}(x_1y_4 + x_4y_1) + 2(x_2y_3 + x_3y_2) - \frac{\sqrt{3}}{3}(3\phi - 1)(x_4y_5 + x_5y_4)] \\ x_1y_5 + x_5y_1 + \frac{1}{3\sqrt{15}}(-11x_1y_1 + 4x_2y_2 + 11x_5y_5) - \frac{4\sqrt{15}}{45}[(2-3\phi)x_3y_3 + (3\phi-1)x_4y_4] \end{pmatrix}_5$$

$$\oplus \begin{pmatrix} -\frac{3\sqrt{5}}{4}(x_1y_1 - x_5y_5) - \frac{\sqrt{3}}{4}(x_1y_5 + x_5y_1) + \sqrt{5}x_2y_2 - \phi^2 x_3y_3 + \frac{1}{\phi^2}x_4y_4 \\ \sqrt{5}(x_1y_2 + x_2y_1) + \sqrt{3}(x_2y_5 + x_5y_2) + x_3y_4 + x_4y_3 \\ -\phi^2(x_1y_3 + x_3y_1) + x_2y_4 + x_4y_2 + \frac{\sqrt{3}}{\phi}(x_3y_5 + x_5y_3) \\ \frac{1}{\phi^2}(x_1y_4 + x_4y_1) + (x_2y_3 + x_3y_2) - \sqrt{3}\phi(x_4y_5 + x_5y_4) \\ -\frac{\sqrt{3}}{4}(x_1y_1 - 4x_2y_2 - x_5y_5) + \frac{3\sqrt{5}}{4}(x_1y_5 + x_5y_1) + \frac{\sqrt{3}}{\phi}x_3y_3 - \sqrt{3}\phi x_4y_4 \end{pmatrix}_5.$$

(D.15)

D.2 Basis II

Here we consider another basis, which is used, e.g., in [2], with the generators a and b satisfying

$$a^2 = b^3 = (ab)^5 = e. \tag{D.16}$$

These generators a and b are given in terms of s and t of Sect. D.1 by

$$a = st^3 st^2 s, \qquad ab = t^4. \tag{D.17}$$

In the above transformation, a is diagonal but ab is not. We diagonalize ab with the unitary transformation U_ϕ:

$$a = U_\phi^\dagger st^3 st^2 s U_\phi, \qquad ab = U_\phi^\dagger t^4 U_\phi. \tag{D.18}$$

These generators are represented as follows:

$$a = \frac{1}{\sqrt{5}}\begin{pmatrix} 1 & -\sqrt{2} & -\sqrt{2} \\ -\sqrt{2} & -\phi & \frac{1}{\phi} \\ -\sqrt{2} & \frac{1}{\phi} & -\phi \end{pmatrix}, \quad ab = \begin{pmatrix} 1 & 0 & 0 \\ 0 & \rho & 0 \\ 0 & 0 & \rho^4 \end{pmatrix}, \quad \text{on } \mathbf{3}, \quad \text{(D.19)}$$

with

$$U_\phi = \frac{1}{\sqrt{25^{1/4}}}\begin{pmatrix} -\sqrt{\frac{2}{\phi}} & -\sqrt{\phi} & -\sqrt{\phi} \\ 0 & i5^{1/4} & -i5^{1/4} \\ -\sqrt{2\phi} & \frac{1}{\sqrt{\phi}} & \frac{1}{\sqrt{\phi}} \end{pmatrix}, \tag{D.20}$$

$$a = \frac{1}{\sqrt{5}}\begin{pmatrix} -1 & \sqrt{2} & \sqrt{2} \\ \sqrt{2} & -\frac{1}{\phi} & \phi \\ \sqrt{2} & \phi & -\frac{1}{\phi} \end{pmatrix}, \quad ab = \begin{pmatrix} 1 & 0 & 0 \\ 0 & \rho^2 & 0 \\ 0 & 0 & \rho^3 \end{pmatrix}, \quad \text{on } \mathbf{3'},$$

(D.21)

D Representations of A_5 in Different Bases

with

$$U_\phi = \frac{1}{\sqrt{2}5^{1/4}}\begin{pmatrix} 0 & i5^{1/4} & -i5^{1/4} \\ \sqrt{2\phi} & -\frac{1}{\sqrt{\phi}} & -\frac{1}{\sqrt{\phi}} \\ \sqrt{\frac{2}{\phi}} & \sqrt{\phi} & \sqrt{\phi} \end{pmatrix},$$ (D.22)

$$a = \frac{1}{\sqrt{5}}\begin{pmatrix} 1 & \frac{1}{\phi} & \phi & -1 \\ \frac{1}{\phi} & -1 & 1 & \phi \\ \phi & 1 & -1 & \frac{1}{\phi} \\ -1 & \phi & \frac{1}{\phi} & 1 \end{pmatrix}, \quad ab = \begin{pmatrix} \rho & 0 & 0 & 0 \\ 0 & \rho^2 & 0 & 0 \\ 0 & 0 & \rho^3 & 0 \\ 0 & 0 & 0 & \rho^4 \end{pmatrix}, \quad \text{on } \mathbf{4},$$ (D.23)

with

$$U_\phi = \frac{1}{2}\begin{pmatrix} 1 & -1 & -1 & 1 \\ \frac{i}{5^{1/4}\phi^{3/2}} & \frac{i\phi^{3/2}}{5^{1/4}} & -\frac{i\phi^{3/2}}{5^{1/4}} & -\frac{i}{5^{1/4}\phi^{3/2}} \\ \frac{i\phi^{3/2}}{5^{1/4}} & -\frac{i}{5^{1/4}\phi^{3/2}} & \frac{i}{5^{1/4}\phi^{3/2}} & -\frac{i\phi^{3/2}}{5^{1/4}} \\ 1 & 1 & 1 & 1 \end{pmatrix},$$ (D.24)

and

$$a = \frac{1}{5}\begin{pmatrix} -1 & \sqrt{6} & \sqrt{6} & \sqrt{6} & \sqrt{6} \\ \sqrt{6} & \frac{1}{\phi^2} & -2\phi & \frac{2}{\phi} & \phi^2 \\ \sqrt{6} & -2\phi & \phi^2 & \frac{1}{\phi^2} & \frac{2}{\phi} \\ \sqrt{6} & \frac{2}{\phi} & \frac{1}{\phi^2} & \phi^2 & -2\phi \\ \sqrt{6} & \phi^2 & \frac{2}{\phi} & -2\phi & \frac{1}{\phi^2} \end{pmatrix},$$

$$ab = \begin{pmatrix} 1 & 0 & 0 & 0 & 0 \\ 0 & \rho & 0 & 0 & 0 \\ 0 & 0 & \rho^2 & 0 & 0 \\ 0 & 0 & 0 & \rho^3 & 0 \\ 0 & 0 & 0 & 0 & \rho^4 \end{pmatrix}, \quad \text{on } \mathbf{5},$$ (D.25)

with

$$U_\phi = \frac{1}{\sqrt{10}}\begin{pmatrix} -\frac{1}{\phi}\sqrt{\frac{3}{2}} & 1 & -\frac{1}{2}\sqrt{\phi^4+8} & -\frac{1}{2}\sqrt{\phi^4+8} & 1 \\ 0 & \frac{i5^{1/4}}{\sqrt{\phi}} & -i5^{1/4}\sqrt{\phi} & i5^{1/4}\sqrt{\phi} & -\frac{i5^{1/4}}{\sqrt{\phi}} \\ 0 & i5^{1/4}\sqrt{\phi} & \frac{i5^{1/4}}{\sqrt{\phi}} & -\frac{i5^{1/4}}{\sqrt{\phi}} & -i5^{1/4}\sqrt{\phi} \\ \sqrt{6} & -1 & -1 & -1 & -1 \\ \frac{\phi^2}{\sqrt{2}} & \sqrt{3} & \frac{\sqrt{3}}{2\phi} & \frac{\sqrt{3}}{2\phi} & \sqrt{3} \end{pmatrix},$$ (D.26)

where $\phi = (1+\sqrt{5})/2$ and $\rho = e^{2i\pi/5}$.

D.2 Basis II

The tensor products are as follows:

$$\begin{pmatrix} x_1 \\ x_2 \\ x_3 \end{pmatrix}_3 \otimes \begin{pmatrix} y_1 \\ y_2 \\ y_3 \end{pmatrix}_3 = (x_1 y_1 + x_2 y_3 + x_3 y_2)_1 \oplus \begin{pmatrix} x_2 y_3 - x_3 y_2 \\ x_1 y_2 - x_2 y_1 \\ x_3 y_1 - x_1 y_3 \end{pmatrix}_3$$

$$\oplus \begin{pmatrix} 2x_1 y_1 - x_2 y_3 - x_3 y_2 \\ -\sqrt{3} x_1 y_2 - \sqrt{3} x_2 y_1 \\ \sqrt{6} x_2 y_2 \\ \sqrt{6} x_3 y_3 \\ -\sqrt{3} x_1 y_3 - \sqrt{3} x_3 y_1 \end{pmatrix}_5 , \qquad (D.27)$$

$$\begin{pmatrix} x_1 \\ x_2 \\ x_3 \end{pmatrix}_{3'} \otimes \begin{pmatrix} y_1 \\ y_2 \\ y_3 \end{pmatrix}_{3'} = (x_1 y_1 + x_2 y_3 + x_3 y_2)_1 \oplus \begin{pmatrix} x_2 y_3 - x_3 y_2 \\ x_1 y_2 - x_2 y_1 \\ x_3 y_1 - x_1 y_3 \end{pmatrix}_{3'}$$

$$\oplus \begin{pmatrix} 2x_1 y_1 - x_2 y_3 - x_3 y_2 \\ \sqrt{6} x_3 y_3 \\ -\sqrt{3} x_1 y_2 - \sqrt{3} x_2 y_1 \\ -\sqrt{3} x_1 y_3 - \sqrt{3} x_3 y_1 \\ \sqrt{6} x_2 y_2 \end{pmatrix}_5 , \qquad (D.28)$$

$$\begin{pmatrix} x_1 \\ x_2 \\ x_3 \end{pmatrix}_3 \otimes \begin{pmatrix} y_1 \\ y_2 \\ y_3 \end{pmatrix}_{3'} = \begin{pmatrix} \sqrt{2} x_2 y_1 + x_3 y_2 \\ -\sqrt{2} x_1 y_2 - x_3 y_3 \\ -\sqrt{2} x_1 y_3 - x_2 y_2 \\ \sqrt{2} x_3 y_1 + x_2 y_3 \end{pmatrix}_4 \oplus \begin{pmatrix} \sqrt{3} x_1 y_1 \\ x_2 y_1 - \sqrt{2} x_3 y_2 \\ x_1 y_2 - \sqrt{2} x_3 y_3 \\ x_1 y_3 - \sqrt{2} x_2 y_2 \\ x_3 y_1 - \sqrt{2} x_2 y_3 \end{pmatrix}_5 , \qquad (D.29)$$

$$\begin{pmatrix} x_1 \\ x_2 \\ x_3 \end{pmatrix}_3 \otimes \begin{pmatrix} y_1 \\ y_2 \\ y_3 \\ y_4 \end{pmatrix}_4 = \begin{pmatrix} -\sqrt{2} x_2 y_4 - \sqrt{2} x_3 y_1 \\ \sqrt{2} x_1 y_2 - x_2 y_1 + x_3 y_3 \\ \sqrt{2} x_1 y_3 + x_2 y_2 - x_3 y_4 \end{pmatrix}_{3'} \oplus \begin{pmatrix} x_1 y_1 - \sqrt{2} x_3 y_2 \\ -x_1 y_2 - \sqrt{2} x_2 y_1 \\ x_1 y_3 + \sqrt{2} x_3 y_4 \\ -x_1 y_4 + \sqrt{2} x_2 y_3 \end{pmatrix}_4$$

$$\oplus \begin{pmatrix} \sqrt{6} x_2 y_4 - \sqrt{6} x_3 y_1 \\ 2\sqrt{2} x_1 y_1 + 2 x_3 y_2 \\ -\sqrt{2} x_1 y_2 + x_2 y_1 + 3 x_3 y_3 \\ \sqrt{2} x_1 y_3 - 3 x_2 y_2 - x_3 y_4 \\ -2\sqrt{2} x_1 y_4 - 2 x_2 y_3 \end{pmatrix}_5 , \qquad (D.30)$$

$$\begin{pmatrix} x_1 \\ x_2 \\ x_3 \end{pmatrix}_{3'} \otimes \begin{pmatrix} y_1 \\ y_2 \\ y_3 \\ y_4 \end{pmatrix}_{4} = \begin{pmatrix} -\sqrt{2}x_2y_3 - \sqrt{2}x_3y_2 \\ \sqrt{2}x_1y_1 + x_2y_4 - x_3y_3 \\ \sqrt{2}x_1y_4 - x_2y_2 + x_3y_1 \end{pmatrix}_{3} \oplus \begin{pmatrix} x_1y_1 + \sqrt{2}x_3y_3 \\ x_1y_2 - \sqrt{2}x_3y_4 \\ -x_1y_3 + \sqrt{2}x_2y_1 \\ -x_1y_4 - \sqrt{2}x_2y_2 \end{pmatrix}_{4}$$

$$\oplus \begin{pmatrix} \sqrt{6}x_2y_3 - \sqrt{6}x_3y_2 \\ \sqrt{2}x_1y_1 - 3x_2y_4 - x_3y_3 \\ 2\sqrt{2}x_1y_2 + 2x_3y_4 \\ -2\sqrt{2}x_1y_3 - 2x_2y_1 \\ -\sqrt{2}x_1y_4 + x_2y_2 + 3x_3y_1 \end{pmatrix}_{5}, \qquad (D.31)$$

$$\begin{pmatrix} x_1 \\ x_2 \\ x_3 \end{pmatrix}_{3} \otimes \begin{pmatrix} y_1 \\ y_2 \\ y_3 \\ y_4 \\ y_5 \end{pmatrix}_{5} = \begin{pmatrix} -2x_1y_1 + \sqrt{3}x_2y_5 + \sqrt{3}x_3y_2 \\ \sqrt{3}x_1y_2 + x_2y_1 - \sqrt{6}x_3y_3 \\ \sqrt{3}x_1y_5 - \sqrt{6}x_2y_4 + x_3y_1 \end{pmatrix}_{3}$$

$$\oplus \begin{pmatrix} \sqrt{3}x_1y_1 + x_2y_5 + x_3y_2 \\ x_1y_3 - \sqrt{2}x_2y_2 - \sqrt{2}x_3y_4 \\ x_1y_4 - \sqrt{2}x_2y_3 - \sqrt{2}x_3y_5 \end{pmatrix}_{3'}$$

$$\oplus \begin{pmatrix} 2\sqrt{2}x_1y_2 - \sqrt{6}x_2y_1 + x_3y_3 \\ -\sqrt{2}x_1y_3 + 2x_2y_2 - 3x_3y_4 \\ \sqrt{2}x_1y_4 + 3x_2y_3 - 2x_3y_5 \\ -2\sqrt{2}x_1y_5 - x_2y_4 + \sqrt{6}x_3y_1 \end{pmatrix}_{4}$$

$$\oplus \begin{pmatrix} \sqrt{3}x_2y_5 - \sqrt{3}x_3y_2 \\ -x_1y_2 - \sqrt{3}x_2y_1 - \sqrt{2}x_3y_3 \\ -2x_1y_3 - \sqrt{2}x_2y_2 \\ 2x_1y_4 + \sqrt{2}x_3y_5 \\ x_1y_5 + \sqrt{2}x_2y_4 + \sqrt{3}x_3y_1 \end{pmatrix}_{5}, \qquad (D.32)$$

$$\begin{pmatrix} x_1 \\ x_2 \\ x_3 \end{pmatrix}_{3'} \otimes \begin{pmatrix} y_1 \\ y_2 \\ y_3 \\ y_4 \\ y_5 \end{pmatrix}_{5} = \begin{pmatrix} \sqrt{3}x_1y_1 + x_2y_4 + x_3y_3 \\ x_1y_2 - \sqrt{2}x_2y_5 - \sqrt{2}x_3y_4 \\ x_1y_5 - \sqrt{2}x_2y_3 - \sqrt{2}x_3y_2 \end{pmatrix}_{3}$$

$$\oplus \begin{pmatrix} -2x_1y_1 + \sqrt{3}x_2y_4 + \sqrt{3}x_3y_3 \\ \sqrt{3}x_1y_3 + x_2y_1 - \sqrt{6}x_3y_5 \\ \sqrt{3}x_1y_4 - \sqrt{6}x_2y_2 + x_3y_1 \end{pmatrix}_{3'}$$

D.2 Basis II

$$\oplus \begin{pmatrix} \sqrt{2}x_1y_2 + 3x_2y_5 - 2x_3y_4 \\ 2\sqrt{2}x_1y_3 - \sqrt{6}x_2y_1 + x_3y_5 \\ -2\sqrt{2}x_1y_4 - x_2y_2 + \sqrt{6}x_3y_1 \\ -\sqrt{2}x_1y_5 + 2x_2y_3 - 3x_3y_2 \end{pmatrix}_4$$

$$\oplus \begin{pmatrix} \sqrt{3}x_2y_4 - \sqrt{3}x_3y_3 \\ 2x_1y_2 + \sqrt{2}x_3y_4 \\ -x_1y_3 - \sqrt{3}x_2y_1 - \sqrt{2}x_3y_5 \\ x_1y_4 + \sqrt{2}x_2y_2 + \sqrt{3}x_3y_1 \\ -2x_1y_5 - \sqrt{2}x_2y_3 \end{pmatrix}_5, \quad (D.33)$$

$$\begin{pmatrix} x_1 \\ x_2 \\ x_3 \\ x_4 \end{pmatrix}_4 \otimes \begin{pmatrix} y_1 \\ y_2 \\ y_3 \\ y_4 \end{pmatrix}_4 = (x_1y_4 + x_2y_3 + x_3y_2 + x_4y_1)_1$$

$$\oplus \begin{pmatrix} -x_1y_4 + x_2y_3 - x_3y_2 + x_4y_1 \\ \sqrt{2}x_2y_4 - \sqrt{2}x_4y_2 \\ \sqrt{2}x_1y_3 - \sqrt{2}x_3y_1 \end{pmatrix}_3$$

$$\oplus \begin{pmatrix} x_1y_4 + x_2y_3 - x_3y_2 - x_4y_1 \\ \sqrt{2}x_3y_4 - \sqrt{2}x_4y_3 \\ \sqrt{2}x_1y_2 - \sqrt{2}x_2y_1 \end{pmatrix}_{3'}$$

$$\oplus \begin{pmatrix} x_2y_4 + x_3y_3 + x_4y_2 \\ x_1y_1 + x_3y_4 + x_4y_3 \\ x_1y_2 + x_2y_1 + x_4y_4 \\ x_1y_3 + x_2y_2 + x_3y_1 \end{pmatrix}_4$$

$$\oplus \begin{pmatrix} \sqrt{3}x_1y_4 - \sqrt{3}x_2y_3 - \sqrt{3}x_3y_2 + \sqrt{3}x_4y_1 \\ -\sqrt{2}x_2y_4 + 2\sqrt{2}x_3y_3 - \sqrt{2}x_4y_2 \\ -2\sqrt{2}x_1y_1 + \sqrt{2}x_3y_4 + \sqrt{2}x_4y_3 \\ \sqrt{2}x_1y_2 + \sqrt{2}x_2y_1 - 2\sqrt{2}x_4y_4 \\ -\sqrt{2}x_1y_3 + 2\sqrt{2}x_2y_2 - \sqrt{2}x_3y_1 \end{pmatrix}_5, \quad (D.34)$$

$$\begin{pmatrix} x_1 \\ x_2 \\ x_3 \\ x_4 \end{pmatrix}_4 \otimes \begin{pmatrix} y_1 \\ y_2 \\ y_3 \\ y_4 \\ y_5 \end{pmatrix}_5 = \begin{pmatrix} 2\sqrt{2}x_1y_5 - \sqrt{2}x_2y_4 + \sqrt{2}x_3y_3 - 2\sqrt{2}x_4y_2 \\ -\sqrt{6}x_1y_1 + 2x_2y_5 + 3x_3y_4 - x_4y_3 \\ x_1y_4 - 3x_2y_3 - 2x_3y_2 + \sqrt{6}x_4y_1 \end{pmatrix}_3$$

$$\oplus \begin{pmatrix} \sqrt{2}x_1y_5 + 2\sqrt{2}x_2y_4 - 2\sqrt{2}x_3y_3 - \sqrt{2}x_4y_2 \\ 3x_1y_2 - \sqrt{6}x_2y_1 - x_3y_5 + 2x_4y_4 \\ -2x_1y_3 + x_2y_2 + \sqrt{6}x_3y_1 - 3x_4y_5 \end{pmatrix}_{3'}$$

$$\oplus \begin{pmatrix} \sqrt{3}x_1y_1 - \sqrt{2}x_2y_5 + \sqrt{2}x_3y_4 - 2\sqrt{2}x_4y_3 \\ -\sqrt{2}x_1y_2 - \sqrt{3}x_2y_1 + 2\sqrt{2}x_3y_5 + \sqrt{2}x_4y_4 \\ \sqrt{2}x_1y_3 + 2\sqrt{2}x_2y_2 - \sqrt{3}x_3y_1 - \sqrt{2}x_4y_5 \\ -2\sqrt{2}x_1y_4 + \sqrt{2}x_2y_3 - \sqrt{2}x_3y_2 + \sqrt{3}x_4y_1 \end{pmatrix}_{4}$$

$$\oplus \begin{pmatrix} \sqrt{2}x_1y_5 - \sqrt{2}x_2y_4 - \sqrt{2}x_3y_3 + \sqrt{2}x_4y_2 \\ -\sqrt{2}x_1y_1 - \sqrt{3}x_3y_4 - \sqrt{3}x_4y_3 \\ \sqrt{3}x_1y_2 + \sqrt{2}x_2y_1 + \sqrt{3}x_3y_5 \\ \sqrt{3}x_2y_2 + \sqrt{2}x_3y_1 + \sqrt{3}x_4y_5 \\ -\sqrt{3}x_1y_4 - \sqrt{3}x_2y_3 - \sqrt{2}x_4y_1 \end{pmatrix}_{5}$$

$$\oplus \begin{pmatrix} 2x_1y_5 + 4x_2y_4 + 4x_3y_3 + 2x_4y_2 \\ 4x_1y_1 + 2\sqrt{6}x_2y_5 \\ -\sqrt{6}x_1y_2 + 2x_2y_1 - \sqrt{6}x_3y_5 + 2\sqrt{6}x_4y_4 \\ 2\sqrt{6}x_1y_3 - \sqrt{6}x_2y_2 + 2x_3y_1 - \sqrt{6}x_4y_5 \\ 2\sqrt{6}x_3y_2 + 4x_4y_1 \end{pmatrix}_{5},$$

(D.35)

$$\begin{pmatrix} x_1 \\ x_2 \\ x_3 \\ x_4 \\ x_5 \end{pmatrix}_{5} \otimes \begin{pmatrix} y_1 \\ y_2 \\ y_3 \\ y_4 \\ y_5 \end{pmatrix}_{5}$$
$$= (x_1y_1 + x_2y_5 + x_3y_4 + x_4y_3 + x_5y_2)_1$$

$$\oplus \begin{pmatrix} x_2y_5 + 2x_3y_4 - 2x_4y_3 - x_5y_2 \\ -\sqrt{3}x_1y_2 + \sqrt{3}x_2y_1 + \sqrt{2}x_3y_5 - \sqrt{2}x_5y_3 \\ \sqrt{3}x_1y_5 + \sqrt{2}x_2y_4 - \sqrt{2}x_4y_2 - \sqrt{3}x_5y_1 \end{pmatrix}_{3}$$

$$\oplus \begin{pmatrix} 2x_2y_5 - x_3y_4 + x_4y_3 - 2x_5y_2 \\ \sqrt{3}x_1y_3 - \sqrt{3}x_3y_1 + \sqrt{2}x_4y_5 - \sqrt{2}x_5y_4 \\ -\sqrt{3}x_1y_4 + \sqrt{2}x_2y_3 - \sqrt{2}x_3y_2 + \sqrt{3}x_4y_1 \end{pmatrix}_{3'}$$

$$\oplus \begin{pmatrix} 3\sqrt{2}x_1y_2 + 3\sqrt{2}x_2y_1 - \sqrt{3}x_3y_5 + 4\sqrt{3}x_4y_4 - \sqrt{3}x_5y_3 \\ 3\sqrt{2}x_1y_3 + 4\sqrt{3}x_2y_2 + 3\sqrt{2}x_3y_1 - \sqrt{3}x_4y_5 - \sqrt{3}x_5y_4 \\ 3\sqrt{2}x_1y_4 - \sqrt{3}x_2y_3 - \sqrt{3}x_3y_2 + 3\sqrt{2}x_4y_1 + 4\sqrt{3}x_5y_5 \\ 3\sqrt{2}x_1y_5 - \sqrt{3}x_2y_4 + 4\sqrt{3}x_3y_3 - \sqrt{3}x_4y_2 + 3\sqrt{2}x_5y_1 \end{pmatrix}_{4}$$

$$\oplus \begin{pmatrix} \sqrt{2}x_1y_2 - \sqrt{2}x_2y_1 + \sqrt{3}x_3y_5 - \sqrt{3}x_5y_3 \\ -\sqrt{2}x_1y_3 + \sqrt{2}x_3y_1 + \sqrt{3}x_4y_5 - \sqrt{3}x_5y_4 \\ -\sqrt{2}x_1y_4 - \sqrt{3}x_2y_3 + \sqrt{3}x_3y_2 + \sqrt{2}x_4y_1 \\ \sqrt{2}x_1y_5 - \sqrt{3}x_2y_4 + \sqrt{3}x_4y_2 - \sqrt{2}x_5y_1 \end{pmatrix}_4$$

$$\oplus \begin{pmatrix} 2x_1y_1 + x_2y_5 - 2x_3y_4 - 2x_4y_3 + x_5y_2 \\ x_1y_2 + x_2y_1 + \sqrt{6}x_3y_5 + \sqrt{6}x_5y_3 \\ -2x_1y_3 + \sqrt{6}x_2y_2 - 2x_3y_1 \\ -2x_1y_4 - 2x_4y_1 + \sqrt{6}x_5y_5 \\ x_1y_5 + \sqrt{6}x_2y_4 + \sqrt{6}x_4y_2 + x_5y_1 \end{pmatrix}_5$$

$$\oplus \begin{pmatrix} 2x_1y_1 - 2x_2y_5 + x_3y_4 + x_4y_3 - 2x_5y_2 \\ -2x_1y_2 - 2x_2y_1 + \sqrt{6}x_4y_4 \\ x_1y_3 + x_3y_1 + \sqrt{6}x_4y_5 + \sqrt{6}x_5y_4 \\ x_1y_4 + \sqrt{6}x_2y_3 + \sqrt{6}x_3y_2 + x_4y_1 \\ -2x_1y_5 + \sqrt{6}x_3y_3 - 2x_5y_1 \end{pmatrix}_5 . \quad (D.36)$$

References

1. Shirai, K.: J. Phys. Soc. Jpn. **61**, 2735 (1992)
2. Ding, G.-J., Everett, L.L., Stuart, A.J.: arXiv:1110.1688 [hep-ph]

Appendix E
Representations of T' in Different Bases

Here we consider another basis for representations of the T' group. All elements of T' can be written as products of the generators s and t, which satisfy

$$s^2 = r, \qquad r^2 = t^3 = (st)^3 = e, \qquad rt = tr. \tag{E.1}$$

In Chap. 5, the doublet and triplet representations were as follows:

$$t = \begin{pmatrix} \omega^2 & 0 \\ 0 & \omega \end{pmatrix}, \quad r = \begin{pmatrix} -1 & 0 \\ 0 & -1 \end{pmatrix}, \quad s = -\frac{1}{\sqrt{3}} \begin{pmatrix} i & \sqrt{2}p \\ -\sqrt{2}\bar{p} & -i \end{pmatrix} \quad \text{on } \mathbf{2}, \tag{E.2}$$

$$t = \begin{pmatrix} 1 & 0 \\ 0 & \omega^2 \end{pmatrix}, \quad r = \begin{pmatrix} -1 & 0 \\ 0 & -1 \end{pmatrix}, \quad s = -\frac{1}{\sqrt{3}} \begin{pmatrix} i & \sqrt{2}p \\ -\sqrt{2}\bar{p} & -i \end{pmatrix} \quad \text{on } \mathbf{2'}, \tag{E.3}$$

$$t = \begin{pmatrix} \omega & 0 \\ 0 & 1 \end{pmatrix}, \quad r = \begin{pmatrix} -1 & 0 \\ 0 & -1 \end{pmatrix}, \quad s = -\frac{1}{\sqrt{3}} \begin{pmatrix} i & \sqrt{2}p \\ -\sqrt{2}\bar{p} & -i \end{pmatrix} \quad \text{on } \mathbf{2''}, \tag{E.4}$$

$$t = \begin{pmatrix} 1 & 0 & 0 \\ 0 & \omega & 0 \\ 0 & 0 & \omega^2 \end{pmatrix}, \quad r = \begin{pmatrix} 1 & 0 & 0 \\ 0 & 1 & 0 \\ 0 & 0 & 1 \end{pmatrix},$$

$$s = \frac{1}{3} \begin{pmatrix} -1 & 2p_1 & 2p_1p_2 \\ 2\bar{p}_1 & -1 & 2p_2 \\ 2\bar{p}_1\bar{p}_2 & 2\bar{p}_2 & -1 \end{pmatrix} \quad \text{on } \mathbf{3}, \tag{E.5}$$

where $p_1 = e^{i\phi_1}$ and $p_2 = e^{i\phi_2}$.

E.1 Basis I

In this section, we recall the basis used in Chap. 5. We take the parameters $p = i$ and $p_1 = p_2 = 1$, whereupon the generator s takes the form

$$s = -\frac{i}{\sqrt{3}}\begin{pmatrix} 1 & \sqrt{2} \\ \sqrt{2} & -1 \end{pmatrix}, \quad \text{on } 2, 2', \text{ and } 2'', \tag{E.6}$$

$$s = \frac{1}{3}\begin{pmatrix} -1 & 2 & 2 \\ 2 & -1 & 2 \\ 2 & 2 & -1 \end{pmatrix}, \quad \text{on } 3. \tag{E.7}$$

The tensor products decompose as follows:

$$\begin{pmatrix} x_1 \\ x_2 \end{pmatrix}_{2(2')} \otimes \begin{pmatrix} y_1 \\ y_2 \end{pmatrix}_{2(2'')} = \begin{pmatrix} \frac{x_1 y_2 - x_2 y_1}{\sqrt{2}} \end{pmatrix}_1 \oplus \begin{pmatrix} \frac{x_1 y_2 + x_2 y_1}{\sqrt{2}} \\ -x_1 y_1 \\ x_2 y_2 \end{pmatrix}_3, \tag{E.8}$$

$$\begin{pmatrix} x_1 \\ x_2 \end{pmatrix}_{2'(2)} \otimes \begin{pmatrix} y_1 \\ y_2 \end{pmatrix}_{2'(2'')} = \begin{pmatrix} \frac{x_1 y_2 - x_2 y_1}{\sqrt{2}} \end{pmatrix}_{1''} \oplus \begin{pmatrix} -x_1 y_1 \\ x_2 y_2 \\ \frac{x_1 y_2 + x_2 y_1}{\sqrt{2}} \end{pmatrix}_3, \tag{E.9}$$

$$\begin{pmatrix} x_1 \\ x_2 \end{pmatrix}_{2''(2)} \otimes \begin{pmatrix} y_1 \\ y_2 \end{pmatrix}_{2''(2')} = \begin{pmatrix} \frac{x_1 y_2 - x_2 y_1}{\sqrt{2}} \end{pmatrix}_{1'} \oplus \begin{pmatrix} x_2 y_2 \\ \frac{x_1 y_2 + x_2 y_1}{\sqrt{2}} \\ -x_1 y_1 \end{pmatrix}_3, \tag{E.10}$$

$$\begin{pmatrix} x_1 \\ x_2 \\ x_3 \end{pmatrix}_3 \otimes \begin{pmatrix} y_1 \\ y_2 \\ y_3 \end{pmatrix}_3 = [x_1 y_1 + x_2 y_3 + x_3 y_2]_1$$

$$\oplus [x_3 y_3 + x_1 y_2 + x_2 y_1]_{1'} \oplus [x_2 y_2 + x_1 y_3 + x_3 y_1]_{1''}$$

$$\oplus \frac{1}{3}\begin{pmatrix} 2x_1 y_1 - x_2 y_3 - x_3 y_2 \\ 2x_3 y_3 - x_1 y_2 - x_2 y_1 \\ 2x_2 y_2 - x_1 y_3 - x_3 y_1 \end{pmatrix}_3$$

$$\oplus \frac{1}{2}\begin{pmatrix} x_2 y_3 - x_3 y_2 \\ x_1 y_2 - x_2 y_1 \\ x_3 y_1 - x_1 y_3 \end{pmatrix}_3, \tag{E.11}$$

$$\begin{pmatrix} x_1 \\ x_2 \end{pmatrix}_{2,2',2''} \otimes \begin{pmatrix} y_1 \\ y_2 \\ y_3 \end{pmatrix}_3 = \begin{pmatrix} \sqrt{2} x_2 y_2 + x_1 y_1 \\ \sqrt{2} x_1 y_3 - x_2 y_1 \end{pmatrix}_{2, 2', 2''} \oplus \begin{pmatrix} \sqrt{2} x_2 y_3 + x_1 y_2 \\ \sqrt{2} x_1 y_1 - x_2 y_2 \end{pmatrix}_{2', 2'', 2}$$

$$\oplus \begin{pmatrix} \sqrt{2} x_2 y_1 + x_1 y_3 \\ \sqrt{2} x_1 y_2 - x_2 y_3 \end{pmatrix}_{2'', 2, 2'}, \tag{E.12}$$

$$(x)_{1'(1'')} \otimes \begin{pmatrix} y_1 \\ y_2 \end{pmatrix}_{2,2',2''} = \begin{pmatrix} x y_1 \\ x y_2 \end{pmatrix}_{2'(2''), 2''(2), 2(2')}, \tag{E.13}$$

$$(x)_{1'} \otimes \begin{pmatrix} y_1 \\ y_2 \\ y_3 \end{pmatrix}_3 = \begin{pmatrix} xy_3 \\ xy_1 \\ xy_2 \end{pmatrix}_3, \quad (x)_{1''} \otimes \begin{pmatrix} y_1 \\ y_2 \\ y_3 \end{pmatrix}_3 = \begin{pmatrix} xy_2 \\ xy_3 \\ xy_1 \end{pmatrix}_3. \quad (E.14)$$

E.2 Basis II

We now consider another basis, which is used, e.g., in [1]. We take the parameters $p = e^{i\pi/12}$ and $p_1 = p_2 = \omega$, whence the generator s takes the form

$$s = -\frac{1}{\sqrt{3}} \begin{pmatrix} i & \sqrt{2}e^{i\pi/12} \\ -\sqrt{2}e^{-i\pi/12} & -i \end{pmatrix}, \quad \text{on } 2, 2', 2'', \quad (E.15)$$

$$s = \frac{1}{3} \begin{pmatrix} -1 & 2\omega & 2\omega^2 \\ 2\omega^2 & -1 & 2\omega \\ 2\omega & 2\omega^2 & -1 \end{pmatrix}, \quad \text{on } 3. \quad (E.16)$$

The tensor products are as follows:

$$\begin{pmatrix} x_1 \\ x_2 \end{pmatrix}_{2(2')} \otimes \begin{pmatrix} y_1 \\ y_2 \end{pmatrix}_{2(2'')} = [x_1 y_2 - x_2 y_1]_1 \oplus \begin{pmatrix} \frac{1-i}{2}(x_1 y_2 + x_2 y_1) \\ i x_1 y_1 \\ x_2 y_2 \end{pmatrix}_3, \quad (E.17)$$

$$\begin{pmatrix} x_1 \\ x_2 \end{pmatrix}_{2'(2)} \otimes \begin{pmatrix} y_1 \\ y_2 \end{pmatrix}_{2'(2'')} = [x_1 y_2 - x_2 y_1]_{1''} \oplus \begin{pmatrix} i x_1 y_1 \\ x_2 y_2 \\ \frac{1-i}{2}(x_1 y_2 + x_2 y_1) \end{pmatrix}_3, \quad (E.18)$$

$$\begin{pmatrix} x_1 \\ x_2 \end{pmatrix}_{2''(2)} \otimes \begin{pmatrix} y_1 \\ y_2 \end{pmatrix}_{2''(2')} = [x_1 y_2 - x_2 y_1]_{1'} \oplus \begin{pmatrix} x_2 y_2 \\ \frac{1-i}{2}(x_1 y_2 + x_2 y_1) \\ i x_1 y_1 \end{pmatrix}_3, \quad (E.19)$$

$$\begin{pmatrix} x_1 \\ x_2 \\ x_3 \end{pmatrix}_3 \otimes \begin{pmatrix} y_1 \\ y_2 \\ y_3 \end{pmatrix}_3 = [x_1 y_1 + x_2 y_3 + x_3 y_2]_1$$

$$\oplus [x_3 y_3 + x_1 y_2 + x_2 y_1]_{1'} \oplus [x_2 y_2 + x_1 y_3 + x_3 y_1]_{1''}$$

$$\oplus \frac{1}{3} \begin{pmatrix} 2x_1 y_1 - x_2 y_3 - x_3 y_2 \\ 2x_3 y_3 - x_1 y_2 - x_2 y_1 \\ 2x_2 y_2 - x_1 y_3 - x_3 y_1 \end{pmatrix}_3$$

$$\oplus \frac{1}{2} \begin{pmatrix} x_2 y_3 - x_3 y_2 \\ x_1 y_2 - x_2 y_1 \\ x_3 y_1 - x_1 y_3 \end{pmatrix}_3, \quad (E.20)$$

$$\begin{pmatrix} x_1 \\ x_2 \end{pmatrix}_{2,2',2''} \otimes \begin{pmatrix} y_1 \\ y_2 \\ y_3 \end{pmatrix}_3 = \begin{pmatrix} (1+i)x_2 y_2 + x_1 y_1 \\ (1-i)x_1 y_3 - x_2 y_1 \end{pmatrix}_{2,2',2''}$$

$$\oplus \begin{pmatrix} (1+i)x_2y_3 + x_1y_2 \\ (1-i)x_1y_1 - x_2y_2 \end{pmatrix}_{2',2'',2}$$

$$\oplus \begin{pmatrix} (1+i)x_2y_1 + x_1y_3 \\ (1-i)x_1y_2 - x_2y_3 \end{pmatrix}_{2'',2,2'}, \tag{E.21}$$

$$(x)_{\mathbf{1}'(\mathbf{1}'')} \otimes \begin{pmatrix} y_1 \\ y_2 \end{pmatrix}_{2,2',2''} = \begin{pmatrix} xy_1 \\ xy_2 \end{pmatrix}_{2'(2''),2''(2),2(2')}, \tag{E.22}$$

$$(x)_{\mathbf{1}'} \otimes \begin{pmatrix} y_1 \\ y_2 \\ y_3 \end{pmatrix}_3 = \begin{pmatrix} xy_3 \\ xy_1 \\ xy_2 \end{pmatrix}_3, \quad (x)_{\mathbf{1}''} \otimes \begin{pmatrix} y_1 \\ y_2 \\ y_3 \end{pmatrix}_3 = \begin{pmatrix} xy_2 \\ xy_3 \\ xy_1 \end{pmatrix}_3. \tag{E.23}$$

References

1. Feruglio, F., Hagedorn, C., Lin, Y., Merlo, L.: Nucl. Phys. B **775**, 120 (2007). arXiv:hep-ph/0702194

Appendix F
Other Smaller Groups

In this appendix, we study finite groups whose orders are less than 31 [1, 2]. Such groups are summarized in Table F.1, where g denotes the order. Here we have omitted the Abelian groups Z_N as well as their direct products. Most of the finite groups in Table F.1 are non-Abelian groups mentioned in the text, and their extensions by direct products with Abelian groups such as $Z_2 \times D_4$, $Z_2 \times Q_4$, etc. However, the table includes non-Abelian groups which are not mentioned in the text, in particular, $Z_4 \rtimes Z_4$, $Z_8 \rtimes Z_2$, $(Z_4 \times Z_2) \rtimes Z_2(I)$, $(Z_4 \times Z_2) \rtimes Z_2(II)$, $Z_3 \rtimes Z_8$, $(Z_6 \times Z_2) \rtimes Z_2$, and $Z_9 \rtimes Z_3$. In this appendix, we shall explicitly discuss these groups.

F.1 $Z_4 \rtimes Z_4$

We denote the first and second Z_4 generators by a and b, respectively. They thus satisfy $a^4 = b^4 = e$. Then we have the relation $ab = ba^m$, which implies $ab^2 = b^2 a^{m^2}$. We require $m \neq 0$ and $m^2 \neq 0$ mod 4. If $m = 1$ mod 4, the generators a and b commute and the group becomes the direct product $Z_4 \times Z_4$. We thus require $m \neq 1$ mod 4. These requirements are then satisfied for $m = 3$ mod 4, i.e.,

$$ab = ba^3. \tag{F.1}$$

Using these generators, all elements of $Z_4 \rtimes Z_4$ can be written in the form

$$g = b^m a^n, \tag{F.2}$$

with $n, m = 0, 1, 2, 3$.

Table F.1 Classification of the non-Abelian groups with $g \leq 30$. Note that there are two finite groups isomorphic to $(Z_4 \times Z_2) \rtimes Z_2$ apart from D_8

g	Groups
6	$S_3 \equiv D_3$
8	D_4, Q_4
10	D_5
12	A_4, D_6, Q_6
14	D_7
16	$D_8, Q_8, QD_{16}, Z_2 \times D_4, Z_2 \times Q_4, Z_4 \rtimes Z_4, Z_8 \rtimes Z_2,$ $(Z_4 \times Z_2) \rtimes Z_2(I), (Z_4 \times Z_2) \rtimes Z_2(II)$
18	$D_9, Z_3 \times D_3, \Sigma(18) \equiv (Z_3 \times Z_3) \rtimes Z_2$
20	D_{10}, Q_{10}
21	$T_7 \equiv Z_7 \rtimes Z_3$
22	D_{11}
24	$D_{12}, S_4, Q_{12}, T' \simeq SL(2,3), Z_2 \times Z_2 \times S_3, Z_4 \times S_3, Z_2 \times Q_6,$ $Z_3 \times D_4, Z_3 \times Q_4, Z_2 \times A_4, Z_3 \rtimes Z_8, (Z_6 \times Z_2) \rtimes Z_2$
26	D_{13}
27	$\Delta(27) \equiv (Z_3 \times Z_3) \rtimes Z_3, Z_9 \rtimes Z_3$
28	Q_{14}, D_{14}
30	$Z_5 \times S_3, Z_3 \times D_5, D_{15}$

All the elements $b^m a^n$ are classified into ten conjugacy classes:

$$
\begin{aligned}
C_1 &: \quad \{e\}, & h &= 1, \\
C_1^{(1)} &: \quad \{a^2\}, & h &= 2, \\
C_1^{(2)} &: \quad \{b^2\}, & h &= 2, \\
C_1^{(3)} &: \quad \{b^2 a^2\}, & h &= 2, \\
C_2^{(1)} &: \quad \{b, ba^2\}, & h &= 4, \\
C_2^{(2)} &: \quad \{ba, ba^3\}, & h &= 4, \\
C_2^{(3)} &: \quad \{b^3, b^3 a^2\}, & h &= 4, \\
C_2^{(4)} &: \quad \{b^3 a, b^3 a^3\}, & h &= 4, \\
C_2'^{(1)} &: \quad \{a, a^3\}, & h &= 4, \\
C_2'^{(2)} &: \quad \{b^2 a, b^2 a^3\}, & h &= 4.
\end{aligned}
\tag{F.3}
$$

The group $Z_4 \rtimes Z_4$ has eight singlets $\mathbf{1}_{\pm,k}$ with $k = 0, 1, 2, 3$, and two doublets $\mathbf{2}_1$ and $\mathbf{2}_2$. The generators b and a are represented by

$$b = \mathrm{i}^k, \qquad a = \pm 1, \quad \text{on } \mathbf{1}_{\pm,k}, \tag{F.4}$$

F.1 $Z_4 \rtimes Z_4$

Table F.2 Characters of $Z_4 \rtimes Z_4$

	h	$\chi_{1_{+,0}}$	$\chi_{1_{+,1}}$	$\chi_{1_{+,2}}$	$\chi_{1_{+,3}}$	$\chi_{1_{-,0}}$	$\chi_{1_{-,1}}$	$\chi_{1_{-,2}}$	$\chi_{1_{-,3}}$	χ_{2_1}	χ_{2_2}
C_1	1	1	1	1	1	1	1	1	1	2	2
$C_1^{(1)}$	2	1	1	1	1	1	1	1	1	-2	-2
$C_1^{(2)}$	2	1	1	1	1	-1	-1	-1	-1	2	-2
$C_1^{(3)}$	2	1	1	1	1	-1	-1	-1	-1	-2	2
$C_2^{(1)}$	4	1	-1	-1	1	i	$-i$	i	$-i$	0	0
$C_2^{(2)}$	4	1	1	-1	-1	$-i$	i	i	$-i$	0	0
$C_2^{(3)}$	4	1	-1	-1	1	$-i$	i	$-i$	i	0	0
$C_2^{(4)}$	4	1	1	-1	-1	i	$-i$	$-i$	i	0	0
$C_2^{\prime(1)}$	4	1	-1	1	-1	-1	-1	1	1	0	0
$C_2^{\prime(2)}$	4	1	-1	1	-1	1	1	-1	-1	0	0

and

$$b = \begin{pmatrix} 0 & 1 \\ 1 & 0 \end{pmatrix}, \quad a = \begin{pmatrix} i & 0 \\ 0 & -i \end{pmatrix}, \quad \text{on } \mathbf{2}_1, \tag{F.5}$$

$$b = \begin{pmatrix} 0 & 1 \\ -1 & 0 \end{pmatrix}, \quad a = \begin{pmatrix} -i & 0 \\ 0 & i \end{pmatrix}, \quad \text{on } \mathbf{2}_2. \tag{F.6}$$

The characters are shown in Table F.2.

The tensor products between doublets are:

$$\begin{pmatrix} x_1 \\ x_3 \end{pmatrix}_{\mathbf{2}_1} \otimes \begin{pmatrix} y_1 \\ y_3 \end{pmatrix}_{\mathbf{2}_1} = \begin{pmatrix} x_3 \\ x_1 \end{pmatrix}_{\mathbf{2}_2} \otimes \begin{pmatrix} y_3 \\ y_1 \end{pmatrix}_{\mathbf{2}_2}$$
$$= (x_1 y_3 + x_3 y_1)_{\mathbf{1}_{+,0}} \oplus (x_1 y_3 - x_3 y_1)_{\mathbf{1}_{+,2}}$$
$$\oplus (x_1 y_1 + x_3 y_3)_{\mathbf{1}_{+,3}} \oplus (x_1 y_1 - x_3 y_3)_{\mathbf{1}_{+,1}}, \tag{F.7}$$

$$\begin{pmatrix} x_1 \\ x_3 \end{pmatrix}_{\mathbf{2}_1} \otimes \begin{pmatrix} y_3 \\ y_1 \end{pmatrix}_{\mathbf{2}_2} = (x_1 y_3 + x_3 y_1)_{\mathbf{1}_{-,2}} \oplus (x_1 y_3 - x_3 y_1)_{\mathbf{1}_{-,0}}$$
$$\oplus (x_1 y_1 + x_3 y_3)_{\mathbf{1}_{-,3}} \oplus (x_1 y_1 - x_3 y_3)_{\mathbf{1}_{-,1}}. \tag{F.8}$$

The tensor products between singlets and doublets are:

$$(x)_{\mathbf{1}_{\pm,0}} \otimes \begin{pmatrix} y_{1(3)} \\ y_{3(1)} \end{pmatrix}_{\mathbf{2}_1(\mathbf{2}_2)} = \begin{pmatrix} x y_{1(3)} \\ x y_{3(1)} \end{pmatrix}_{\mathbf{2}_1(\mathbf{2}_2)}, \tag{F.9}$$

$$(x)_{\mathbf{1}_{\pm,2}} \otimes \begin{pmatrix} y_1 \\ y_3 \end{pmatrix}_{\mathbf{2}_1} = \begin{pmatrix} x y_1 \\ x y_3 \end{pmatrix}_{\mathbf{2}_2}, \tag{F.10}$$

$$(x)_{\mathbf{1}_{\pm,2}} \otimes \begin{pmatrix} y_3 \\ y_1 \end{pmatrix}_{\mathbf{2}_2} = \begin{pmatrix} x y_3 \\ x y_1 \end{pmatrix}_{\mathbf{2}_1}, \tag{F.11}$$

$$(x)_{\mathbf{1}_{\pm,1}} \otimes \begin{pmatrix} y_{1(3)} \\ y_{3(1)} \end{pmatrix}_{\mathbf{2}_1(\mathbf{2}_2)} = (xy_1)_{\mathbf{1}_{\pm,2}} \oplus (xy_3)_{\mathbf{1}_{\pm,0}}, \tag{F.12}$$

$$(x)_{\mathbf{1}_{\pm,3}} \otimes \begin{pmatrix} y_{1(3)} \\ y_{3(1)} \end{pmatrix}_{\mathbf{2}_1(\mathbf{2}_2)} = (xy_1)_{\mathbf{1}_{\pm,0}} \oplus (xy_3)_{\mathbf{1}_{\pm,2}}. \tag{F.13}$$

The tensor products between singlets are:

$$\mathbf{1}_{\pm,i} \otimes \mathbf{1}_{\pm,j} = \mathbf{1}_{\pm,i+j \,(\mathrm{mod}\,4)}, \qquad \mathbf{1}_{\pm,i} \otimes \mathbf{1}_{\mp,j} = \mathbf{1}_{\mp,i+j \,(\mathrm{mod}\,4)}, \tag{F.14}$$

where $i, j = 0, 1, 2, 3$.

F.2 $Z_8 \rtimes Z_2$

Here we study the group $Z_8 \rtimes Z_2$ other than D_8 and QD_{16}. We denote the generators of Z_8 and Z_2 by a and b, respectively. That is, they satisfy

$$a^8 = e, \qquad b^2 = e. \tag{F.15}$$

In addition, we require

$$bab = a^m, \tag{F.16}$$

where $m \neq 0$. This leads to $b^2 a b^2 = a^{m^2}$. Since $b^2 = e$, consistency requires $m^2 = 1$ mod 8. Then the possible values are found to be $m = 1, 3, 5, 7$. However, the groups with $m = 7$ and 3 correspond to D_8 and QD_{16}, respectively, while the group with $m = 1$ is just the direct product $Z_8 \times Z_2$. Therefore, we shall focus here on the group with $m = 5$.

All elements of the group can be written in the form $b^k a^\ell$ with $k = 0, 1$, and $\ell = 0, \ldots, 7$. These elements are classified into ten conjugacy classes:

$$\begin{aligned}
C_1 &: \quad \{e\}, & h &= 1, \\
C_2^{(1)} &: \quad \{a, a^5\}, & h &= 8, \\
C_2^{(2)} &: \quad \{a^3, a^7\}, & h &= 8, \\
C_1^{(1)} &: \quad \{a^2\}, & h &= 4, \\
C_1^{(2)} &: \quad \{a^4\}, & h &= 2, \\
C_1^{(3)} &: \quad \{a^6\}, & h &= 4, \\
C_2'^{(1)} &: \quad \{b, ba^4\}, & h &= 2, \\
C_2'^{(2)} &: \quad \{ba, ba^5\}, & h &= 2, \\
C_2'^{(3)} &: \quad \{ba^2, ba^6\}, & h &= 2, \\
C_2'^{(4)} &: \quad \{ba^3, ba^7\}, & h &= 2.
\end{aligned} \tag{F.17}$$

F.2 $Z_8 \rtimes Z_2$

Table F.3 Characters of $Z_8 \rtimes Z_2$

	h	$\chi_{1_{+0}}$	$\chi_{1_{-0}}$	$\chi_{1_{+1}}$	$\chi_{1_{-1}}$	$\chi_{1_{+2}}$	$\chi_{1_{-2}}$	$\chi_{1_{+3}}$	$\chi_{1_{-3}}$	χ_{2_1}	χ_{2_2}
C_1	1	1	1	1	1	1	1	1	1	2	2
$C_2^{(1)}$	8	1	1	i	i	-1	-1	$-i$	$-i$	0	0
$C_2^{(2)}$	8	1	1	$-i$	$-i$	-1	-1	i	i	0	0
$C_1^{(1)}$	4	1	1	-1	-1	1	1	-1	-1	$2i$	$-2i$
$C_1^{(2)}$	2	1	1	1	1	1	1	1	1	-2	-2
$C_1^{(3)}$	4	1	1	-1	-1	1	1	i	i	$-2i$	$2i$
$C_2'^{(1)}$	2	1	-1	1	-1	1	-1	1	-1	0	0
$C_2'^{(2)}$	2	1	-1	i	$-i$	-1	1	$-i$	i	0	0
$C_2'^{(3)}$	2	1	-1	-1	1	1	-1	-1	1	0	0
$C_2'^{(4)}$	2	1	-1	$-i$	i	-1	1	i	$-i$	0	0

The group $Z_8 \rtimes Z_2$ has eight singlets $\mathbf{1}_{\pm,k}$ with $k = 0, 1, 2, 3$, and two doublets $\mathbf{2}_1$ and $\mathbf{2}_2$. The characters are shown in Table F.3.

The generators a and b can be represented by

$$a = i^k, \quad b = \pm 1, \quad \text{on } \mathbf{1}_{\pm,k}. \tag{F.18}$$

They can also be represented by the following matrices:

$$a = \begin{pmatrix} \rho & 0 \\ 0 & \rho^5 \end{pmatrix}, \quad b = \begin{pmatrix} 0 & 1 \\ 1 & 0 \end{pmatrix}, \quad \text{on } \mathbf{2}_1, \tag{F.19}$$

$$a = \begin{pmatrix} \rho^7 & 0 \\ 0 & \rho^3 \end{pmatrix}, \quad b = \begin{pmatrix} 0 & 1 \\ 1 & 0 \end{pmatrix}, \quad \text{on } \mathbf{2}_2. \tag{F.20}$$

The tensor products between doublets are:

$$\begin{pmatrix} x_1 \\ x_2 \end{pmatrix}_{\mathbf{2}_1} \otimes \begin{pmatrix} y_1 \\ y_2 \end{pmatrix}_{\mathbf{2}_1} = (x_1 y_1 + x_2 y_2)_{\mathbf{1}_{+,1}} \oplus (x_1 y_1 - x_2 y_2)_{\mathbf{1}_{-,1}}$$
$$\oplus (x_1 y_2 + x_2 y_1)_{\mathbf{1}_{+,3}} \oplus (x_1 y_2 - x_2 y_1)_{\mathbf{1}_{-,3}}, \tag{F.21}$$

$$\begin{pmatrix} x_1 \\ x_2 \end{pmatrix}_{\mathbf{2}_1} \otimes \begin{pmatrix} y_1 \\ y_2 \end{pmatrix}_{\mathbf{2}_2} = (x_1 y_1 + x_2 y_2)_{\mathbf{1}_{+,0}} \oplus (x_1 y_1 - x_2 y_2)_{\mathbf{1}_{-,0}}$$
$$\oplus (x_1 y_2 + x_2 y_1)_{\mathbf{1}_{+,2}} \oplus (x_1 y_2 - x_2 y_1)_{\mathbf{1}_{-,2}}, \tag{F.22}$$

$$\begin{pmatrix} x_1 \\ x_2 \end{pmatrix}_{\mathbf{2}_2} \otimes \begin{pmatrix} y_1 \\ y_2 \end{pmatrix}_{\mathbf{2}_2} = (x_1 y_1 + x_2 y_2)_{\mathbf{1}_{+,3}} \oplus (x_1 y_1 - x_2 y_2)_{\mathbf{1}_{-,3}}$$
$$\oplus (x_1 y_2 + x_2 y_1)_{\mathbf{1}_{+,1}} \oplus (x_1 y_2 - x_2 y_1)_{\mathbf{1}_{-,1}}. \tag{F.23}$$

The tensor products between doublets and singlets are:

$$(x)1_{\pm,0} \otimes \begin{pmatrix} y_1 \\ y_2 \end{pmatrix}_{2_i} = \begin{pmatrix} xy_1 \\ \pm xy_2 \end{pmatrix}_{2_i}, \tag{F.24}$$

$$(x)1_{\pm,1} \otimes \begin{pmatrix} y_1 \\ y_2 \end{pmatrix}_{2_1} = \begin{pmatrix} xy_2 \\ \pm xy_1 \end{pmatrix}_{2_2}, \qquad (x)1_{\pm,1} \otimes \begin{pmatrix} y_1 \\ y_2 \end{pmatrix}_{2_2} = \begin{pmatrix} xy_1 \\ \pm xy_2 \end{pmatrix}_{2_1}, \tag{F.25}$$

$$(x)1_{\pm,2} \otimes \begin{pmatrix} y_1 \\ y_2 \end{pmatrix}_{2_1} = \begin{pmatrix} xy_2 \\ \pm xy_1 \end{pmatrix}_{2_1}, \qquad (x)1_{\pm,2} \otimes \begin{pmatrix} y_1 \\ y_2 \end{pmatrix}_{2_2} = \begin{pmatrix} xy_2 \\ \pm xy_1 \end{pmatrix}_{2_2}, \tag{F.26}$$

$$(x)1_{\pm,3} \otimes \begin{pmatrix} y_1 \\ y_2 \end{pmatrix}_{2_1} = \begin{pmatrix} xy_1 \\ \pm xy_2 \end{pmatrix}_{2_2}, \qquad (x)1_{\pm,3} \otimes \begin{pmatrix} y_1 \\ y_2 \end{pmatrix}_{2_2} = \begin{pmatrix} xy_2 \\ \pm xy_1 \end{pmatrix}_{2_1}. \tag{F.27}$$

The tensor products between singlets are

$$\mathbf{1}_{s,k} \otimes \mathbf{1}_{s',k'} = \mathbf{1}_{s'',k+k'}, \tag{F.28}$$

where $s'' = ss'$.

F.3 $(Z_2 \times Z_4) \rtimes Z_2$ (I)

Here we discuss the group $(Z_2 \times Z_4) \rtimes Z_2(I)$. We denote the first and second Z_2 generators by a and b, respectively, while the generator of Z_4 is written $\tilde{a}$. The generators a, $\tilde{a}$, and b satisfy the conditions

$$a^2 = e, \qquad \tilde{a}^4 = e, \qquad b^2 = e, \qquad bab = a\tilde{a}^2, \qquad a\tilde{a} = \tilde{a}a, \qquad \tilde{a}b = b\tilde{a}. \tag{F.29}$$

All elements can written $b^k a^\ell \tilde{a}^m$ with $k = 0, 1$, $\ell = 0, 1$, and $m = 0, 1, 2, 3$. These elements are classified into ten conjugacy classes:

$$\begin{aligned}
C_1: & \quad \{e\}, & h &= 1, \\
C_1^{(1)}: & \quad \{\tilde{a}\}, & h &= 4, \\
C_1^{(2)}: & \quad \{\tilde{a}^2\}, & h &= 2, \\
C_1^{(3)}: & \quad \{\tilde{a}^3\}, & h &= 4, \\
C_2^{(1)}: & \quad \{a, a\tilde{a}^2\}, & h &= 2, \\
C_2^{(2)}: & \quad \{a\tilde{a}, a\tilde{a}^3\}, & h &= 4, \\
C_2^{(3)}: & \quad \{b, \tilde{a}^2 b\}, & h &= 2, \\
C_2^{(4)}: & \quad \{ab, a\tilde{a}^2 b\}, & h &= 4, \\
C_2^{(5)}: & \quad \{\tilde{a}b, \tilde{a}^3 b\}, & h &= 4, \\
C_2^{(6)}: & \quad \{a\tilde{a}b, a\tilde{a}^3 b\}, & h &= 2.
\end{aligned} \tag{F.30}$$

F.3 $(Z_2 \times Z_4) \rtimes Z_2$ (I)

Table F.4 Characters of $(Z_2 \times Z_4) \rtimes Z_2(I)$

h	χ_{1+++}	χ_{1++-}	χ_{1+-+}	χ_{1+--}	χ_{1-++}	χ_{1-+-}	χ_{1--+}	χ_{1---}	χ_{2_1}	χ_{2_2}
C_1	1	1	1	1	1	1	1	1	1	2	2
$C_1^{(1)}$	4	1	1	−1	−1	1	1	−1	−1	2i	−2i
$C_1^{(2)}$	2	1	1	1	1	1	1	1	1	−2	−2
$C_1^{(3)}$	4	1	1	−1	−1	1	1	−1	−1	−2i	2i
$C_2^{(1)}$	2	1	1	1	1	−1	−1	−1	−1	0	0
$C_2^{(2)}$	4	1	1	−1	−1	−1	−1	1	1	0	0
$C_2^{(3)}$	2	1	−1	1	−1	1	−1	1	−1	0	0
$C_2^{(4)}$	4	1	−1	1	−1	−1	1	−1	1	0	0
$C_2^{(5)}$	4	1	−1	−1	1	1	−1	−1	1	0	0
$C_2^{(6)}$	2	1	−1	−1	1	−1	1	1	−1	0	0

The group $(Z_2 \rtimes Z_4) \rtimes Z_2(I)$ has eight singlets $\mathbf{1}_{\pm\pm\pm}$ and two doublets $\mathbf{2}_1$ and $\mathbf{2}_2$. The characters are shown in Table F.4.

Regarding the singlets, the generators a, $\tilde{a}$, and b can be represented by

$$a = \pm 1, \quad \text{on } \mathbf{1}_{\pm ss'}, \tag{F.31}$$

for any s and s',

$$\tilde{a} = \pm 1, \quad \text{on } \mathbf{1}_{s \pm s'}, \tag{F.32}$$

for any s and s', and

$$b = \pm 1, \quad \text{on } \mathbf{1}_{ss' \pm}, \tag{F.33}$$

for any s and s'. For the doublets, the generator $\tilde{a}$ can be represented by

$$\tilde{a} = \begin{pmatrix} i & 0 \\ 0 & i \end{pmatrix}, \quad \text{on } \mathbf{2}_1, \tag{F.34}$$

$$\tilde{a} = \begin{pmatrix} -i & 0 \\ 0 & -i \end{pmatrix}, \quad \text{on } \mathbf{2}_2. \tag{F.35}$$

The generators a and b can be represented by

$$a = \begin{pmatrix} 1 & 0 \\ 0 & -1 \end{pmatrix}, \quad b = \begin{pmatrix} 0 & 1 \\ 1 & 0 \end{pmatrix}, \tag{F.36}$$

on both doublets.

The tensor products between doublets are:

$$\begin{pmatrix} x_1 \\ x_2 \end{pmatrix}_{\mathbf{2}_1} \otimes \begin{pmatrix} y_1 \\ y_2 \end{pmatrix}_{\mathbf{2}_1} = (x_1 y_1 + x_2 y_2)_{\mathbf{1}_{+-+}} \oplus (x_1 y_1 - x_2 y_2)_{\mathbf{1}_{+--}}$$

$$\oplus (x_1y_2 + x_2y_1)\mathbf{1}_{--+} \oplus (x_1y_2 - x_2y_1)\mathbf{1}_{---}, \quad \text{(F.37)}$$

$$\begin{pmatrix} x_1 \\ x_2 \end{pmatrix}_{2_1} \otimes \begin{pmatrix} y_1 \\ y_2 \end{pmatrix}_{2_2} = (x_1y_1 + x_2y_2)\mathbf{1}_{+++} \oplus (x_1y_1 - x_2y_2)\mathbf{1}_{++-}$$

$$\oplus (x_1y_2 + x_2y_1)\mathbf{1}_{-++} \oplus (x_1y_2 - x_2y_1)\mathbf{1}_{-+-}, \quad \text{(F.38)}$$

$$\begin{pmatrix} x_1 \\ x_2 \end{pmatrix}_{2_2} \otimes \begin{pmatrix} y_1 \\ y_2 \end{pmatrix}_{2_2} = (x_1y_1 + x_2y_2)\mathbf{1}_{+-+} \oplus (x_1y_1 - x_2y_2)\mathbf{1}_{+--}$$

$$\oplus (x_1y_2 + x_2y_1)\mathbf{1}_{--+} \oplus (x_1y_2 - x_2y_1)\mathbf{1}_{---}. \quad \text{(F.39)}$$

The tensor products between doublets and singlets are:

$$(x)_{\mathbf{1}_{++\pm}} \otimes \begin{pmatrix} y_1 \\ y_2 \end{pmatrix}_{2_i} = \begin{pmatrix} xy_1 \\ \pm xy_2 \end{pmatrix}_{2_i}, \quad \text{(F.40)}$$

$$(x)_{\mathbf{1}_{+-\pm}} \otimes \begin{pmatrix} y_1 \\ y_2 \end{pmatrix}_{2_1} = \begin{pmatrix} xy_1 \\ \pm xy_2 \end{pmatrix}_{2_2}, \quad (x)_{\mathbf{1}_{+-\pm}} \otimes \begin{pmatrix} y_1 \\ y_2 \end{pmatrix}_{2_2} = \begin{pmatrix} xy_1 \\ \pm xy_2 \end{pmatrix}_{2_1}, \quad \text{(F.41)}$$

$$(x)_{\mathbf{1}_{-+\pm}} \otimes \begin{pmatrix} y_1 \\ y_2 \end{pmatrix}_{2_i} = \begin{pmatrix} xy_2 \\ \pm xy_1 \end{pmatrix}_{2_i}, \quad \text{(F.42)}$$

$$(x)_{\mathbf{1}_{--\pm}} \otimes \begin{pmatrix} y_1 \\ y_2 \end{pmatrix}_{2_1} = \begin{pmatrix} xy_2 \\ \pm xy_1 \end{pmatrix}_{2_2}, \quad (x)_{\mathbf{1}_{--\pm}} \otimes \begin{pmatrix} y_1 \\ y_2 \end{pmatrix}_{2_2} = \begin{pmatrix} xy_2 \\ \pm xy_1 \end{pmatrix}_{2_1}. \quad \text{(F.43)}$$

The tensor products between singlets are

$$\mathbf{1}_{s_1 s_2 s_3} \otimes \mathbf{1}_{s'_1 s'_2 s'_3} = \mathbf{1}_{s''_1 s''_2 s''_3}, \quad \text{(F.44)}$$

where $s''_1 = s_1 s'_1$, $s''_2 = s_2 s'_2$, and $s''_3 = s_3 s'_3$.

F.4 $(Z_2 \times Z_4) \rtimes Z_2$ (II)

We now turn to $(Z_2 \times Z_4) \rtimes Z_2(\text{II})$. We denote the first and second Z_2 generators by a and b, while the generator of Z_4 is written $\tilde{a}$. The generators a, $\tilde{a}$, and b satisfy the conditions

$$a^2 = e, \quad \tilde{a}^4 = e, \quad b^2 = e, \quad b\tilde{a}b = a\tilde{a}, \quad a\tilde{a} = \tilde{a}a, \quad ab = ba. \quad \text{(F.45)}$$

F.4 $(Z_2 \times Z_4) \rtimes Z_2$ (II)

Table F.5 Characters of $(Z_2 \times Z_4) \rtimes Z_2$(II)

	h	$\chi 1_{+0}$	$\chi 1_{+1}$	$\chi 1_{+2}$	$\chi 1_{+3}$	$\chi 1_{-0}$	$\chi 1_{-1}$	$\chi 1_{-2}$	$\chi 1_{-3}$	$\chi 2_1$	$\chi 2_2$
C_1	1	1	1	1	1	1	1	1	1	2	2
$C_1^{(1)}$	2	1	1	1	1	1	1	1	1	-2	-2
$C_1^{(2)}$	2	1	-1	1	-1	1	-1	1	-1	2	-2
$C_1^{(3)}$	2	1	-1	1	-1	1	-1	1	-1	-2	2
$C_2^{(1)}$	4	1	i	-1	$-i$	1	i	-1	$-i$	0	0
$C_2^{(2)}$	4	1	$-i$	-1	i	1	$-i$	-1	i	0	0
$C_2^{(3)}$	2	1	1	1	1	-1	-1	-1	-1	0	0
$C_2^{(4)}$	4	1	i	-1	$-i$	-1	$-i$	1	i	0	0
$C_2^{(5)}$	4	1	-1	1	-1	-1	1	-1	1	0	0
$C_2^{(6)}$	4	1	$-i$	-1	i	-1	i	1	$-i$	0	0

All elements can be written in the form $b^k a^\ell \tilde{a}^m$ with $k, \ell = 0, 1$, and $m = 0, 1, 2, 3$. These elements are classified into ten conjugacy classes:

$$\begin{aligned}
C_1: & \quad \{e\}, & h &= 1, \\
C_1^{(1)}: & \quad \{a\}, & h &= 2, \\
C_1^{(2)}: & \quad \{\tilde{a}^2\}, & h &= 2, \\
C_1^{(3)}: & \quad \{a\tilde{a}^2\}, & h &= 2, \\
C_2^{(1)}: & \quad \{\tilde{a}, a\tilde{a}\}, & h &= 4, \\
C_2^{(2)}: & \quad \{\tilde{a}^3, a\tilde{a}^3\}, & h &= 4, \\
C_2^{(3)}: & \quad \{b, ab\}, & h &= 2, \\
C_2^{(4)}: & \quad \{\tilde{a}b, a\tilde{a}b\}, & h &= 4, \\
C_2^{(5)}: & \quad \{\tilde{a}^2 b, a\tilde{a}^2 b\}, & h &= 2, \\
C_2^{(6)}: & \quad \{\tilde{a}^3 b, a\tilde{a}^3 b\}, & h &= 2.
\end{aligned} \quad (\text{F.46})$$

The group $(Z_2 \times Z_4) \rtimes Z_2$(II) has eight singlets $1_{\pm,k}$ with $k = 0, 1, 2, 3$, and two doublets 2_1 and 2_2. The characters are shown in Table F.5.

The generators, a, $\tilde{a}$ and b, can be represented by

$$a = 1, \quad \tilde{a} = i^k, \quad b = \pm 1, \quad \text{on } 1_{\pm,k}. \quad (\text{F.47})$$

In addition, the generator $\tilde{a}$ can be represented by

$$\tilde{a} = \begin{pmatrix} 1 & 0 \\ 0 & -1 \end{pmatrix}, \quad \text{on } 2_1, \quad (\text{F.48})$$

$$\tilde{a} = \begin{pmatrix} i & 0 \\ 0 & -i \end{pmatrix}, \quad \text{on } \mathbf{2}_2, \tag{F.49}$$

and for both doublets the generators a and b can be represented by

$$a = \begin{pmatrix} -1 & 0 \\ 0 & -1 \end{pmatrix}, \quad b = \begin{pmatrix} 0 & 1 \\ 1 & 0 \end{pmatrix}. \tag{F.50}$$

The tensor products between doublets are:

$$\begin{pmatrix} x_1 \\ x_2 \end{pmatrix}_{\mathbf{2}_1} \otimes \begin{pmatrix} y_1 \\ y_2 \end{pmatrix}_{\mathbf{2}_1} = (x_1 y_1 + x_2 y_2)\mathbf{1}_{+,0} \oplus (x_1 y_1 - x_2 y_2)\mathbf{1}_{-,0}$$

$$\oplus (x_1 y_2 + x_2 y_1)\mathbf{1}_{+,2} \oplus (x_1 y_2 - x_2 y_1)\mathbf{1}_{-,2}, \tag{F.51}$$

$$\begin{pmatrix} x_1 \\ x_2 \end{pmatrix}_{\mathbf{2}_1} \otimes \begin{pmatrix} y_1 \\ y_2 \end{pmatrix}_{\mathbf{2}_2} = (x_1 y_1 + x_2 y_2)\mathbf{1}_{+,1} \oplus (x_1 y_1 - x_2 y_2)\mathbf{1}_{-,1}$$

$$\oplus (x_1 y_2 + x_2 y_1)\mathbf{1}_{+,3} \oplus (x_1 y_2 - x_2 y_1)\mathbf{1}_{-,3}, \tag{F.52}$$

$$\begin{pmatrix} x_1 \\ x_2 \end{pmatrix}_{\mathbf{2}_2} \otimes \begin{pmatrix} y_1 \\ y_2 \end{pmatrix}_{\mathbf{2}_2} = (x_1 y_1 + x_2 y_2)\mathbf{1}_{+,2} \oplus (x_1 y_1 - x_2 y_2)\mathbf{1}_{-,2}$$

$$\oplus (x_1 y_2 + x_2 y_1)\mathbf{1}_{+,0} \oplus (x_1 y_2 - x_2 y_1)\mathbf{1}_{-,0}. \tag{F.53}$$

The tensor products between doublets and singlets are:

$$(x)\mathbf{1}_{\pm,0} \otimes \begin{pmatrix} y_1 \\ y_2 \end{pmatrix}_{\mathbf{2}_i} = \begin{pmatrix} xy_1 \\ \pm xy_2 \end{pmatrix}_{\mathbf{2}_i}, \tag{F.54}$$

$$(x)\mathbf{1}_{\pm,1} \otimes \begin{pmatrix} y_1 \\ y_2 \end{pmatrix}_{\mathbf{2}_1} = \begin{pmatrix} xy_1 \\ \pm xy_2 \end{pmatrix}_{\mathbf{2}_2}, \quad (x)\mathbf{1}_{\pm,1} \otimes \begin{pmatrix} y_1 \\ y_2 \end{pmatrix}_{\mathbf{2}_2} = \begin{pmatrix} xy_2 \\ \pm xy_1 \end{pmatrix}_{\mathbf{2}_1},$$

$$\tag{F.55}$$

$$(x)\mathbf{1}_{\pm,2} \otimes \begin{pmatrix} y_1 \\ y_2 \end{pmatrix}_{\mathbf{2}_i} = \begin{pmatrix} xy_2 \\ \pm xy_1 \end{pmatrix}_{\mathbf{2}_i}, \tag{F.56}$$

$$(x)\mathbf{1}_{\pm,3} \otimes \begin{pmatrix} y_1 \\ y_2 \end{pmatrix}_{\mathbf{2}_1} = \begin{pmatrix} xy_2 \\ \pm xy_1 \end{pmatrix}_{\mathbf{2}_2}, \quad (x)\mathbf{1}_{\pm,3} \otimes \begin{pmatrix} y_1 \\ y_2 \end{pmatrix}_{\mathbf{2}_2} = \begin{pmatrix} xy_1 \\ \pm xy_2 \end{pmatrix}_{\mathbf{2}_1}.$$

$$\tag{F.57}$$

The tensor products between singlets are

$$\mathbf{1}_{s,k} \otimes \mathbf{1}_{s',k'} = \mathbf{1}_{s'',k+k'}, \tag{F.58}$$

where $s'' = ss'$.

F.5 $Z_3 \rtimes Z_8$

Here we denote the Z_3 and Z_8 generators by a and b, respectively. These generators satisfy the conditions

$$a^3 = e, \qquad b^8 = e, \qquad b^{-1}ab = a^2. \tag{F.59}$$

All elements can be written $b^k a^\ell$ with $k = 0, \ldots, 7$, and $\ell = 0, 1, 2$. These elements are classified into twelve conjugacy classes:

$$\begin{aligned}
C_1 &: \{e\}, & h &= 1, \\
C_1^{(1)} &: \{b^2\}, & h &= 4, \\
C_1^{(2)} &: \{b^4\}, & h &= 2, \\
C_1^{(3)} &: \{b^6\}, & h &= 4, \\
C_2^{(1)} &: \{a, a^2\}, & h &= 3, \\
C_2^{(2)} &: \{b^2 a, b^2 a^2\}, & h &= 12, \\
C_2^{(3)} &: \{b^4 a, b^4 a^2\}, & h &= 3, \\
C_2^{(4)} &: \{b^6 a, b^6 a^2\}, & h &= 3, \\
C_3^{(1)} &: \{b, ba, ba^2\}, & h &= 8, \\
C_3^{(2)} &: \{b^3, b^3 a, b^3 a^2\}, & h &= 8, \\
C_3^{(3)} &: \{b^5, b^5 a, b^5 a^2\}, & h &= 8, \\
C_3^{(4)} &: \{b^7, b^7 a, b^7 a^2\}, & h &= 8.
\end{aligned} \tag{F.60}$$

The group $Z_3 \rtimes Z_8$ has eight singlets $\mathbf{1}_r$ with $r = 0, \ldots, 7$, and four doublets $\mathbf{2}_k$ with $k = 1, 2, 3, 4$. The characters are shown in Table F.6.

The generators a and b can be represented by

$$a = 1, \qquad b = \rho^k, \qquad \text{on } \mathbf{1}_k, \tag{F.61}$$

where $\rho = e^{\pi i/4}$. The generators a and b are also represented by the following matrices:

$$a = \begin{pmatrix} \omega & 0 \\ 0 & \omega^2 \end{pmatrix}, \qquad b = \begin{pmatrix} 0 & 1 \\ 1 & 0 \end{pmatrix}, \qquad \text{on } \mathbf{2}_1, \tag{F.62}$$

$$a = \begin{pmatrix} \omega & 0 \\ 0 & \omega^2 \end{pmatrix}, \qquad b = \begin{pmatrix} 0 & 1 \\ i & 0 \end{pmatrix}, \qquad \text{on } \mathbf{2}_2, \tag{F.63}$$

$$a = \begin{pmatrix} \omega & 0 \\ 0 & \omega^2 \end{pmatrix}, \qquad b = \begin{pmatrix} 0 & 1 \\ -1 & 0 \end{pmatrix}, \qquad \text{on } \mathbf{2}_3, \tag{F.64}$$

$$a = \begin{pmatrix} \omega & 0 \\ 0 & \omega^2 \end{pmatrix}, \qquad b = \begin{pmatrix} 0 & 1 \\ -i & 0 \end{pmatrix}, \qquad \text{on } \mathbf{2}_4, \tag{F.65}$$

Table F.6 Characters of $Z_3 \rtimes Z_8$

	h	χ_{1_0}	χ_{1_1}	χ_{1_2}	χ_{1_3}	χ_{1_4}	χ_{1_5}	χ_{1_6}	χ_{1_7}	χ_{2_1}	χ_{2_2}	χ_{2_3}	χ_{2_4}
C_1	1	1	1	1	1	1	1	1	1	2	2	2	2
$C_1^{(1)}$	8	1	i	-1	$-i$	1	i	-1	$-i$	2	$2i$	-2	$-2i$
$C_1^{(2)}$	8	1	-1	1	-1	1	-1	1	-1	2	-2	2	-2
$C_1^{(3)}$	8	1	$-i$	-1	i	1	$-i$	-1	i	2	$-2i$	-2	$2i$
$C_2^{(1)}$	3	1	1	1	1	1	1	1	1	-1	-1	-1	-1
$C_2^{(2)}$	12	1	i	-1	$-i$	1	i	-1	$-i$	-2	$-2i$	2	$2i$
$C_2^{(3)}$	3	1	-1	1	-1	1	-1	1	-1	-2	2	-2	2
$C_2^{(4)}$	3	1	$-i$	-1	i	1	$-i$	-1	i	-2	$2i$	2	$-2i$
$C_3^{(1)}$	8	1	ρ	i	$i\rho$	-1	$-\rho$	$-i$	$-i\rho$	0	0	0	0
$C_3^{(2)}$	12	1	$i\rho$	$-i$	ρ	-1	$-i\rho$	i	$-\rho$	0	0	0	0
$C_3^{(3)}$	8	1	$-\rho$	i	$-i\rho$	-1	ρ	$-i$	$i\rho$	0	0	0	0
$C_3^{(4)}$	8	1	$-i\rho$	$-i$	$-\rho$	-1	$i\rho$	i	ρ	0	0	0	0

where $\omega = e^{2\pi i/3}$.

The tensor products between doublets are:

$$\begin{pmatrix} x_1 \\ x_2 \end{pmatrix}_{2_1} \otimes \begin{pmatrix} y_1 \\ y_2 \end{pmatrix}_{2_1} = \begin{pmatrix} x_2 y_2 \\ x_1 y_1 \end{pmatrix}_{2_1} \oplus (x_1 y_2 + x_2 y_1)_{1_0} \oplus (x_1 y_2 - x_2 y_1)_{1_4}, \quad (F.66)$$

$$\begin{pmatrix} x_1 \\ x_2 \end{pmatrix}_{2_1} \otimes \begin{pmatrix} y_1 \\ y_2 \end{pmatrix}_{2_2} = \begin{pmatrix} x_2 y_1 \\ x_1 y_2 \end{pmatrix}_{2_2} \oplus (x_1 y_1 + i\rho x_2 y_2)_{1_1} \oplus (x_1 y_1 - i\rho x_2 y_2)_{1_5},$$

(F.67)

$$\begin{pmatrix} x_1 \\ x_2 \end{pmatrix}_{2_1} \otimes \begin{pmatrix} y_1 \\ y_2 \end{pmatrix}_{2_3} = \begin{pmatrix} x_2 y_2 \\ -x_1 y_1 \end{pmatrix}_{2_3} \oplus (x_1 y_2 + i x_2 y_1)_{1_2} \oplus (x_1 y_2 - i x_2 y_1)_{1_6},$$

(F.68)

$$\begin{pmatrix} x_1 \\ x_2 \end{pmatrix}_{2_1} \otimes \begin{pmatrix} y_1 \\ y_2 \end{pmatrix}_{2_4} = \begin{pmatrix} x_2 y_1 \\ x_1 y_2 \end{pmatrix}_{2_1} \oplus (x_1 y_2 + i\rho x_2 y_1)_{1_3} \oplus (x_1 y_2 - i\rho x_2 y_1)_{1_7},$$

(F.69)

$$\begin{pmatrix} x_1 \\ x_2 \end{pmatrix}_{2_2} \otimes \begin{pmatrix} y_1 \\ y_2 \end{pmatrix}_{2_2} = \begin{pmatrix} x_2 y_2 \\ -x_1 y_1 \end{pmatrix}_{2_3} \oplus (x_1 y_2 + x_2 y_1)_{1_2} \oplus (x_1 y_2 - x_2 y_1)_{1_6},$$

(F.70)

$$\begin{pmatrix} x_1 \\ x_2 \end{pmatrix}_{2_2} \otimes \begin{pmatrix} y_1 \\ y_2 \end{pmatrix}_{2_3} = \begin{pmatrix} x_2 y_2 \\ -i x_1 y_1 \end{pmatrix}_{2_4} \oplus (x_1 y_2 + i\rho x_2 y_1)_{1_3} \oplus (x_1 y_2 - i\rho x_2 y_1)_{1_7},$$

(F.71)

F.6 $(Z_6 \times Z_2) \rtimes Z_2$

$$\begin{pmatrix} x_1 \\ x_2 \end{pmatrix}_{2_2} \otimes \begin{pmatrix} y_1 \\ y_2 \end{pmatrix}_{2_4} = \begin{pmatrix} x_2 y_2 \\ x_1 y_1 \end{pmatrix}_{2_1} \oplus (x_1 y_2 + i x_2 y_1)_{1_0} \oplus (x_1 y_2 - i x_2 y_1)_{1_4},$$
(F.72)

$$\begin{pmatrix} x_1 \\ x_2 \end{pmatrix}_{2_3} \otimes \begin{pmatrix} y_1 \\ y_2 \end{pmatrix}_{2_3} = \begin{pmatrix} x_2 y_2 \\ x_1 y_1 \end{pmatrix}_{2_1} \oplus (x_1 y_2 + x_2 y_1)_{1_0} \oplus (x_1 y_2 - x_2 y_1)_{1_4},$$
(F.73)

$$\begin{pmatrix} x_1 \\ x_2 \end{pmatrix}_{2_3} \otimes \begin{pmatrix} y_1 \\ y_2 \end{pmatrix}_{2_4} = \begin{pmatrix} x_2 y_2 \\ i x_1 y_1 \end{pmatrix}_{2_2} \oplus (x_1 y_2 + \rho x_2 y_1)_{1_1} \oplus (x_1 y_2 - \rho x_2 y_1)_{1_5},$$
(F.74)

$$\begin{pmatrix} x_1 \\ x_2 \end{pmatrix}_{2_4} \otimes \begin{pmatrix} y_1 \\ y_2 \end{pmatrix}_{2_4} = \begin{pmatrix} x_2 y_2 \\ -x_1 y_1 \end{pmatrix}_{2_3} \oplus (x_1 y_2 + x_2 y_1)_{1_2} \oplus (x_1 y_2 - x_2 y_1)_{1_6}.$$
(F.75)

The tensor products between doublets and singlets are

$$(x)_{1_i} \otimes \begin{pmatrix} y_1 \\ y_2 \end{pmatrix}_{2_j} = \begin{pmatrix} x y_1 \\ \rho^i x y_2 \end{pmatrix}_{2_{i+j} \bmod 4}.$$
(F.76)

The tensor products between singlets are

$$\mathbf{1}_k \otimes \mathbf{1}_{k'} = \mathbf{1}_{k+k' \bmod 8}.$$
(F.77)

F.6 $(Z_6 \times Z_2) \rtimes Z_2$

Now we consider the group $(Z_6 \times Z_2) \rtimes Z_2$. We denote the generator of Z_6 by a, while the first and second Z_2 generators are written b and c, respectively. The generators a, b, and c satisfy the conditions

$$\begin{gathered} a^6 = e, \quad b^2 = e, \quad c^2 = e, \\ c^{-1} ab = a^5, \quad c^{-1} bc = a^3 b, \quad ab = ba. \end{gathered}$$
(F.78)

Table F.7 Characters of $(Z_6 \times Z_2) \rtimes Z_2$

	h	$\chi_{1_{++}}$	$\chi_{1_{+-}}$	$\chi_{1_{-+}}$	$\chi_{1_{--}}$	χ_{2_1}	χ_{2_2}	χ_{2_3}	χ_{2_4}	χ_{2_5}
C_1	1	1	1	1	1	2	2	2	2	2
$C_1^{(1)}$	2	1	1	1	1	2	2	-2	-2	-2
$C_2^{(1)}$	6	1	1	1	1	-1	-1	1	1	-2
$C_2^{(2)}$	3	1	1	1	1	-1	-1	-1	-1	2
$C_2^{(3)}$	2	1	1	-1	-1	2	-2	0	0	0
$C_2^{(4)}$	6	1	1	-1	-1	-1	1	$\sqrt{3}i$	$-\sqrt{3}i$	0
$C_2^{(5)}$	6	1	1	-1	-1	-1	1	$-\sqrt{3}i$	$\sqrt{3}i$	0
$C_6^{(1)}$	2	1	-1	1	-1	0	0	0	0	0
$C_6^{(2)}$	6	1	-1	-1	1	0	0	0	0	0

All elements can be written in the form $a^k b^\ell c^m$ with $k = 0, \ldots, 5$, and $\ell, m = 0, 1$. These elements are classified into nine conjugacy classes:

$$
\begin{aligned}
C_1: & \quad \{e\}, & h &= 1, \\
C_1^{(1)}: & \quad \{a^3\}, & h &= 2, \\
C_2^{(1)}: & \quad \{a, a^5\}, & h &= 2, \\
C_2^{(2)}: & \quad \{a^2, a^4\}, & h &= 2, \\
C_2^{(3)}: & \quad \{b, a^3 b\}, & h &= 2, \\
C_2^{(4)}: & \quad \{ab, a^2 b\}, & h &= 6, \\
C_2^{(5)}: & \quad \{a^4 b, a^5 b\}, & h &= 6, \\
C_6^{(1)}: & \quad \{c, ac, a^2 c, a^3 c, a^4 c, a^5 c\}, & h &= 2, \\
C_6^{(2)}: & \quad \{bc, abc, a^2 bc, a^3 bc, a^4 bc, a^5 bc\}, & h &= 6.
\end{aligned}
\tag{F.79}
$$

The group $(Z_6 \times Z_2) \rtimes Z_2$ has four singlets denoted by $1_{\pm\pm}$ and five doublets 2_k with $k = 1, \ldots, 5$. The characters are shown in Table F.7.

For singlets, the generators b and c can be represented by

$$b = \pm 1 \quad \text{on } 1_{\pm s}, \tag{F.80}$$

for both $s = \pm$, and

$$c = \pm 1 \quad \text{on } 1_{s\pm}, \tag{F.81}$$

for both $s = \pm$, while the generator a is represented by $a = 1$ on all the singlets. For doublets, we use the following representations for the generators a and b:

$$a = \begin{pmatrix} \omega & 0 \\ 0 & \omega^2 \end{pmatrix}, \quad b = \begin{pmatrix} 1 & 0 \\ 0 & 1 \end{pmatrix}, \quad \text{on } 2_1, \tag{F.82}$$

F.6 $(Z_6 \times Z_2) \rtimes Z_2$

$$a = \begin{pmatrix} \omega & 0 \\ 0 & \omega^2 \end{pmatrix}, \quad b = \begin{pmatrix} -1 & 0 \\ 0 & -1 \end{pmatrix}, \quad \text{on } \mathbf{2}_2, \tag{F.83}$$

$$a = \begin{pmatrix} -\omega^2 & 0 \\ 0 & -\omega \end{pmatrix}, \quad b = \begin{pmatrix} 1 & 0 \\ 0 & -1 \end{pmatrix}, \quad \text{on } \mathbf{2}_3, \tag{F.84}$$

$$a = \begin{pmatrix} -\omega & 0 \\ 0 & -\omega^2 \end{pmatrix}, \quad b = \begin{pmatrix} 1 & 0 \\ 0 & -1 \end{pmatrix}, \quad \text{on } \mathbf{2}_4, \tag{F.85}$$

$$a = \begin{pmatrix} -1 & 0 \\ 0 & -1 \end{pmatrix}, \quad b = \begin{pmatrix} 1 & 0 \\ 0 & -1 \end{pmatrix}, \quad \text{on } \mathbf{2}_5, \tag{F.86}$$

where $\omega = e^{2\pi i/3}$, while c is represented by

$$c = \begin{pmatrix} 0 & 1 \\ 1 & 0 \end{pmatrix}, \tag{F.87}$$

on all the doublets.

The tensor products between doublets are:

$$\begin{pmatrix} x_1 \\ x_2 \end{pmatrix}_{\mathbf{2}_1} \otimes \begin{pmatrix} y_1 \\ y_2 \end{pmatrix}_{\mathbf{2}_1} = \begin{pmatrix} x_2 y_2 \\ x_1 y_1 \end{pmatrix}_{\mathbf{2}_1} \oplus (x_1 y_2 + x_2 y_1) \mathbf{1}_{++} \oplus (x_1 y_2 - x_2 y_1) \mathbf{1}_{+-},$$
$$\tag{F.88}$$

$$\begin{pmatrix} x_1 \\ x_2 \end{pmatrix}_{\mathbf{2}_1} \otimes \begin{pmatrix} y_1 \\ y_2 \end{pmatrix}_{\mathbf{2}_2} = \begin{pmatrix} x_2 y_2 \\ x_1 y_1 \end{pmatrix}_{\mathbf{2}_2} \oplus (x_1 y_2 + x_2 y_1) \mathbf{1}_{-+} \oplus (x_1 y_2 - x_2 y_1) \mathbf{1}_{--},$$
$$\tag{F.89}$$

$$\begin{pmatrix} x_1 \\ x_2 \end{pmatrix}_{\mathbf{2}_1} \otimes \begin{pmatrix} y_1 \\ y_2 \end{pmatrix}_{\mathbf{2}_3} = \begin{pmatrix} x_2 y_1 \\ x_1 y_2 \end{pmatrix}_{\mathbf{2}_4} \oplus \begin{pmatrix} x_1 y_1 \\ x_2 y_2 \end{pmatrix}_{\mathbf{2}_5}, \tag{F.90}$$

$$\begin{pmatrix} x_1 \\ x_2 \end{pmatrix}_{\mathbf{2}_1} \otimes \begin{pmatrix} y_1 \\ y_2 \end{pmatrix}_{\mathbf{2}_4} = \begin{pmatrix} x_1 y_1 \\ x_2 y_2 \end{pmatrix}_{\mathbf{2}_3} \oplus \begin{pmatrix} x_2 y_1 \\ x_1 y_2 \end{pmatrix}_{\mathbf{2}_5}, \tag{F.91}$$

$$\begin{pmatrix} x_1 \\ x_2 \end{pmatrix}_{\mathbf{2}_1} \otimes \begin{pmatrix} y_1 \\ y_2 \end{pmatrix}_{\mathbf{2}_5} = \begin{pmatrix} x_2 y_1 \\ x_1 y_2 \end{pmatrix}_{\mathbf{2}_3} \oplus \begin{pmatrix} x_1 y_1 \\ x_2 y_2 \end{pmatrix}_{\mathbf{2}_4}, \tag{F.92}$$

$$\begin{pmatrix} x_1 \\ x_2 \end{pmatrix}_{\mathbf{2}_2} \otimes \begin{pmatrix} y_1 \\ y_2 \end{pmatrix}_{\mathbf{2}_2} = \begin{pmatrix} x_2 y_2 \\ x_1 y_1 \end{pmatrix}_{\mathbf{2}_1} \oplus (x_1 y_2 + x_2 y_1) \mathbf{1}_{++} \oplus (x_1 y_2 - x_2 y_1) \mathbf{1}_{+-},$$
$$\tag{F.93}$$

$$\begin{pmatrix} x_1 \\ x_2 \end{pmatrix}_{\mathbf{2}_2} \otimes \begin{pmatrix} y_1 \\ y_2 \end{pmatrix}_{\mathbf{2}_3} = \begin{pmatrix} x_1 y_2 \\ x_2 y_1 \end{pmatrix}_{\mathbf{2}_3} \oplus \begin{pmatrix} x_2 y_2 \\ x_1 y_1 \end{pmatrix}_{\mathbf{2}_5}, \tag{F.94}$$

$$\begin{pmatrix} x_1 \\ x_2 \end{pmatrix}_{\mathbf{2}_2} \otimes \begin{pmatrix} y_1 \\ y_2 \end{pmatrix}_{\mathbf{2}_4} = \begin{pmatrix} x_2 y_2 \\ x_1 y_1 \end{pmatrix}_{\mathbf{2}_4} \oplus \begin{pmatrix} x_1 y_2 \\ x_2 y_1 \end{pmatrix}_{\mathbf{2}_5}, \tag{F.95}$$

$$\begin{pmatrix} x_1 \\ x_2 \end{pmatrix}_{2_2} \otimes \begin{pmatrix} y_1 \\ y_2 \end{pmatrix}_{2_5} = \begin{pmatrix} x_2 y_2 \\ x_1 y_1 \end{pmatrix}_{2_3} \oplus \begin{pmatrix} x_1 y_2 \\ x_2 y_1 \end{pmatrix}_{2_4}, \tag{F.96}$$

$$\begin{pmatrix} x_1 \\ x_2 \end{pmatrix}_{2_3} \otimes \begin{pmatrix} y_1 \\ y_2 \end{pmatrix}_{2_3} = \begin{pmatrix} x_1 y_1 \\ x_2 y_2 \end{pmatrix}_{2_1} \oplus (x_1 y_2 + x_2 y_1)_{1_{-+}} \oplus (x_1 y_2 - x_2 y_1)_{1_{--}}, \tag{F.97}$$

$$\begin{pmatrix} x_1 \\ x_2 \end{pmatrix}_{2_3} \otimes \begin{pmatrix} y_1 \\ y_2 \end{pmatrix}_{2_4} = \begin{pmatrix} x_1 y_2 \\ x_2 y_1 \end{pmatrix}_{2_2} \oplus (x_1 y_1 + x_2 y_2)_{1_{++}} \oplus (x_1 y_1 - x_2 y_2)_{1_{+-}}, \tag{F.98}$$

$$\begin{pmatrix} x_1 \\ x_2 \end{pmatrix}_{2_3} \otimes \begin{pmatrix} y_1 \\ y_2 \end{pmatrix}_{2_5} = \begin{pmatrix} x_2 y_2 \\ x_1 y_1 \end{pmatrix}_{2_1} \oplus \begin{pmatrix} x_2 y_1 \\ x_1 y_2 \end{pmatrix}_{2_2}, \tag{F.99}$$

$$\begin{pmatrix} x_1 \\ x_2 \end{pmatrix}_{2_4} \otimes \begin{pmatrix} y_1 \\ y_2 \end{pmatrix}_{2_4} = \begin{pmatrix} x_2 y_1 \\ x_1 y_1 \end{pmatrix}_{2_1} \oplus (x_1 y_2 + x_2 y_1)_{1_{-+}} \oplus (x_1 y_2 - x_2 y_1)_{1_{--}}, \tag{F.100}$$

$$\begin{pmatrix} x_1 \\ x_2 \end{pmatrix}_{2_4} \otimes \begin{pmatrix} y_1 \\ y_2 \end{pmatrix}_{2_5} = \begin{pmatrix} x_1 y_1 \\ x_2 y_2 \end{pmatrix}_{2_1} \oplus \begin{pmatrix} x_1 y_2 \\ x_2 y_1 \end{pmatrix}_{2_2}, \tag{F.101}$$

$$\begin{pmatrix} x_1 \\ x_2 \end{pmatrix}_{2_5} \otimes \begin{pmatrix} y_1 \\ y_2 \end{pmatrix}_{2_5} = (x_1 y_1 + x_2 y_2)_{1_{++}} \oplus (x_1 y_1 - x_2 y_2)_{1_{+-}}$$
$$\oplus (x_1 y_2 + x_2 y_1)_{1_{-+}} \oplus (x_1 y_2 - x_2 y_1)_{1_{--}}. \tag{F.102}$$

The tensor products between doublets and singlets are:

$$(x)_{1_{++}} \otimes \begin{pmatrix} y_1 \\ y_2 \end{pmatrix}_{2_i} = \begin{pmatrix} x y_1 \\ \pm x y_2 \end{pmatrix}_{2_i}, \tag{F.103}$$

$$(x)_{1_{-\pm}} \otimes \begin{pmatrix} y_1 \\ y_2 \end{pmatrix}_{2_1} = \begin{pmatrix} x y_1 \\ \pm x y_2 \end{pmatrix}_{2_2}, \quad (x)_{1_{-\pm}} \otimes \begin{pmatrix} y_1 \\ y_2 \end{pmatrix}_{2_2} = \begin{pmatrix} x y_1 \\ \pm x y_2 \end{pmatrix}_{2_1}, \tag{F.104}$$

$$(x)_{1_{-\pm}} \otimes \begin{pmatrix} y_1 \\ y_2 \end{pmatrix}_{2_3} = \begin{pmatrix} x y_2 \\ \pm x y_1 \end{pmatrix}_{2_4}, \quad (x)_{1_{-\pm}} \otimes \begin{pmatrix} y_1 \\ y_2 \end{pmatrix}_{2_4} = \begin{pmatrix} x y_2 \\ \pm x y_1 \end{pmatrix}_{2_3}, \tag{F.105}$$

$$(x)_{1_{-\pm}} \otimes \begin{pmatrix} y_1 \\ y_2 \end{pmatrix}_{2_5} = \begin{pmatrix} x y_2 \\ \pm x y_1 \end{pmatrix}_{2_5}. \tag{F.106}$$

The tensor products between singlets are

$$\mathbf{1}_{s_1 s_2} \otimes \mathbf{1}_{s'_1 s'_2} = \mathbf{1}_{s''_1 s''_2}, \tag{F.107}$$

where $s''_1 = s_1 s'_1$ and $s''_2 = s_2 s'_2$.

F.7 $Z_9 \rtimes Z_3$

We denote the Z_9 and Z_3 generators by a and b, respectively. They satisfy

$$a^9 = 1, \qquad ab = ba^7. \tag{F.108}$$

Using these, all elements of $Z_9 \rtimes Z_3$ can be written in the form

$$g = b^m a^n, \tag{F.109}$$

with $m = 0, 1, 2$, and $n = 0, \ldots, 8$.

These elements are classified into eleven conjugacy classes:

$$\begin{aligned}
C_1 &: \{e\}, & h &= 1, \\
C_1^{(2)} &: \{a^3\}, & h &= 3, \\
C_1^{(3)} &: \{a^6\}, & h &= 3, \\
C_3^{(1)} &: \{b, ba^3, ba^6\}, & h &= 3, \\
C_3^{(2)} &: \{ba, ba^4, ba^7\}, & h &= 9, \\
C_3^{(3)} &: \{ba^2, ba^5, ba^8\}, & h &= 9, \\
C_3^{(4)} &: \{b^2, b^2a^3, b^2a^6\}, & h &= 3, \\
C_3^{(5)} &: \{b^2a, b^2a^4, b^2a^7\}, & h &= 9, \\
C_3^{(6)} &: \{b^2a^2, b^2a^5, b^2a^8\}, & h &= 9, \\
C_3^{(7)} &: \{a, a^4, a^7\}, & h &= 3, \\
C_3^{(8)} &: \{a^2, a^5, a^8\}, & h &= 3.
\end{aligned} \tag{F.110}$$

This group has nine singlets $\mathbf{1}_{n,k}$ with $n, k = 0, 1, 2$, and two triplets $\mathbf{3}_1$ and $\mathbf{3}_2$. The characters are shown in Table F.8, where $\omega = e^{2\pi i/3}$.

The generators a and b are represented by

$$a = \omega^k, \qquad b = \omega^\ell, \qquad \text{on } \mathbf{1}_{k,\ell}. \tag{F.111}$$

We use the following representations for the generator a:

$$a = \begin{pmatrix} \rho & 0 & 0 \\ 0 & \rho^4 & 0 \\ 0 & 0 & \rho^7 \end{pmatrix}, \text{ on } \mathbf{3}_1, \qquad a = \begin{pmatrix} \rho^2 & 0 & 0 \\ 0 & \rho^8 & 0 \\ 0 & 0 & \rho^5 \end{pmatrix}, \text{ on } \mathbf{3}_2, \tag{F.112}$$

where $\rho = e^{2\pi i/9}$, while b is represented by

$$b = \begin{pmatrix} 0 & 1 & 0 \\ 0 & 0 & 1 \\ 1 & 0 & 0 \end{pmatrix}, \tag{F.113}$$

Table F.8 Characters of T_9

h	$\chi 1_{00}$	$\chi 1_{01}$	$\chi 1_{02}$	$\chi 1_{10}$	$\chi 1_{11}$	$\chi 1_{12}$	$\chi 1_{20}$	$\chi 1_{21}$	$\chi 1_{22}$	$\chi 3_1$	$\chi 3_2$
C_1	1	1	1	1	1	1	1	1	1	3	3
$C_1^{(2)}$	3	1	1	1	1	1	1	1	1	3ω	$3\omega^2$
$C_1^{(2)}$	3	1	1	1	1	1	1	1	1	$3\omega^2$	3ω
$C_3^{(1)}$	9	1	ω	ω^2	1	ω	ω^2	1	ω	ω^2	0	0
$C_3^{(2)}$	9	1	ω	ω^2	ω	ω^2	1	ω^2	1	ω	0	0
$C_3^{(3)}$	3	1	ω	ω^2	ω^2	1	ω^2	ω	ω^2	1	0	0
$C_3^{(4)}$	9	1	ω^2	ω	1	ω^2	ω	1	ω^2	ω	0	0
$C_3^{(5)}$	9	1	ω^2	ω	ω	1	ω^2	ω^2	ω	1	0	0
$C_3^{(6)}$	3	1	ω^2	ω	ω^2	ω	1	ω	1	ω^2	0	0
$C_3^{(7)}$	3	1	1	1	ω	ω	ω	ω^2	ω^2	ω^2	0	0
$C_3^{(8)}$	3	1	1	1	ω^2	ω^2	ω^2	ω	ω	ω	0	0

on both triplets.

The tensor products between triplets are:

$$\begin{pmatrix} x_1 \\ x_2 \\ x_3 \end{pmatrix}_{3_1} \otimes \begin{pmatrix} y_1 \\ y_2 \\ y_3 \end{pmatrix}_{3_1} = \begin{pmatrix} x_1 y_1 \\ x_2 y_2 \\ x_3 y_3 \end{pmatrix}_{3_2} \oplus \begin{pmatrix} x_2 y_3 \\ x_3 y_1 \\ x_1 y_2 \end{pmatrix}_{3_2} \oplus \begin{pmatrix} x_3 y_2 \\ x_1 y_3 \\ x_2 y_1 \end{pmatrix}_{3_2}, \quad (F.114)$$

$$\begin{pmatrix} x_1 \\ x_2 \\ x_3 \end{pmatrix}_{3_2} \otimes \begin{pmatrix} y_1 \\ y_2 \\ y_3 \end{pmatrix}_{3_2} = \begin{pmatrix} x_3 y_3 \\ x_1 y_1 \\ x_2 y_2 \end{pmatrix}_{3_1} \oplus \begin{pmatrix} x_1 y_2 \\ x_2 y_3 \\ x_3 y_1 \end{pmatrix}_{3_1} \oplus \begin{pmatrix} x_2 y_1 \\ x_3 y_2 \\ x_1 y_3 \end{pmatrix}_{3_1}, \quad (F.115)$$

$$\begin{pmatrix} x_1 \\ x_2 \\ x_3 \end{pmatrix}_{3_1} \otimes \begin{pmatrix} y_1 \\ y_2 \\ y_3 \end{pmatrix}_{3_2} = \sum_{k=0,1,2} (x_1 y_2 + \omega^{2k} x_2 y_3 + \omega^k x_3 y_1)_{1_{0k}}$$

$$\oplus \sum_{k=0,1,2} (x_1 y_1 + \omega^{2k} x_2 y_2 + \omega^k x_3 y_3)_{1_{1k}}$$

$$\oplus \sum_{k=0,1,2} (x_1 y_3 + \omega^{2k} x_2 y_1 + \omega^k x_3 y_2)_{1_{2k}}. \quad (F.116)$$

The tensor products between triplets and singlets are:

$$(x)_{1_{0k}} \otimes \begin{pmatrix} y_1 \\ y_2 \\ y_3 \end{pmatrix}_{3_1} = \begin{pmatrix} x y_1 \\ \omega^k x y_2 \\ \omega^{2k} x y_3 \end{pmatrix}_{3_1}, \quad (x)_{1_{0k}} \otimes \begin{pmatrix} y_1 \\ y_2 \\ y_3 \end{pmatrix}_{3_2} = \begin{pmatrix} x y_1 \\ \omega^k x y_2 \\ \omega^{2k} x y_3 \end{pmatrix}_{3_2},$$

(F.117)

$$(x)_{1_{1k}} \otimes \begin{pmatrix} y_1 \\ y_2 \\ y_3 \end{pmatrix}_{3_1} = \begin{pmatrix} xy_3 \\ \omega^k xy_1 \\ \omega^{2k} xy_2 \end{pmatrix}_{3_1}, \quad (x)_{1_{1k}} \otimes \begin{pmatrix} y_1 \\ y_2 \\ y_3 \end{pmatrix}_{3_2} = \begin{pmatrix} xy_2 \\ \omega^k xy_3 \\ \omega^{2k} xy_1 \end{pmatrix}_{3_2},$$

(F.118)

$$(x)_{1_{2k}} \otimes \begin{pmatrix} y_1 \\ y_2 \\ y_3 \end{pmatrix}_{3_1} = \begin{pmatrix} xy_2 \\ \omega^k xy_3 \\ \omega^{2k} xy_1 \end{pmatrix}_{3_1}, \quad (x)_{1_{2k}} \otimes \begin{pmatrix} y_1 \\ y_2 \\ y_3 \end{pmatrix}_{3_2} = \begin{pmatrix} xy_3 \\ \omega^k xy_1 \\ \omega^{2k} xy_2 \end{pmatrix}_{3_2}.$$

(F.119)

The tensor products between singlets are

$$\mathbf{1}_{n,k} \otimes \mathbf{1}_{n',k'} = \mathbf{1}_{n+n',k+k'}.$$

(F.120)

References

1. Frampton, P.H., Kephart, T.W.: Int. J. Mod. Phys. A **10**, 4689 (1995). arXiv:hep-ph/9409330
2. Frampton, P.H., Kephart, T.W., Rohm, R.M.: Phys. Lett. B **679**, 478 (2009). arXiv:0904.0420 [hep-ph]

Index

A
Abelian, 14
A_4, 31
A_4 flavor model, 207
A_5, 34
Alternating group, 31
Anomaly of discrete symmetry, 187
Associativity, 13

B
Bimaximal mixing, 219, 221
Binary dihedral group, 61
Breaking pattern, 147

C
Character, 16
Closure, 13
Completely reducible, 16
Conjugacy class, 16
Cube, 25
Cummins-Patera's basis, 35
Cyclic group, 13

D
D_4, 58
D_5, 59
$\Delta(27)$, 94
$\Delta(54)$, 138
Dihedral group, 51
Dimension, 16
Dodecahedron, 34
Double covering group, 43

F
Flavor mixing, 205
Fujikawa's method, 185

G
Golden ratio, 219, 220
Gravitational anomaly, 186

H
Homomorphic, 16

I
Icosahedron, 34
Invariant subspace, 16
Irreducible, 16
Isomorphic, 16

L
Lagrange's theorem, 14, 229

N
Non-Abelian, 14
Normal subgroup, 14

O
Octahedron, 25
Order, 14
Origin of flavor symmetries, 223

P
Pentagon, 59
Polygon, 51

Q
Q_4, 66
Q_6, 67
QD_{16}, 72

R
Reducible, 16
Representation, 16

S
Schur's lemma, 230
S_3, 21
S_4, 25
S_4 flavor model, 211
Semi-direct product, 19
Shirai's basis, 35, 37
Square, 58
Subgroup, 15, 147
Symmetric group, 14, 21
$\Sigma(18)$, 78
$\Sigma(32)$, 80
$\Sigma(50)$, 84
$\Sigma(81)$, 113

T
Tetrahedron, 31, 32
T', 43
T_7, 100
T_{13}, 102
T_{19}, 104
Tri-bimaximal mixing, 205, 206